# ADVANCED AIRBREATHING PROPULSION

by
Y. M. Timnat
Department of Aerospace Engineering
and Space Research Institute,
Technion—Israel Institute of Technology
Haifa, Israel

AN ORBIT SERIES BOOK

**KRIEGER PUBLISHING COMPANY**

MALABAR, FLORIDA
1996

Original Edition 1996


Printed and Published by
**KRIEGER PUBLISHING COMPANY**
**KRIEGER DRIVE**
**MALABAR, FLORIDA 32950**






**Library of Congress Cataloging-In-Publication Data**

Timnat, Y. M.
    Advanced airbreathing propulsion / Y. M. Timnat.
      p.  cm. — (Orbit)
    Includes bibliographical references and index.
    ISBN 0-89464-049-6 (acid-free paper)
    1. Airplanes—Jet propulsion. 2. Airplanes—Ramjet engines.
  3. Airplanes—Scramjet engines. 4. High-Speed aeronautics.
  I. Title. II. Series.
TL709.T54  1996
629.134′353—dc20          94-36668
                               CIP

10 9 8 7 6 5 4 3 2

*Series editors*
Edwin F. Strother, Ph.D.
Donald M. Waltz

# Contents

# Preface

In this book I have tried to give a general picture of modern air breathing propulsion. In the chapters dedicated to engines that do not include rotating parts the aim is to present a comprehensive review of the various systems in use or in advanced stages of development. Their principles are described and a rather detailed account of applications is given in the relevant cases. Attention is also devoted to various measuring methods, in particular to those specific to this type of propulsion. This is of some importance, in my opinion, since much of this information was not available in book form at the time of writing.

While there are a number of text books dealing with jet engines, the aim in the chapters treating this subject is to outline trends in the development of powerplants, for both civil and military applications. In the civil engine field the accent is on energy efficient engines, the increasing trend towards international cooperation, the prospects for propfan propulsion, and the progress in supersonic propulsion. With regard to military applications, the emphasis is on low observability requirements and stealth technology, which have proven their importance in the 1991 Gulf War, on thrust vector control, and on V/STOL propulsion.

An extensive chapter is dedicated to combustion chambers, reflecting both the increasing importance of pollution control and my personal interest and involvement in the subject. Another chapter is dedicated to noise problems, a subject which has come to the forefront in the last decade.

Chapter 15 deals with modern developments in components and materials for turbine engines, ramjets, scramjets, and combined propulsion, pointing out the latest advances and future trends.

The last chapter treats briefly maintenance and reliability in the context of turbine engines. This subject will certainly be important in the future for scramjets and combined propulsion, but at the present moment the state of the art is such that only experimental powerplants are under consideration.

# Acknowledgments

I want to thank my colleagues Prof. A. Gany and Dr. Y. Goldman for helpful discussions, Miss O. Segal, who contributed to the drawings, and Mrs. H. Burcat, who typed the manuscript. Particular thanks are due to my wife Shoshana, who gave me both technical and moral support throughout the work on this monograph.

I would also like to state that the manuscript was closed at the end of 1992.

# Chapter 1

# The Importance of Airbreathing Propulsion

Airbreathing propulsion is playing an ever increasing role in modern society. In civil applications it makes possible high speed leisure and business travel and provides transportation for much high value commercial cargo. Its military importance has been vividly demonstrated by the Falkland Island conflict and the 1991 Gulf War. Until now the major role has been played by aircraft and missiles powered by turboengines or rockets, but the prospects for ramjets and scramjets are also bright, as shown by the Soviet SA-6 used successfully in the 1973 Arab-Israeli War. Our treatment shall distinguish between civil applications, which are almost exclusively dominated by turbojets, and military uses, for which the ramjet and the scramjet hold great promise. The pulse jet which has potential for both civil and military uses, will also be treated.

## 1.1 Classification of Airbreathing Propulsion

A propulsion device, or an engine, is designed to generate thrust, which will overcome the aerodynamic drag on the vehicle, applying enough net thrust to overcome vehicle inertia and gain speed and altitude as required. Airbreathing devices are limited to that portion of the atmosphere, where air is dense enough to provide sufficient thrust; today, this limits the height to about 100 km.

In our treatment of modern airbreathing propulsion we shall consider only jet propulsion and shall divide it into two main groups: turbojets, which include rotating machinery, and ramjets, which are devoid of it. In the first category, the types treated include turboprops, propfans, turbofans, and pure jets; in the second, we shall deal with ramjets proper and scramjets; these can be fueled by hydrocarbons (preferred for ramjets) and by hydrogen (preferred for scramjets). Turboprops are limited to speeds of about Mach 0.7, while propfans can reach Mach 0.8; turbofans and pure jets are usually limited to Mach 3. Ramjets operate between Mach 2 and 5, while for higher speeds scramjets are employed.

An important characteristic of the various modes of propulsion is the specific impulse $I_{sp}$ which denotes the impulse per unit mass and is expressed in units of N-sec/kg which is the same as m/sec. $I_{sp}$ may also be defined as impulse per unit weight in which case the units are seconds. For turbojets the specific impulse can reach over 4000 sec at low speeds, it is somewhat lower for scramjets (around 3000 sec), while the value for ramjets is around 2000 sec. This is shown in Figure 1.1, which also presents for comparison the specific impulse of rocket engines. $I_{sp}$ is close to 500 sec for liquid oxygen-liquid hydrogen rocket engines in vacuo and around 300 for LOX-hydrocarbon or solid propellants.

It is interesting to note that the speed of sound in the earth's atmosphere is remarkably constant, at a value of approximately $300 \pm 5$ m/sec, while density and pressure drop very steeply, by about a factor of 1000 for every 50 km rise in altitude. Generally one speaks of subsonic speed up to Mach 0.92, transonic speed in the range of Mach numbers between 0.92 and 1.2, then supersonic speed up to Mach 5, and hypersonic speed above that. Not all speeds can be achieved at every altitude in the atmosphere. Since aerodynamic drag and heating increase with Mach number as $M^2$ and $M^3$, respectively, for hypersonic flight at a fixed altitude, it can be considered only for low atmospheric pressures and densities. For instance, cruising at $M = 5$ to 6 is feasible only for altitudes of 30–40 km, where the density ratio $\rho/\rho_{SL} < 10^{-3}$. At higher Mach numbers, say, between 15 and 20, hypersonic flight will be attractive only at altitudes exceeding 50 km when $\rho/\rho_{SL} < 10^{-6}$ (see Figure 1.2).

### 1.1.1 Civil Applications

In the civil applications of turbojets one can distinguish three types of engines—the large-size powerplants, which are employed mainly on long haul/high density routes, the medium-size engines, used for smaller aircraft and shorter routes, and the regional transport and business aviation powerplants.

The large-size engine market is dominated by the big three: General Electric (GE), Pratt and Whitney (PW), and Rolls-Royce (RR). The trend is towards increasingly powerful engines, going towards a thrust of 400,000 N which allows direct flights from the Western United States to Japan or from Southeast Asia to Europe (using for example the

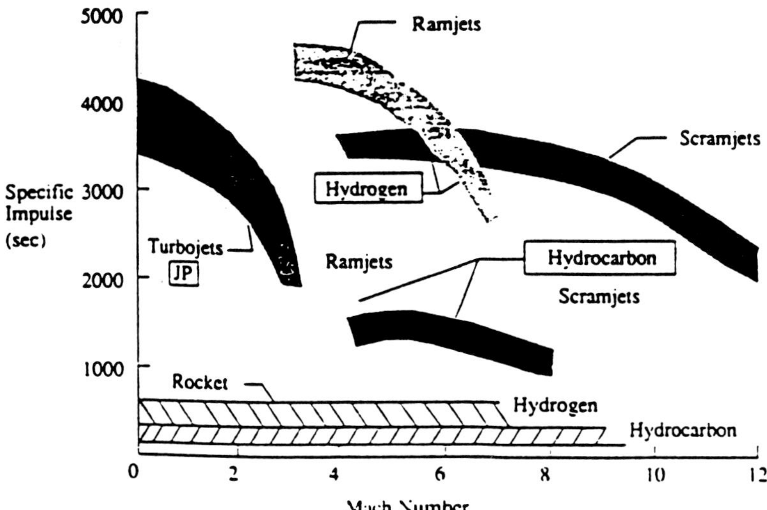


**Figure 1.1**  Operating Mach number ranges and specific impulse of various propulsion alternatives (Cheng 1989). *Reprinted with permission from Pergamon Press Ltd.*

Boeing 747-400). Other manufacturers tend to form consortia with one of the big three, i.e. SNECMA with GE, the Japanese engine manufacturers with PW, and so on.

Notwithstanding the high cost of developing new engines, all three manufacturers are engaged in the process, with GE working on the 90 series, PW developing new models in the 4000 series, and RR passing from the RB 211 to the Trent. Generally the early models have a thrust around 275,000 N, but the derivatives should reach 400,000 N. All the big engines are bypass engines with a bypass ratio of 6 or more; also, their architecture is modular, since this facilitates maintenance and improves reliability. The Russians are far behind in the area of very big civil powerplants.

In the field of medium-sized engines, the situation is similar; GE-SNECMA with the CFM series and PW with associates (2000 and 7000 series, V2500). In this context one should also mention the probable development of propfans (PW 5000 series, GE-SNECMA UDF). Only toward the lower end of the market does one find other suppliers, like Allison and Garrett in the United States, Turbomeca in France, and the Brazilian CASA.

With regard to civil engines for supersonic aircraft, these will probably operate without afterburners and will certainly profit from the great effort dedicated to the powerplants of fighter planes. Coming to hypersonic flight, the American national aerospace plane (NASP) is a joint effort of NASA and the Department of Defense (DOD), with the bulk of the money likely to come from the military; it could possibly develop sometime in the twenty-first century into an "Orient Express" (Europe or America to the Far East in a couple of hours). The Japanese are studying a hydrogen ramjet/scramjet, that could reach Mach 25 and be applied to a single-stage to orbit vehicle; the British Hotol and the

German Sänger are also aimed at hypersonic flight, perhaps in a joint European development.

### 1.1.2 Military Uses

To evaluate the developments in the military field, one can look at the American program of integrated high performance turbine engine technology (IHPTET), initiated in 1988 by NASA and DOD as a clear indicator for this type of powerplant. The goal of the program is to double propulsion capability for fighter, attack, and transport aircraft, as well as for vertical take-off and landing (VTOL), and rotorcraft vehicles, including also tactical and strategic cruise missiles. This program treats turbofans, turbojets, turboshafts, turboprops, and expendable missile engines.

The specific goals of IHPTET are a 100% increase in thrust-to-weight ratio, together with a 40% reduction in fuel consumption for turbofan and turbojet engines powering fighter and attack aircraft; a 40% drop in fuel consumption and 120% increase in power-weight ratio for turboshaft and turboprop engines, while for tactical missiles the aim is a 40% decrease in fuel consumption for supersonic flight, a 100% increase in thrust to airflow ratio and a 60% cost reduction. The means to reach the above goals include raising ignition temperature in order to increase efficiency, decrease fuel consumption, and expand the flight envelope, increasing the turbine inlet temperature, thus augmenting output per unit airflow, and reducing weight per unit airflow to obtain a higher output per unit weight.

The results expected are short takeoff and landing (STOVL) capability for the fighter, together with improved performance and mission capability, while lowering the takeoff gross weight by 35%. It is expected that the new fighter, the ATF (advanced tactical fighter), will be able to cruise at Mach 3 and above with an 1800 km mission radius.

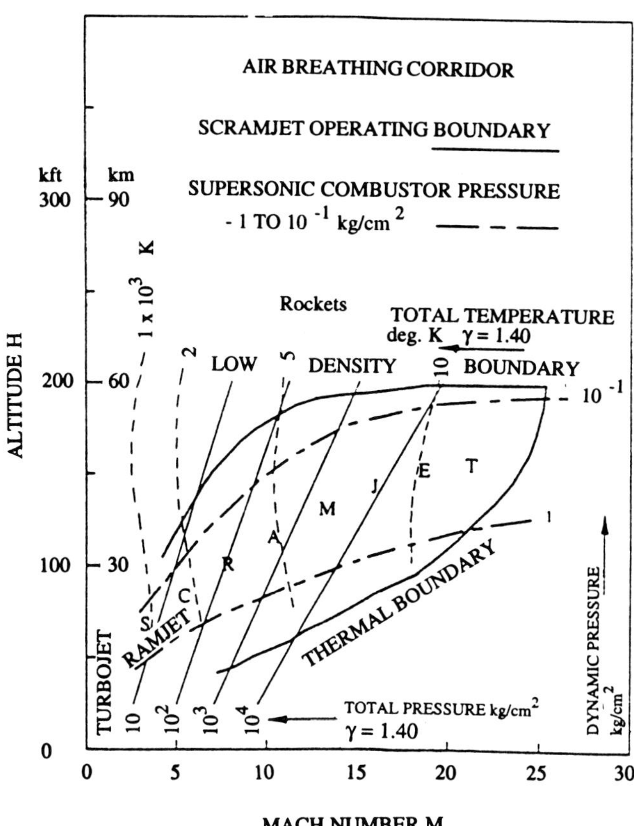



**Figure 1.2**  Altitude and Mach number map of the corridor for airbreathing engines (Cheng 1989). *Reprinted with permission from Pergamon Press Ltd.*

Helicopters are expected to double their range-payload capabilities, while transport aircraft will increase their range by 50% without increase in size, improving at the same time fuel economy.

In order to accomplish the tasks outlined, better components and technological progress are required. Two examples of this approach are the development of a four-stage compressor and the introduction of segmented combustors. The new compressor will have four stages and weigh one-third that of the J79 compressor, which has 17 stages, while attaining the same performance and efficiency. To achieve this one must increase, by a large factor, the pressure ratio per stage; this implies raising the blade tip speed. The resulting increased centrifugal forces will be handled by better materials as they become available, while possible aerodynamic losses will be offset by using swept blades. In the second example, segmented double wall lined combustors can eliminate fatigue cracking and have already been introduced using metallic components; new composite ceramic units are being developed, which will further enhance performance.

Similar advanced technological research programs are going on in Europe for the EJ200 engine, which is being developed by a consortium of RR, Deutsche Aerospace, FIAT, and CASA for the Eurofighter.

The French are already flight-testing the SNECMA M-88 engine for their Rafale; naturally the advances are not so sweeping, but due to the modular configuration, they should be able to introduce more advanced materials and manufacturing techniques in future, as has been the practice in many derivative engines (Lenorowitz 1991).

In the ramjet field, six missiles were in operational status at the beginning of the 1990s: two first generation missiles, the British Bloodhound and its Chinese version, the HY-3/C101, which are large and heavy, and therefore limited to surface to air operations. Two second generation ramjets, the British Sea Dart and the Soviet SA-4, have been developed since the 1960s. Although smaller and lighter, they are also used only in surface to air missiles. The Sea Dart proved very effective against Argentinean aircraft during the Falkland conflict.

The two operational third generation missiles are the Soviet SA-6, an integral rocket ramjet (IRR), which was used with great success against Israeli aircraft in 1973 and the French ASMP (Air Sol Moyenne Portée) missile, a liquid fuel IRR. This is the first air to surface high performance missile using ramjet propulsion. Other French ramjet missiles (Rustic, ANS) are entering the inventory, while it is expected that the United States will apply integral rocket ramjet propulsion, both liquid and solid, to some of their new missiles that are now in the development stage. Japan, Taiwan, China, and South Korea also have programs in this area.

## 1.2 Outline of the Book

After this introductory chapter, the book contains two main parts; the first deals with means of propulsion that do not include rotating machinery, while the second deals with modern turbojets.

The topics treated in Chapter 2 comprise all types of state-of-the-art ramjets, starting with a short review of the ideal ramjet and taking into account also aerodynamic effects. There is a discussion of the various types (liquid and solid fuel ramjets as well as ducted rockets), followed by a description of modern applications in Russia, France, Germany, the United States, and China. Particular attention is devoted to thrust vector control.

The next chapter deals with scramjets. After an analysis of the theoretical cycle, the principles of supersonic and hypersonic combustion, including the problems of aerodynamic losses and integration, are treated. Preliminary design, performance calculations, and scramjet research and applications are discussed. The chapter concludes with reviews of ramjet-scramjet transition and current aerospace plane projects, such as NASP, Sänger, Hotol, Japanese efforts, including the ATR, as well as studies of Western-Russian cooperation programs.

Chapter 4 deals with the pulsejet, a propulsion device invented at the beginning of the century which has recently received renewed interest for different applications. In this chapter the theory of these engines is also treated, since it is often not included in undergraduate courses or texts. All the means of propulsion mentioned require extensive testing. This is discussed in some detail in Chapter 5, which describes different test facilities for ramjets, scramjets, and ducted rockets. The discussion of the current status of hypersonic combustion technology in Chapter 6 concludes Part I.

Part II on turbojet engines treats a very broad subject and therefore the choice of the material is somewhat personal. I have put emphasis on development trends in general, treating then particular examples, like the energy efficient engine. The next chapter treats civil engines for subsonic aircraft, according to their size and special features and is followed by a description of civil supersonic powerplants. We pass then to military engines, with attention to stealth technology and vectored thrust. This is followed by an exhaustive treatment of combustion chambers, including discussion of pollutant formation and methods of reducing it, various types of measurements and diagnostics as well as special types of combustion. This part is concluded with shorter overviews of other components and materials as well as noise problems and questions of maintenance and reliability.

# Part I

# Engines Without Rotating Machinery

# Chapter 2

# *Ramjets*

## *2.1 Introduction*

The advantage of airbreathing engines over rockets, as has been pointed out in the first chapter, stems from the fact that they do not carry the oxidizer, but rather utilize atmospheric oxygen, thus increasing considerably their efficiency. Up to Mach 3 the turbine engine has a clear economic advantage. This is particularly true for the bypass engine using an appropriate bypass ratio. For higher speeds it is necessary to adopt a different approach; the ramjet is especially attractive due to its simplicity, stemming from the absence of rotating components. This makes it possible to use higher combustion temperatures in the combustor and in the nozzle, by keeping their walls cooler than the main fuel stream, thanks to the utilization of a fuel-injection pattern, which leaves a shielding layer of relatively cool air near the walls.

Zucrow (1958) gives an account of the conception of the ramjet engine, which is attributed to the Frenchman Lorin, who described such a device in 1913. In 1926 Carter described the application of two ramjet-like devices for propelling shells. A. Fono holds the first patent for the use of a ramjet engine as a propulsion means for supersonic flight, issued in Germany in 1928. The first tests of a ramjet engine were performed by Leduc in France during the late thirties.

But the real beginning of the ramjet era occurred during the second world war, when their development was pursued in the United States, Germany, and England. The first flight tests were performed in 1945. Since then a number of flight vehicles were developed in the U.S., U.K., USSR, and France in the period up to 1965. Of these one should mention the U.S. Talos and Bomarc missiles, the British Bloodhound and Sea Dart, the Russian SA-4, and the French CT-41. Although most of them reached operational status, their success was limited, since they required two stages (the first generally a rocket booster) and their length was considerable. These vehicles used liquid fuels and their speed was limited to Mach 4.

In the sixties attention was switched to solid fuel ramjets and the concept of the ram rocket evolved, which was first deployed successfully by the Russians in their SA-6 missile. The United States, France, and Germany also started vigorous programs at that time and have since developed both operational and experimental vehicles such as the American ALVRJ; the French ASMP (air ground), ANS (antiship), Rustic (antiaircraft); and various German boron-fueled vehicles. Some variants, called ducted rockets and hybrid rockets, although they are airbreathing, were developed in parallel.

Scramjet activity (above Mach 5) has been going on in France (Scorpion, Esope experimental vehicles up to $M = 7$), the U.S. (NASP), the U.K. (Hotol), Germany (Sänger), and Japan. After treating in detail the principles of the scramjet these projects will be described briefly.

The last type of propulsion we shall discuss in Part I is the pulsejet, a related form of airbreathing propulsion, which has the advantage of also allowing takeoff, but has a much lower specific impulse.

This is followed by a brief description of test facilities that are associated with ramjet and scramjet R&D. Part I concludes with a review of the status of hypersonic combustion technology at the beginning of the 1990s.

## *2.2 The Liquid Fuel Ramjet*

Figure 2.1 shows schematically a liquid fuel ramjet which consists of a diffuser, a combustor and an exhaust nozzle. The air that enters the diffuser is compressed there, mixed with the fuel that is introduced by suitable injectors, and is then burned in the combustor. The hot gases are expelled through the nozzle by the pressure rise in the diffuser as the incoming air decelerates from flight speed to a relatively low velocity in the combustor.

Although the ramjet can operate at subsonic flight speeds, it is most suitable for supersonic flight, due to the higher pressure rise accompanying higher flight speeds. The figure relates to a supersonic ramjet employing partially supersonic diffusion through a system of shocks. The combustor requires an inlet Mach number of 0.3 to 0.4, entailing a substantial pressure rise for supersonic flight speeds. For isentropic deceleration from Mach 3 to 0.3 this would amount to 34. One must, however, remember that pressure losses as-

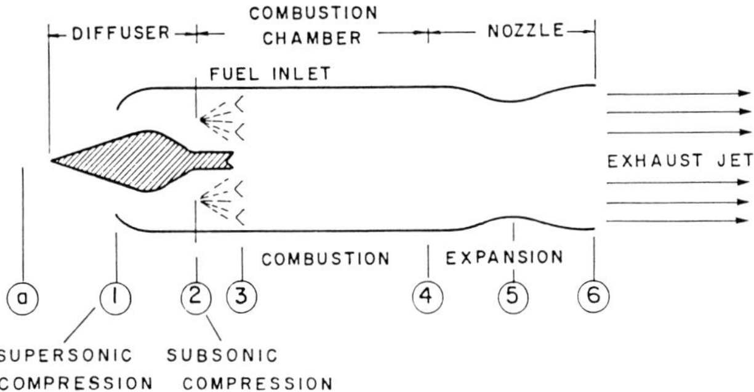


**Figure 2.1**   Schematic diagram of a ramjet engine (Hill and Peterson 1992). *Reprinted by permission of Addison-Wesley Publishing Co., Inc., Reading, MA.*

sociated with shocks are high and thus the rise is much smaller. The compressed air flows past the fuel injectors which spray a stream of fine fuel droplets producing rapid mixing with the air. The mixture is often stabilized by flameholders and the combustion raises the temperature to over 2000 K before the products expand in the nozzle.

The reaction to the propellant momentum is the thrust, $T$, in the engine according to

$$T = m_a[(1 + f)u_e - u] + (p_e - p_a)A_e. \qquad (2.1)$$

Here $m$ is the air mass flow, $f$ the fuel-air ratio, $u$ and $u_e$ the flow and exit speed, $p_a$ and $p_e$ the ambient and exit pressure respectively, and $A_e$ the nozzle exit area. The thrust is developed by pressure and shear forces distributed over the engine surface.

The relatively high peak temperature limit of the ramjet allows it to operate at flight Mach numbers higher than that of turbine engines. These are, however, limited by wall materials and cooling methods; at Mach 6.7, for instance, assuming an outside temperature of $-50°C$, the stagnation temperature is about 2000 K and dissociation may become significant. In order to avoid these effects, which can penalize severely the performance, one must switch to supersonic combustion if higher flight speeds are required. The vehicle becomes then a scramjet that will be treated later (see Chapter 3).

Finally Figure 2.2 depicts a typical ramjet, showing the geometry of the supersonic and subsonic portions of the diffuser, the fuel injector, the flame holder, the combustor and the nozzle.

### 2.2.1 The Ideal Ramjet

It is useful to perform a thermodynamic analysis of a simplified model, which is applicable with suitable modifications also to a solid fuel ramjet, in order to understand the operation of the device. The assumptions are that the compression and expansion processes are reversible and adiabatic and that combustion takes place at constant pressure.

Figure 2.3 is a temperature-entropy diagram, representing the processes, through which the fluid goes, using the nomenclature of Figure 2.1. The compression process takes the air from its initial state, $a$, isentropically to the stagnation state 02 at station 2. The combustion process, in which mass and heat are added at constant pressure, brings us to station 4, where the stagnation temperature reaches its

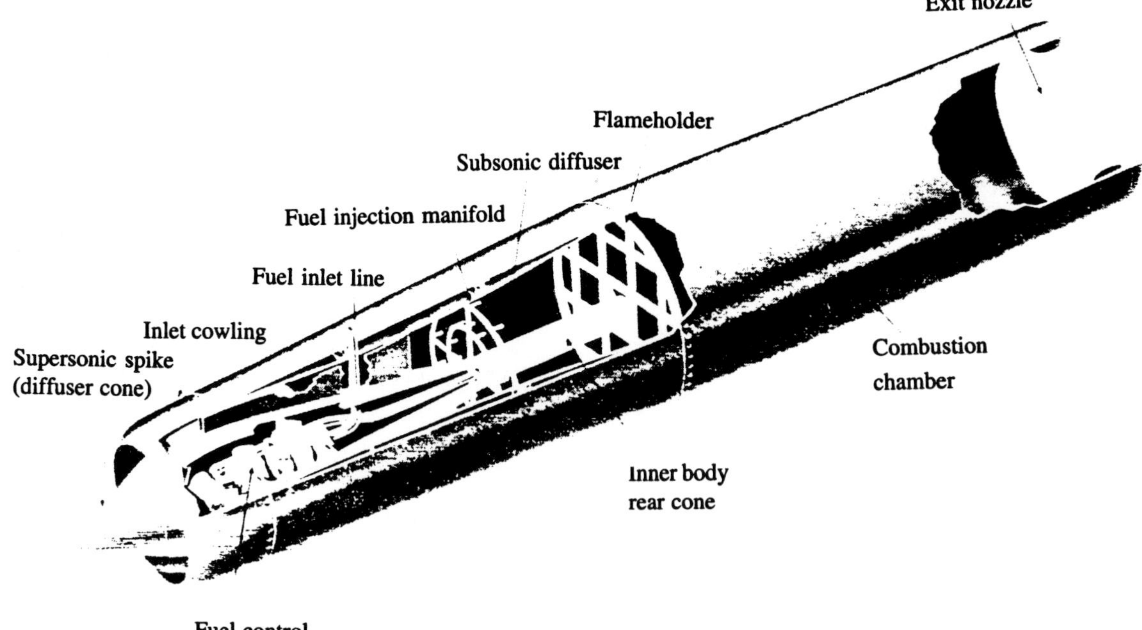


**Figure 2.2**   Cutaway view of a typical ramjet engine (Hill and Peterson 1992). *Reprinted by permission of Addison-Wesley Publishing Co., Inc., Reading, MA.*

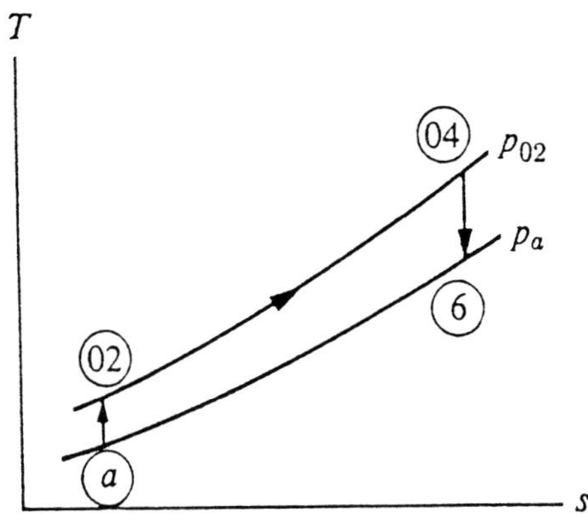

**Figure 2.3** Thermodynamic path of the fluid in an ideal ramjet (Hill and Peterson 1992). *Reprinted by permission of Addison-Wesley Publishing Co., Inc., Reading, MA.*

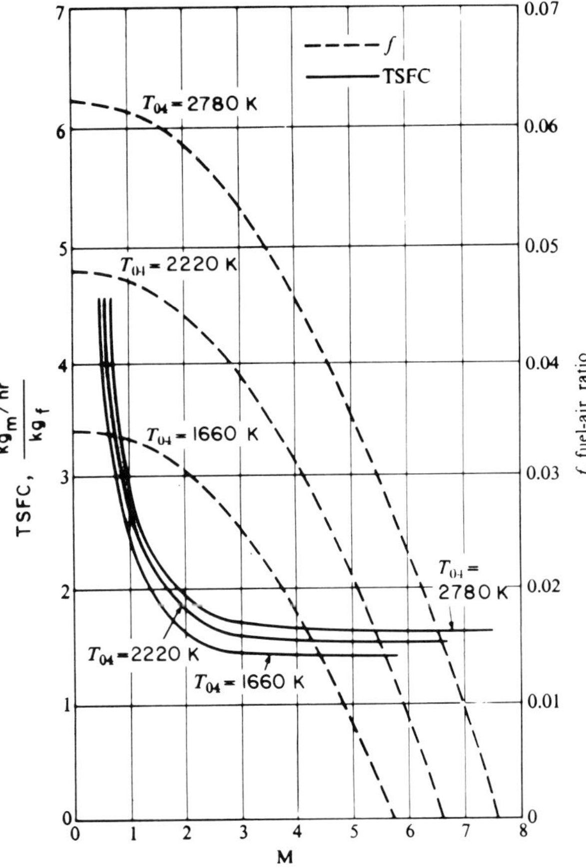

**Figure 2.4** Performance of three ideal ramjets as a function of flight Mach number and peak temperature limit for $T_a = 220$ K, $Q = 10,550$ Kcal/kg, $C_p = 0.24$ Kcal/Kg·k and $\gamma = 1.4$ (Hill and Peterson 1992). *Reprinted by permission of Addison-Wesley Publishing Co., Inc., Reading, MA.*

maximum value. In the nozzle the gases expand isentropically to the ambient pressure $p_a$ at station 6. The ideal engine thrust is then obtained from Eq. (2.1) as:

$$T = m_a[(1 + f)u_e - u] \tag{2.2}$$

According to the assumptions the stagnation pressure must remain constant through the engine; that is, $p_{0a} = p_{06}$. In the ideal case one may also assume that the fluid properties (gas constant $R$ and heat capacity ratio $\gamma$) remain constant.

From the ideal gas equation of state and the definition of the Mach number, $M$, it can be shown (see Hill and Peterson, 1965 and 1993) that:

$$p_{0a}/p_a = \left(1 + \frac{\gamma - 1}{2} M_a^2\right)^{\gamma/\gamma-1} \tag{2.3a}$$

and

$$p_{06}/p_e = \left(1 + \frac{\gamma - 1}{2} M_e^2\right)^{\gamma/\gamma-1} \tag{2.3b}$$

where $M_a$ is the flight Mach number and $M_e$ the fluid Mach number in the exit plane. Since $p_e = p_a$ this results in

$$p_{0a}/p_a = p_{06}/p_e \quad \text{and} \quad M_e = M_a. \tag{2.4}$$

As a consequence, one has for the exhaust velocity

$$u_e = (a_e/a_a)u. \tag{2.5}$$

Here $a$ is the speed of sound and from the relation $a = [\gamma RT]^{1/2}$ one obtains $a_e/a_a = [T_e/T_a]^{1/2}$; using Eq. (2.4) one gets $T_e/T_a = T_{06}/T_{0a}$ and since $T_{04} = T_{06}$ one finally obtains

$$u_e = [T_{04}/T_{0a}]^{1/2} u_a. \tag{2.6}$$

Neglecting the enthalpy $h$ of the incoming fuel, which is low compared to its heating value $Q$, conservation of energy gives

$$(1 + f)h_{04} = h_{02} + fQ. \tag{2.7}$$

Introducing the specific heat $c_p$, which is also assumed constant in the ideal case, one obtains for $f$:

$$f = \frac{(T_{04}/T_{0a}) - 1}{(Q/c_p T_{0a}) - (T_{04}/T_{0a})}. \tag{2.8}$$

Combining (2.7) and (2.8) one obtains an expression for the thrust per unit mass flow of air as follows:

$$\frac{T}{m_a} = M(\gamma RT_a)^{1/2}\left[(1 + f)(T_{04}/T_a)^{1/2}\left(1 + \frac{\gamma - 1}{2} M^2\right)^{-1/2} - 1\right] \tag{2.9}$$

The thrust specific fuel consumption (TSFC), which is an important figure of merit for airbreathing engines, is:

$$\text{TSFC} = m_f/T = f/(T/m_a) \tag{2.10}$$

Here $m_f$ is the fuel flow rate.

Figure 2.4 presents the TSFC and the corresponding fuel/air ratio of an ideal ramjet as a function of flight Mach number and peak temperature limit for 2800 K, 2250 K and 1700 K when $T = 222$ K, $Q = 10,550$ Kcal/Kg, $c_p = 0.24$ Kcal/Kg·K, and $\gamma = 1.4$. The figure shows that for each temperature there is a maximum flight Mach number, above which no fuel may be burned in air.

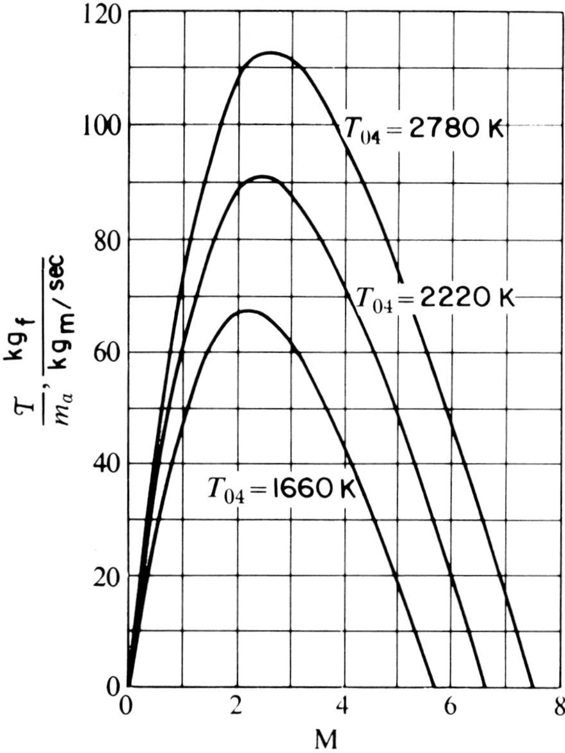

**Figure 2.5** Thrust per unit airflow rate for the ideal ramjets of Fig. 2.4 (Hill and Peterson 1992). *Reprinted by permission of Addison-Wesley Publishing Co., Inc., Reading, MA.*

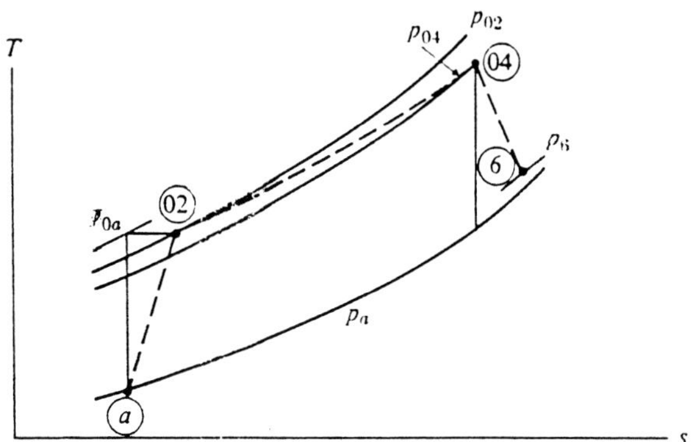

**Figure 2.6** A T-s diagram with aerodynamic losses (Hill and Peterson 1992). *Reprinted by permission of Addison-Wesley Publishing Co., Inc., Reading, MA.*

Figure 2.5 presents the thrust per unit airflow rate for the same conditions and shows that it reaches a maximum close to $M = 2.5$. In any real design one must take into account aerodynamic losses.

## 2.2.2 The Effect of Aerodynamics

In a real engine the propellant suffers stagnation pressure losses, while flowing through the engine. As a result the thrust goes to zero long before the fuel consumption does; thus TSFC first decreases, then goes up with increasing flight Mach number, as we shall see in Figure 2.7.

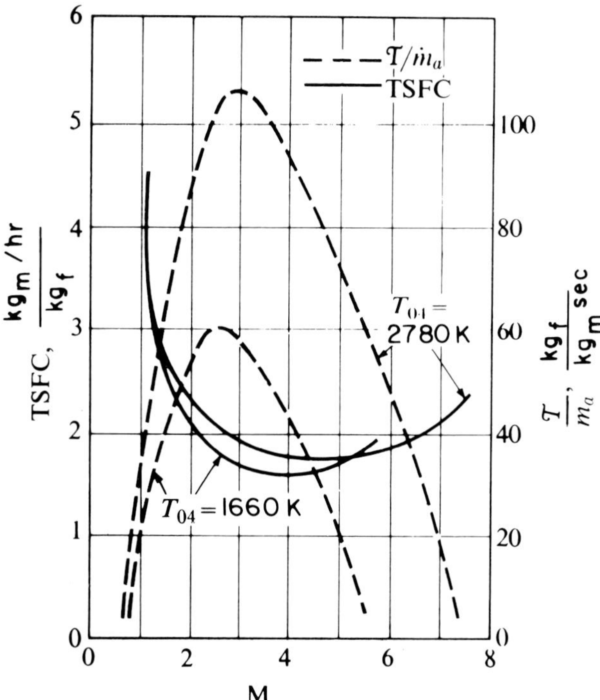

**Figure 2.7** Estimated ramjet performance with $r_d = 0.7$, $r_c = 0.97$, $r_n = 0.96$; $Q = 10{,}550$ Kcal/Kg and $\eta_b = 1.0$ (Hill and Peterson 1992). *Reprinted by permission of Addison-Wesley Publishing Co., Inc., Reading, MA.*

A T-S diagram, which takes into account the irreversible aerodynamic losses occurring during compression, burning, and expansion is shown in Figure 2.6. The compression process from "a" to point 02 is no longer isentropic (for comparison the figure also shows the isentropic process). Thus one can characterize the diffuser performance by a stagnation pressure ratio $r_d$ defined as

$$r_d = p_{02}/p_{0a}. \tag{2.11}$$

Stagnation pressure ratios can also be defined for the combustor, $r_c$, and the nozzle, $r_n$ as follows:

$$r_c = p_{04}/p_{02} \quad \text{and} \quad r_n = p_{06}/p_{04}. \tag{2.12}$$

The overall stagnation pressure will be $p_{06}/p_{0a} = r_d r_c r_n$. Since Eq. (2.4) remains true regardless of losses, one can write

$$M_e^2 = \frac{2}{\gamma - 1}\left[\left(1 + \frac{\gamma - 1}{2} M^2\right)\left(\frac{p_{06}}{p_{0a}} \frac{p_a}{p_e}\right)^{\gamma-1/\gamma} - 1\right] \tag{2.13}$$

or

$$M_e^2 = \frac{2}{\gamma - 1}\left[\left(1 + \frac{\gamma - 1}{2} M^2\right)(r_d r_c r_n p_a/p_e)^{\gamma-1/\gamma} - 1\right]$$

$$\tag{2.13a}$$

Here we have taken into account the possibility that $p_6$ may be different from $p_a$, as shown in Fig. 2.6.

Assuming negligible heat transfer from the engine one obtains for the exhaust velocity

$$u_e = M_e (\gamma R T_e)^{1/2} \quad \text{or}$$

$$u_e = M_e \left[ \gamma R T_{04} \Big/ \left( 1 + \frac{\gamma - 1}{2} M_e^2 \right) \right]^{1/2} \tag{2.14}$$

Since the stagnation temperature is not affected by losses, one has from (2.8)

$$f = \frac{(T_{04}/T_{0a}) - 1}{(\eta_b \, Q/c_p T_{0a}) - (T_{04}/T_{0a})} \tag{2.15}$$

where $\eta_b$ is the combustion efficiency. One then has

$$T/m_a = [(1 + f)u_e - u_a)] + (1/m_a)(p_e - p_a)A_e , \tag{2.16}$$

or, using equations 2-13 and 2-14

$$\begin{aligned}
T/m_a = {} & (1 + f)\left[ \frac{2\gamma R T_{04}(c - 1)}{(\gamma - 1)c} \right]^{1/2} \\
& - M[\gamma R T_a]^{1/2} + \frac{p_c A_e}{m_a}\left( 1 - \frac{p_a}{p_e} \right)
\end{aligned} \tag{2.17}$$

where $c$ stands for

$$\left( 1 + \frac{\gamma - 1}{2} M^2 \right)(r_d \, r_c \, r_n \, p_e)^{\gamma - 1/\gamma} . \tag{2.18}$$

The thrust specific fuel consumption will be given, as before by

$$\text{TSFC} = f/(T/m_a) . \tag{2.19}$$

Figure 2.7 presents results for a ramjet having $r_d = 0.7$, $r_c = 0.97$, $r_n = 0.96$ with $\gamma = 1.4$ and $Q = 10{,}550$ Kcal/Kg as in Figure 2.4, and assuming $\eta_b = 1.0$. Comparing the two figures one sees that for a given peak temperature there is a well-defined minimum TSFC which is now close to Mach 3.

## 2.3 Solid Fuel Ramjets and Ducted Rockets

### 2.3.1 The Solid Fuel Ramjet

The solid fuel ramjet (SFRJ) has the advantage that it is possible to insulate the wall of the combustor from the gases and to achieve high temperatures. The fuels employed are generally less energetic than those used in the liquid fuel ramjet; the most common are polymers such as polyether, polyester, polyurethane and polymethilmetacrylate (PMMA). As a consequence there has been research aimed at incorporating more energetic materials (e.g., boron) in the fuel matrix.

One of the important tasks in the design of an SFRJ is the design of an appropriate solid fuel combustion chamber. Research performed in many countries (U.S., France, Germany, USSR, Israel, Holland, Sweden, China and Japan) has shown that it is necessary to use a configuration comprising an inlet, followed by a diaphragm, to which the solid fuel is

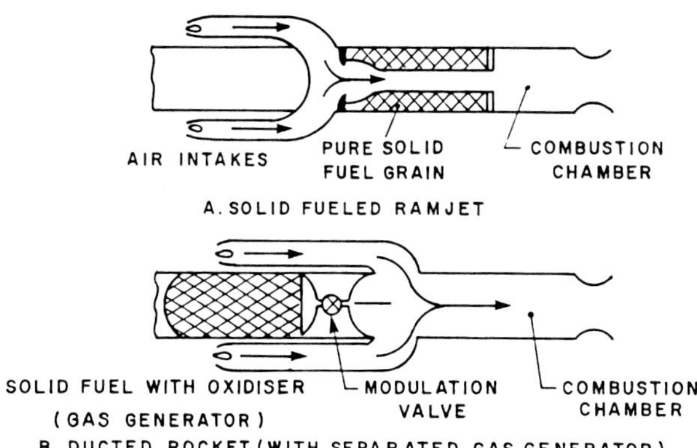


**Figure 2.8**   Solid fuel ramjet variations.

attached, and terminating with a secondary combustion chamber, which may be separated from the fuel by a rear diaphragm (see Figure 2.8).

The main results obtained (Mady et al. 1978) are that the regression rate $r$ depends on the mass flow $m$, the pressure $p$ and the inlet temperature $T_{in}$, according to

$$r \propto p^a m^b T_{in}^c \tag{2.20}$$

with representative values $a = 0.25 - 0.5$, $b = 0.4$, and $c = 0.4$.

Schadow et al. (1978) measured temperature profiles in the solid fuel region and in the secondary combustion chamber as a function of the mass flow and the fuel/air ratio. They found there is a fuel rich zone close to the wall and an air-rich core in the center. Schulte et al. (1987), who measured temperature and concentration in an SFRJ combustor, confirm that for low air mass flow the high temperature zone over the fuel surface is broader and hotter than for high air mass flow. This is a consequence of the fact that the regression rate increases more slowly than the air mass flow, causing the creation of a thicker boundary layer and favoring better mixing and a higher fuel/air ratio. The measured fuel temperature was 900 K, while the core reached 2000 K. In the secondary chamber a temperature of 2100 K was measured with thermocouples, and they estimate it may be 10% lower than the actual value because of losses. The concentration measurements showed that near the wall the amount of oxygen was very small, while significant quantities of carbon monoxide were observed. In the secondary chamber the products were mainly carbon dioxide and water; they found that a minimum length is required to obtain complete combustion. Netzer and Gany (1991) showed that the inlet temperature of the air has a great influence on both ignition and regression rate; they, as well as Ben-Arosh and Gany (1989), treated scale effects.

The main theoretical models were developed by Netzer and coworkers (1977, 1978, and 1981), who succeeded in obtaining good agreements with experimental results even for complicated geometrical configurations, and by Vos (1987), who obtained reasonable agreement with experiment (Korting and coworkers 1986 and 1988), apart from the recirculation zone. Gany and Netzer (1986) investigated the behavior of highly metallized boron containing solid fuels in an SFRJ, using high speed photography with a 2D combustor. They found in the boundary layer above the fuel surface a gas phase diffusion flame of the volatile fuel ingredients. In some cases large pieces and glowing segments of materials were emitted from the surface.

Netzer developed two models for SFRJ combustion. In the first he applied a finite-difference solution to the governing equations in terms of vorticity and stream function (Netzer 1977). Although it provided reasonable agreement with experimental data taken inside the fuel grain, the predicted pressure distributions were inaccurate and it was difficult to specify boundary conditions. Therefore he decided to try a primitive-variable (velocity-pressure) model to increase the accuracy (Stevenson and Netzer 1981). He assumed steady, two dimensional subsonic flow, taking for simplicity constant $c_p$. The effective viscosity was calculated by a modified Jones-Launder turbulence model (Jones and Launder 1972), using the general form of the conservation equations for axisymmetric flow (3); namely,

$$\frac{\partial}{\partial x}(\rho u \phi) + \frac{1}{r}\frac{\partial}{\partial r}(\rho r v \phi) - \frac{\partial}{\partial x}\left(\Gamma_\phi \frac{\partial \phi}{\partial x}\right)$$
$$\text{convection terms} \qquad\qquad \text{diffusion terms}$$
$$(2.21)$$
$$-\frac{1}{r}\frac{\partial}{\partial r}\left(r\Gamma_\phi \frac{\partial \phi}{\partial r}\right) = \rho_\phi$$
$$\text{source terms}$$

where $\phi$ stands for the dependent variables ($u$, $v$, $\varepsilon$, $h$, etc.) considered ($\phi = 1$ for the continuity equation), $\Gamma_\phi$ is the appropriate effective exchange coefficient for turbulent flow, while $\rho_\phi$ is the source term. The reader is referred to Stevenson and Netzer, 1981 for more detail. The energy equation, expressed in terms of stagnation enthalpy, has no source terms, since the turbulent Prandtl and Schmidt numbers were chosen as unity and radiative transport was neglected. The stagnation enthalpy is given by

$$h = h_o + (u^2 + v^2)/2 + k \qquad (2.22)$$

where for non reactive flows

$$h_o = c_p T \qquad (2.23)$$

and for reacting flows

$$h_o = m_{ox}\,\Delta H/i + c_p(T - T_o)\,. \qquad (2.24)$$

The temperature was calculated using Eqs. (2.22) to (2.24), while the density was calculated from the ideal gas law. Fixed boundary conditions were specified (if desired experimentally determined values could be chosen). In the numerical procedure all quantities were non-dimensionalized, plug flow was assumed at the inlet, radial and axial gradients were set equal to zero on the centerline and exit, re-

spectively; the nonreacting solid boundaries were taken as adiabatic and the no slip condition was assumed.

The results obtained allow one to predict the fuel flow inside the fuel grain and in the aft-mixing region as well as the fuel regression rate. The primitive-variable model gave better results, in particular, for the aft-mixing region; the agreement with experimental data was reasonable. Additional experimental work continues with a view of improving the available empirical data and thus the applicability of the model.

Gany and coworkers (Ben-Arosh and Gany 1989; Netzer and Gany 1991) investigated small solid fuel ramjet combustors with an internal diameter between 10 and 14 mm. In the earlier work (Zvuloni, Levy, and Gany 1989) they employed a static test system with a 25 kW electrical heater, which enabled them to simulate the air temperature and pressure encountered for a flight Mach number of 3 at sea level. They measured pressure, air temperature, motor thrust and exhaust gas composition. The transparent polymethilmetacrilate (PMMA) fuel used in the tests permitted continuous video photography, recording the local fuel regression rate behaviour and the instantaneous ignition and combustion phenomena. The results demonstrated high combustion efficiency and indicated peculiar local and average fuel regression rate correlations. Analytical studies showed that the specific conditions resulting from the low Reynolds number range in small SFRJ motors, in contrast to larger combustors, enhance the effect of the sudden expansion heat transfer regime, compared with the boundary layer regime. This might cause the fuel regression rate behaviour to be dominated by the sudden expansion characteristics. It was also demonstrated that the specific fuel regression mechanism in small SFRJ motors may lead to an increase in the apparent exponent in the regression rate law.

In the later work three different fuel types were tested: PMMA, polyethylene, and polybutadiene. The air inlet temperature was also varied, using 850 K, 520 K and room temperature, to simulate flight Mach numbers of 3, 2 and low subsonic speed. All three fuels exhibited very good flameholding capability at the upper air temperature (800 K). The flammability limits decreased at 520 K and became very narrow at room temperature. The flameholding capability also deteriorated with decreasing combustor size. The flow reattachment distance, coinciding with the maximum fuel regression rate location, was shown to depend on the inlet step height and was independent of fuel type and port-flow Reynolds number. Small-scale SFRJ combustors may be considered for specific missions, especially for self propelled projectiles.

### 2.3.2. Ducted rockets

A ducted rocket is a particularly simple ramjet variant, sometimes called an air augmented rocket, shown schematically in Figure 2.9. It consists of a solid propellant

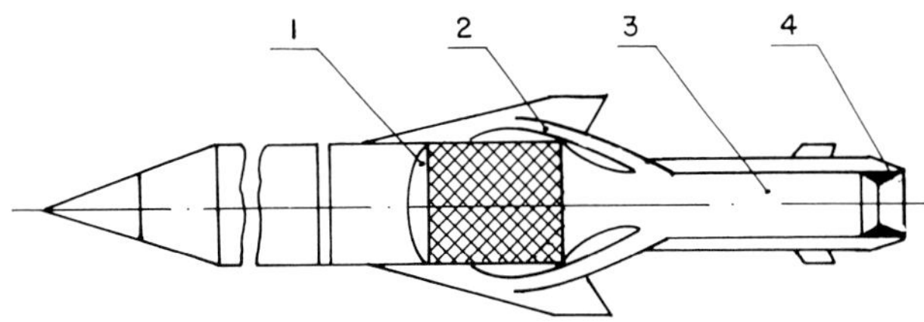

**Figure 2.9** Model of air-augmented rocket: (1) solid propellant rocket, (2) air inlet, (3) ram chamber, and (4) nozzle (Zhongqin et al. 1986). *Copyright © AIAA 1986. Used with permission.*

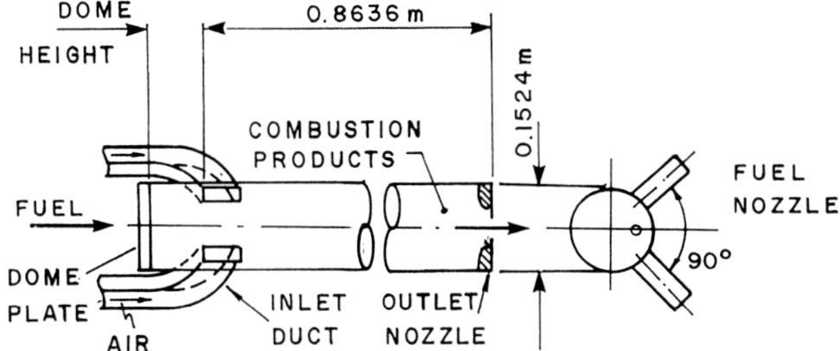

**Figure 2.10** Ducted rocket combustor configuration (Vanka et al. 1986). *Copyright © AIAA 1986. Used with permission.*

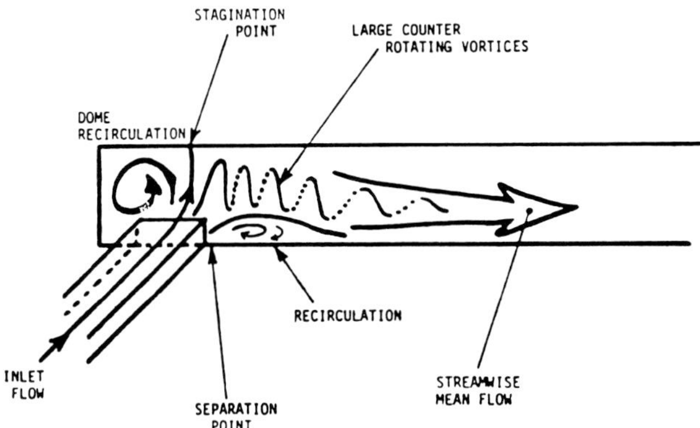

**Figure 2.11** Major combustor flow regions (Stull et al. 1985).

rocket, acting as a gas generator, a number of air inlets (at least two), a ram chamber, and an outlet nozzle. The configuration of the ram chamber is shown in Figure 2.10. Gaseous fuel from the gas generator is injected through the dome plate and the air is supplied through two side arms attached to the combustor periphery. The side arms are inclined (in the figure by 45°) with the duct axis located symmetrically in the azimuthal direction. The mixing of the fuel and the air streams occurs in the complex flowfield formed by the two flow streams. The complex flow recirculation patterns in the dome region and behind the air stream aid in stabilizing the combustion process. A detailed understanding of the aerodynamics and fuel-air mixing processes in this type of configuration is necessary for improving combustion efficiency and increasing the thrust produced. One can then optimize the geometrical variables (the number and angle of the side arms, the distance be-

tween them and the dome plate and the fuel injector location) to obtain maximum efficiency and thrust.

The flow in the ducted rocket configuration is very complex; the side entry of the airstream sets up a complex 3D flow pattern, consisting of a pair of vortices in the cross-sectional plane and a complex recirculation pattern in the dome region. The turbulence-chemistry interactions and the complex composition of the flow from the gas generator complicate the situation further and can influence ignition and blow-down.

Studies have been conducted since the 1970s by Sosounov (1974) in Russia, Schadow (1969 and 1981) at China Lake, a group at Wright-Patterson (Stull and coworkers 1974, 1985; Vanka et al. 1983, 1986), Chen and Tao (1984) in Taiwan, and a group in China (Zhongqin et al. 1986). Work took place also at ONERA, DLR, and Technion. Sosounov indicated the problems concerning optimal ducted rockets; one of his important conclusions concerns the choice of an optimal chamber length. Schadow performed windowed combustion and water tunnel tests to study the effect of air injection, fuel injection momentum, air inlet angle and fuel composition on the combustion efficiency. The Wright-Patterson group started with water-flow visualization, continuing with experiments using JP-4 fuel since it allowed longer runs than the solid fuel, reducing the cost and the hazards. They later used gaseous ethylene as a fuel and developed a model which allowed comparison with experiments. Chen and Tao simplified the ducted rocket geometry to be axisymmetric and numerically solved the 2D reacting flow equations making a number of approximations. Zhongqin and coworkers performed a mainly experimental investigation, starting also with cold flow tests and going on to a gas generator employing metallized propellant. They used an axisymmetric configuration and varied the geometry, arriving at interesting conclusions.

We shall give a brief description of the Wright-Patterson experiments first. The water tunnel combustor model was built of plexiglass and gave information on the streamline pattern using photography (1/4 sec shutter speed) as well as motion pictures. As shown in Figure 2.11 the flow appears to consist of two primary zones: (1) a recirculating flow in the dome region of the combustor and (2) two counter-rotating helical vortices trailing down stream from the inlet. In addition, there are other secondary flow zones; for example, a pair of small, counter-rotating vortices immediately downstream of the inlet in the bottom side of the combustor. This recirculation zone is the result of flow separation due to the inlet flow configuration. These tests provided insight into the mixing-controlled processes of the flow and showed that variations in dome height affected strongly the head end flowfield, but had little influence on the downstream flow. The results were in good agreement with calculations, except in the dome region.

In their analytical study the Wright-Patterson group

**Table 2.1** Conditions for base calculations of a ducted rocket (Vanka et al. 1986) *Copyright © AIAA 1986. Used with permission.*

| | |
|---|---|
| Diameter of combustor | 0.1524 m |
| Length of combustor | 0.8636 m |
| Dome height | 0.0508 m |
| Angle of side arms | 45° |
| Temperature of inlet air | 556 K |
| Airflow rate (both arms) | 1.814 kg/s |
| Fuel flow rate (F/A = 0.06) | 0.1088 kg/s |

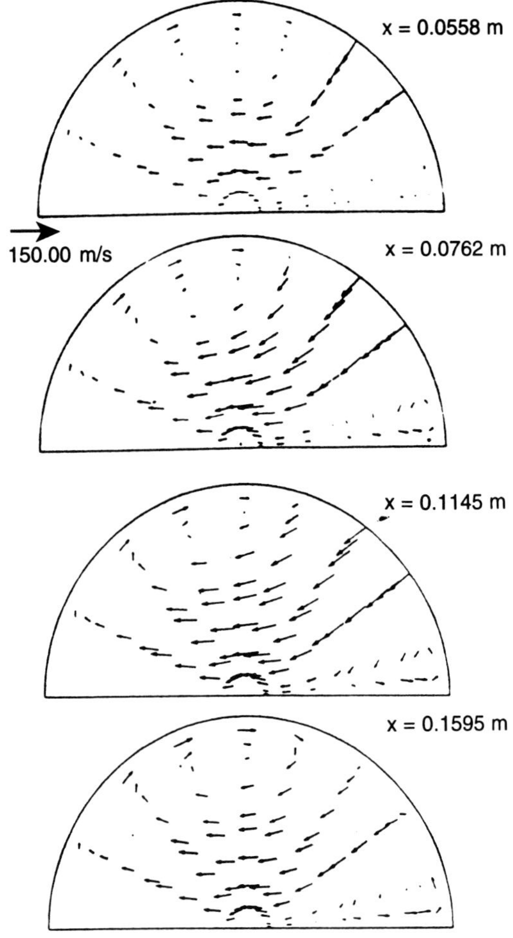


**Figure 2.12** Cross-stream flow pattern for base configuration (Vanka et al. 1986). *Copyright © AIAA 1986. Used with permission.*

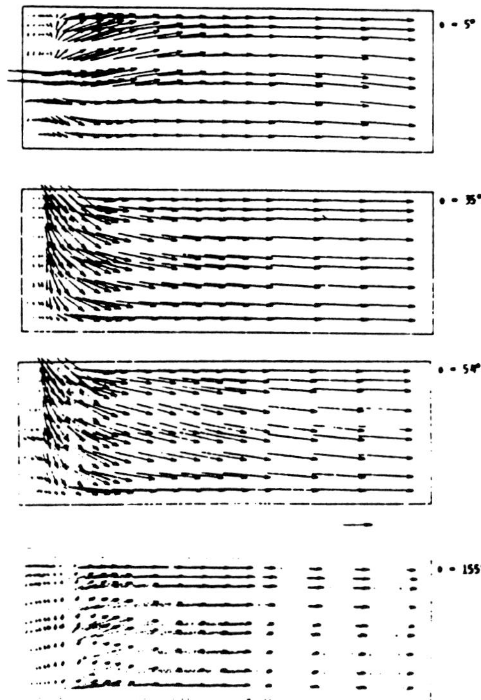


**Figure 2.13** Flow pattern in azimuthal planes for base configuration (Vanka et al. 1986). *Copyright © AIAA 1986. Used with permission.*

solved numerically the equations governing the fully elliptical 3D reacting flow; this was accomplished by an iterative finite-difference algorithm. For the combustion process they assumed the following one-step, infinitely fast, chemical reaction:

$$1 \text{ kg fuel} + i \text{ kg oxidant} = (1+i) \text{ kg products}$$

where $i$ is the stoichiometric oxidizer fuel ratio. They assumed a mixing-limited instantaneous chemical reaction.

Wall heat losses were neglected and the flow was taken as adiabatic. The turbulence was represented by the $k$-$\varepsilon$ model (see Launder and Spalding, 1972). Radiation was neglected. The computations were performed with different geometries. The interested reader can find the governing equations in the original paper (Vanka et al. 1986) include mass continuity, three momentum equations $(x, r, \vartheta)$, and two equations for the turbulence model. They also used the ideal gas equations and obtained the initial and boundary conditions from the experiments.

These were performed with ethylene-oxide, injected through a single port in the dome plate. The heated air (566 K) rate was 1.814 kg/sec and the fuel flow rate 0.101 kg/sec. Two combustor lengths were checked (0.712 and 0.457 m). The conditions for the basic calculations are given in Table 1). The calculated results are shown in a series of graphs: for the flow pattern in the cross stream (Figure 2.12) two symmetrical pairs of vortices—similar to the cold flow results. In the azimuthal plane (Figure 2.13) one sees two regions —in the dome region there are low-velocity recirculating eddies, while downstream of the air inlet the flow is helical (superposition of a vortex pattern on a unidirectional flow). This is similar to the cold flow, but no secondary recirculation zones appear. The temperature distribution (Figure 2.14) shows an increase from the dome region (1300 K maximum) downwards (up to 1900 K), while the unburnt fuel fraction diminishes rapidly (see Figure 2.15), but some

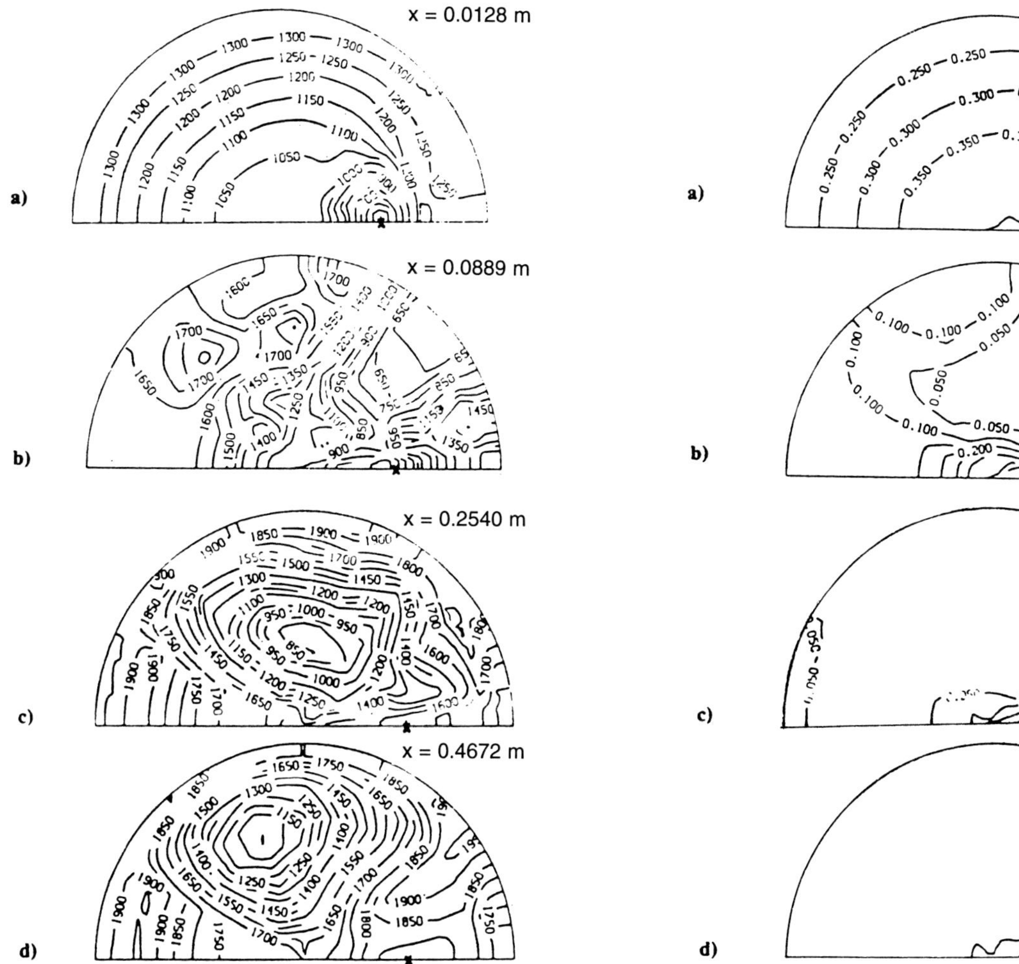

**Figure 2.14**  Cross-stream temperature for base configuration; $x$ = location of fuel injection (Vanka et al. 1986). *Copyright © AIAA 1986. Used with permission.*

**Figure 2.15**  Cross-stream fuel fraction distribution for base configuration; $x$ = location of fuel injection (Vanka et al. 1986). *Copyright © AIAA 1986. Used with permission.*

unburnt fuel is still present after 25 cm. Increasing the dome height improves the burning up to a height of 2″. This gives also the best combustion efficiency and is in good agreement with experiments (see Figure 2.16). The side arm angle has also an effect on the combustion efficiency, with a steeper angle having a slightly higher efficiency (see Figure 2.17). Finally, it is apparent that an eccentric fuel injection, as used in the base configuration, is more efficient than a concentric one (see Figure 2.18); this can be explained by observations on the airflow in the dome region.

Zhongqin and coworkers simulated a typical tactical missile using a combustor assembly consisting of a primary solid propellant rocket and a secondary constant area combustion chamber (see Figure 2.9). They knew from tests performed by others that the metallized fuel-side propellant affects the performance of the secondary chamber which also depends on the speed and altitude of the missile, which, in turn, influences the air/fuel ratio and the mixing characteristics. Furthermore, the secondary duct length and the

injection angle and location also influence the mixing and combustion performance. Therefore, their aim was to achieve improved combustion efficiency and extend the operational range.

The experimental apparatus, shown in Figure 2.19, employed air preheated to 500 K in a kerosene burner (2). The air flow was measured by a sonic throat (10), above which a large plenum chamber stabilized the flow. The assembly was mounted vertically and was separated by a gap (9) from the air pipe, so that the test rig was independent of the forces and movement of the air supply. The air flows into an annular manifold and is injected into the ram chamber (12) through four inlet ports at injection angles of 40, 60 or 90°. The air temperature was measured before entering the ram chamber, which could have different lengths, giving an L/D ratio between 2 and 12. Figure 2.20 shows the solid propellant fuel rich motor; the end burning grain was 150 mm long with a diameter of 212 mm. The properties of the propellant and the combustion products are given in Table 2.2.

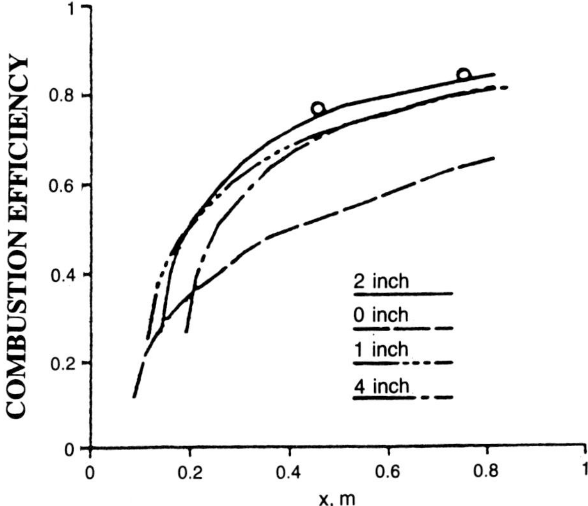

**Figure 2.16** Axial variation of combustion efficiency for different side-arm lengths. Data from Stull et al. 1985 (Vanka et al. 1986). *Copyright © AIAA 1986. Used with permission.*

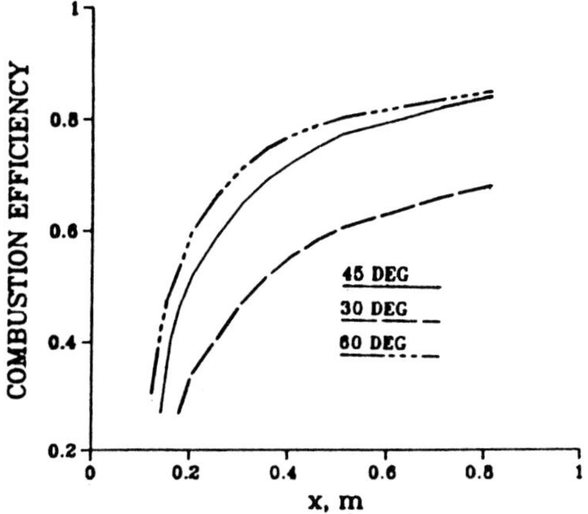

**Figure 2.17** Axial variation of combustion efficiency for different side-arm angles (Vanka et al. 1986). *Copyright © AIAA 1986. Used with permission.*

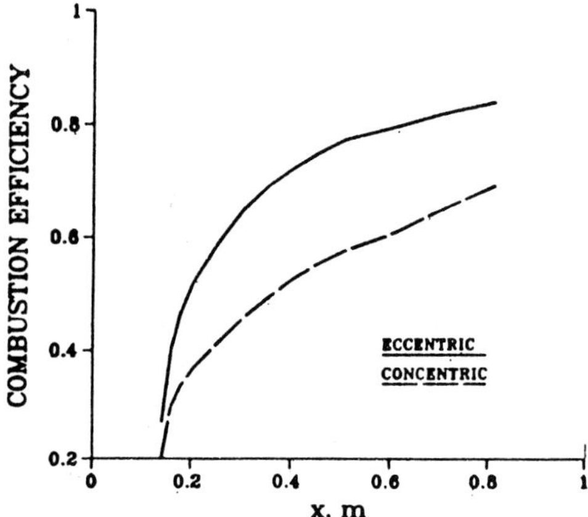

**Figure 2.18** Axial variation of combustion efficiency for concentric fuel injection (Vanka et al. 1986). *Copyright © AIAA 1986. Used with permission.*

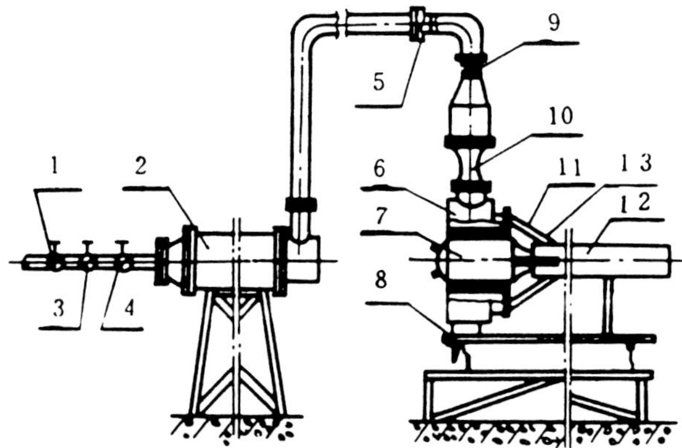

**Figure 2.19** Schematic diagram of test facility (Zhongqin et al. 1986). *Copyright © AIAA 1986. Used with permission.*

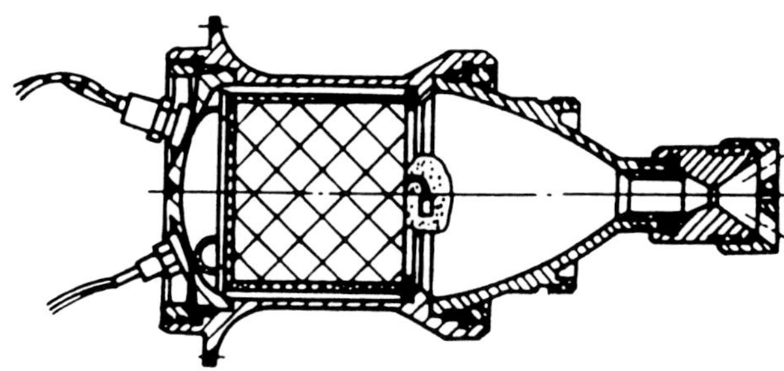

**Figure 2.20** Solid propellant motor (Zhongqin et al. 1986). *Copyright © AIAA 1986. Used with permission.*

**Table 2.2** Properties of the propellant and the combustion products (Zhongqin et al. 1986). *Copyright © AIAA 1986. Used with permission.*

| | |
|---|---|
| *Propellant* | |
| HTPB, % by wt. | 17 |
| AP, % by wt. | 50 |
| Al, % by wt. | 20 |
| Mg, % by wt. | 3 |
| Heating value $Hu$ (of complete combustion), KJ/kg | 18,225 |
| Heating value $Q_p$ (in primary rocket), KJ/kg | 3,638 |
| Burn rate $r$, mm/s | 5–7 |
| *Gas generator* | |
| Run duration $t$, seconds | 19–24 |
| Pressure $P_p$, MPa | 7.0924 |
| Flow rate $m_p$, kg/s | 0.40 |
| *Combustion chemistry* | |
| Flame temperature $T_p$*, K | 2400 |
| Specific heat $c_p$, KJ/kgK | 1.795 |
| Specific heat ratio $k$ | 1.24 |
| Gas constant $R$, KJ/kg K | 0.3506 |
| Characteristic velocity $C^*$, m/s | 1400 |
| Stoichiometric A/F ratio $L_0$ | 2.812 |

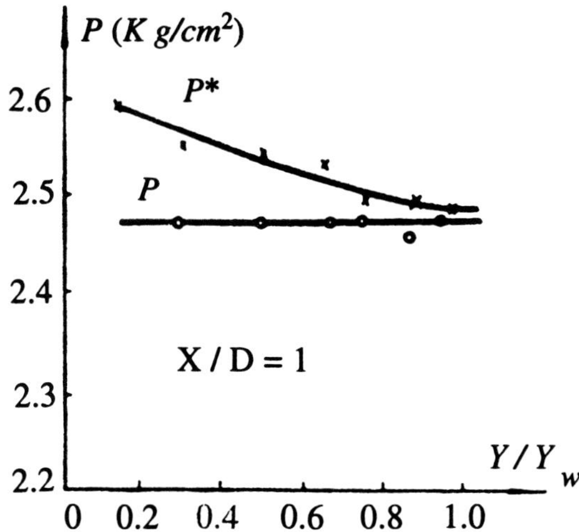

**Figure 2.21**  Radial secondary flow profiles of static and total pressure (Zhongqin et al. 1986). *Copyright © AIAA 1986. Used with permission.*

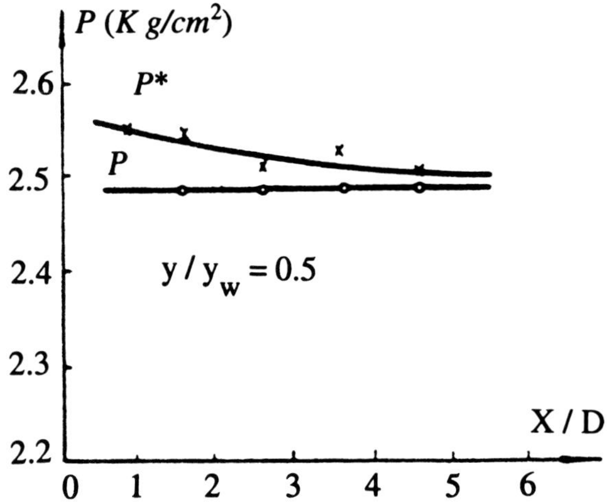

**Figure 2.22**  Axial secondary flow profile of static and total pressure (Zhongqin et al. 1986). *Copyright © AIAA 1986. Used with permission.*

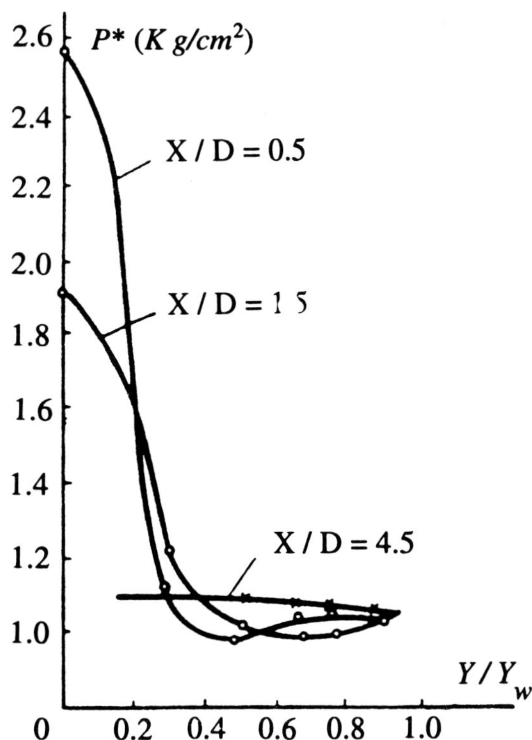

**Figure 2.23**  Static pressure profiles of primary air stream (Zhongqin et al. 1986). *Copyright © AIAA 1986. Used with permission.*

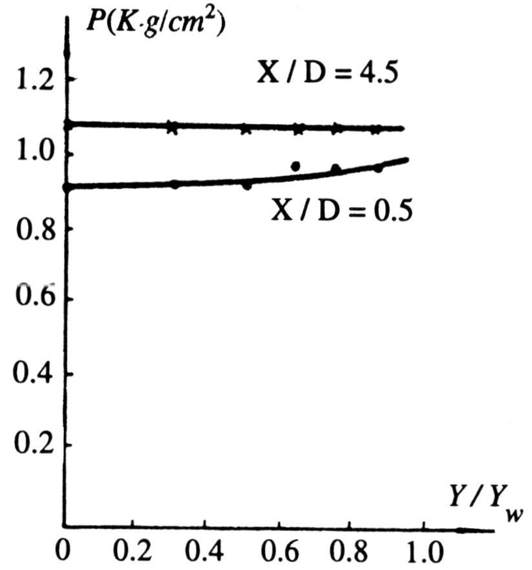

**Figure 2.24**  Total axial pressure profiles of primary air stream (Zhongqin et al. 1986). *Copyright © AIAA 1986. Used with permission.*

Note the large amount of metal, the high pressure of the gas generator, and the relatively long burning time. Cold flow tests were first performed to determine mixing characteristics for a primary air jet and a nonparallel (40° injector angle) four-inlet, secondary air stream. Four primary air nozzles were used: namely, a convergent (sonic) nozzle, two supersonic (15° and 30°) nozzles, and one having thirteen orifices. Figures 2.21 and 2.22 show radial and axial secondary pressure profiles. Without primary flow the static pressure remains constant, so it appears that the secondary flow is parallel near the exit plane. Pressure measurements in the primary flow showed that there too the static pressure was constant near the exit and increased slightly with radial distance near the front end. The total pressure decreased rapidly near the front and was lower along the chamber, becoming constant near the exit (see Figures 2.23 and 2.24).

The effect of the nozzle configuration on the mixing is shown in Figures 2.25 and 2.26. The core velocity at X/D = 1 was much higher (350 m/s) than near the well (80 m/s). At X/D = 5.8 the flow was uniform. Neither a supersonic central flow nor a nozzle with four 8 mm parallel jets improved the performance. With four parallel jets the central flow ve-

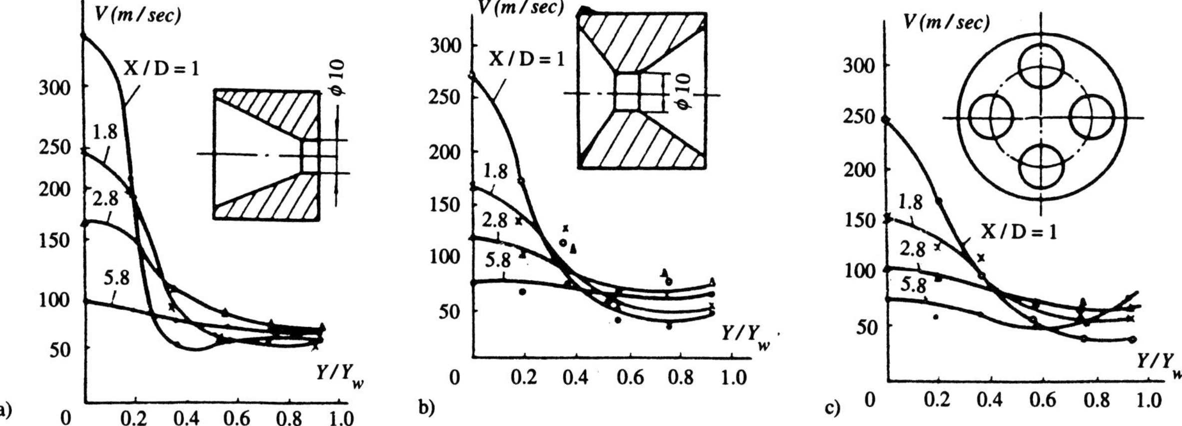

**Figure 2.25**  Velocity profiles of (a) sonic nozzle, (b) supersonic nozzle, and (c) nozzle with four parallel jets (Zhongqin et al. 1986). *Copyright © AIAA 1986. Used with permission.*

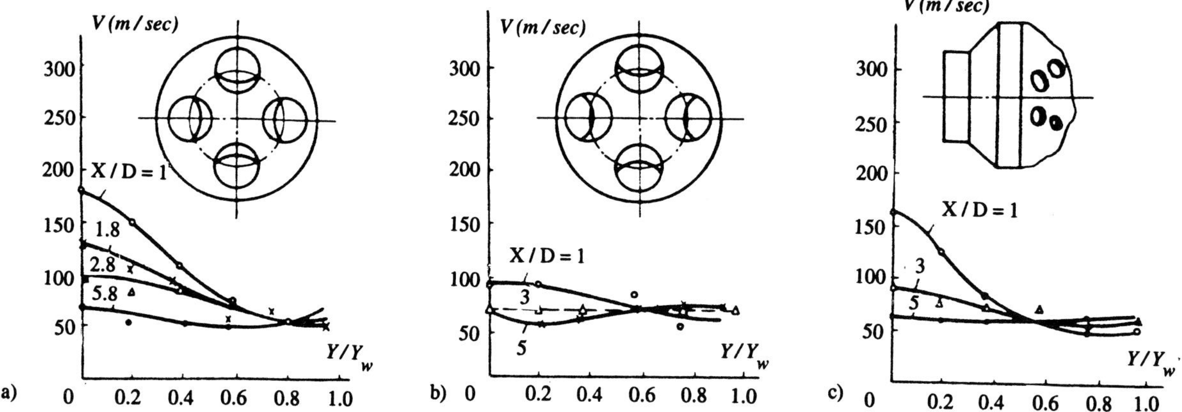

**Figure 2.26**  Velocity of nozzles with (a) four jets at 15° angle, (b) four jets at 30° angle, and (c) 13 orifice injection head (Zhongqin et al. 1986). *Copyright © AIAA 1986. Used with permission.*

locity was 250 m/s and the outer flow 60 m/s. Figure 2.26 shows that four jets at 15° and thirteen jets have similar velocity distributions; at X/D = 5.8 and 5 the flow is almost uniform. From the standpoint of mixing rate, the 30° injector is the best with very uniform flow between 50 and 100 m/sec.

During hot runs it was found that a large injection angle brings the hot gases into direct contact with the duct wall. Hence an angle of 15° is a good tradeoff. It is also important to allow enough time for the reaction between the fuel rich gases and the air to go to completion; therefore, the primary flow velocity should be moderate.

In order to have efficient combustion, the aluminium agglomerates present in the exhaust of this primary flow must be burned in the secondary chamber. A temperature above 2300 K is also required to ensure the melting of the $Al_2O_3$ particles. The ignition delay time should be shorter than the particle residence time; this is achieved by using a subsonic nozzle. The experiments also showed that using a multijet

nozzle raised the combustion efficiency by about 30%, probably because less slag remains in the chamber, while the $I_{sp}$ increases by as much as 8%. It is, however, known that the $I_{sp}$ of the SAM 6, the only known ducted rocket employed operationally, is between 400 and 500 sec. This rocket burns a mixture of magnesium and sodium nitrate, which is easy to ignite and produces relatively high thrust. This depends on the amount of heat evolved, $fQ_r$ (where $f$ is the stoichiometric ratio and $Q_r$ the heat content). As is shown below, the amount of heat evolved is higher for Mg than for hydrocarbons (HC):

For HC: $Q_r \sim 10$ Kcal/g    $f \sim 0.067$,   so $fQ_r \sim 670$ cal/g

For Mg: $Q_r \sim 5.7$ Kcal/g    $f \sim 0.35$,    so $fQ_r \sim 2100$ cal/g

Magnesium is also easy to ignite. With optimal composition the $I_{sp}$ could be much higher (close to 1000 sec).

Another advantage of ducted rockets is the ease of controlling the thrust, either by using a pintle or by secondary injection.

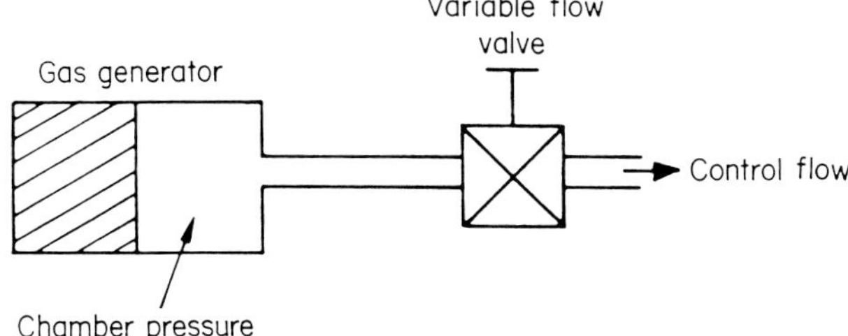



**Figure 2.27**  Schematic of a modulation system with mechanical valves (Timnat 1987).

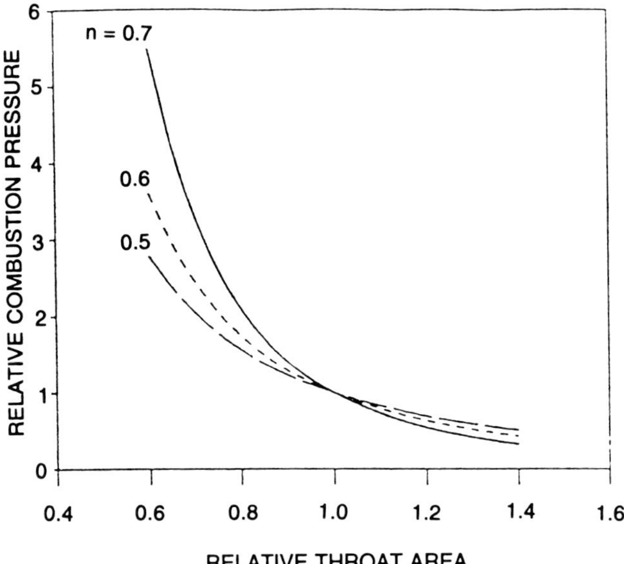



**Figure 2.28**  Relative pressure vs. relative exit area with $n$ as a parameter (Goldman 1991).

### 2.3.2.1. Thrust Magnitude Control

We shall describe first a method for controlling the fuel flow rate by changing the nozzle throat area. A schematic of such a method is shown in Figure 2.27. As one diminishes the throat area of the gas generator, the pressure goes up and since the burning rate depends on the pressure, the fuel flow rate also augments. This goes on until a new equilibrium is reached. From the basic equations one obtains

$$\frac{p_{c2}}{p_{c1}} = \left(\frac{A_{t_1}}{A_{t_2}}\right)^{\frac{1}{1-n}} \quad \text{and} \quad \frac{m_2}{m_1} = \left(\frac{A_{t_1}}{A_{t_2}}\right)^{\frac{n}{1-n}} \quad (2.25)$$

These results are shown in Figures 2.28 and 2.29, with $n$ as a parameter. Clearly a higher exponent $n$ has a stronger effect on the pressure and the mass flow rate, as shown.

Sayles (1975) developed a controllable solid rocket motor using a pintle. This is also described by Timnat (1987). Although he worked with a solid rocket motor, his results are applicable to a gas generator.

It is clear that a pintle arrangement has many drawbacks,

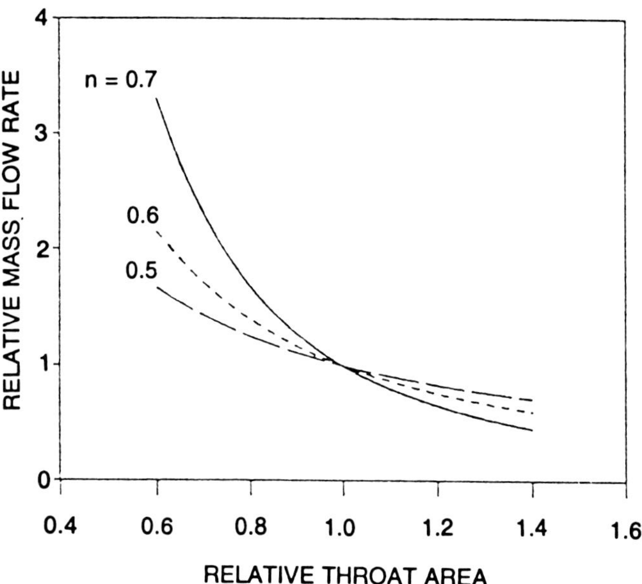



**Figure 2.29**  Relative mass flow rate exhausted from the gas generator vs. relative exit area with $n$ as a parameter (Goldman 1991).

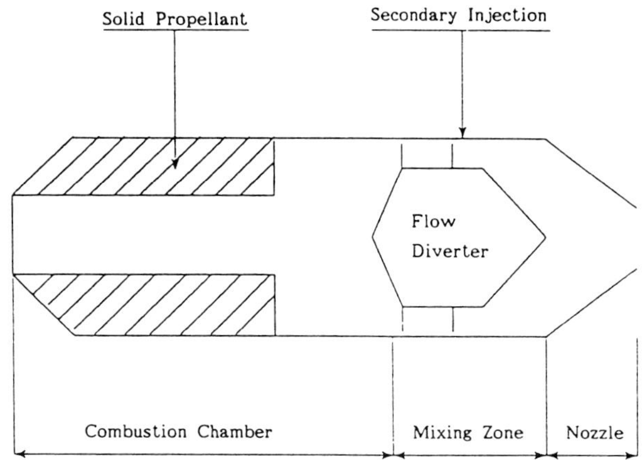



**Figure 2.30**  Vortex valve (Goldman 1991).

the principal being the high temperature that the pintle and the control system must withstand. Therefore, it is desirable to have an alternative method for thrust modulation such as the introduction of a secondary swirling flow, in which the change of the effective throat is achieved by aerodynamic and not mechanical means. This is accomplished by introducing a vortex valve, shown schematically in Figure 2.30. This was studied in the 1960s by Mager (1961), King (1967), and Norton et al., (1969). At the Technion, Greenberg and Wolff (1975) started the study in the mid 1970s, while Gany and coworkers (Natan et al. 1982; Goldman and Gany 1992) continued it.

I want to dwell in particular on Goldman's work (1991 and 1992), in which a new model for the flow rate modulation in a ram-rocket was developed and experiments were performed to validate it. In the experiments a gas generator

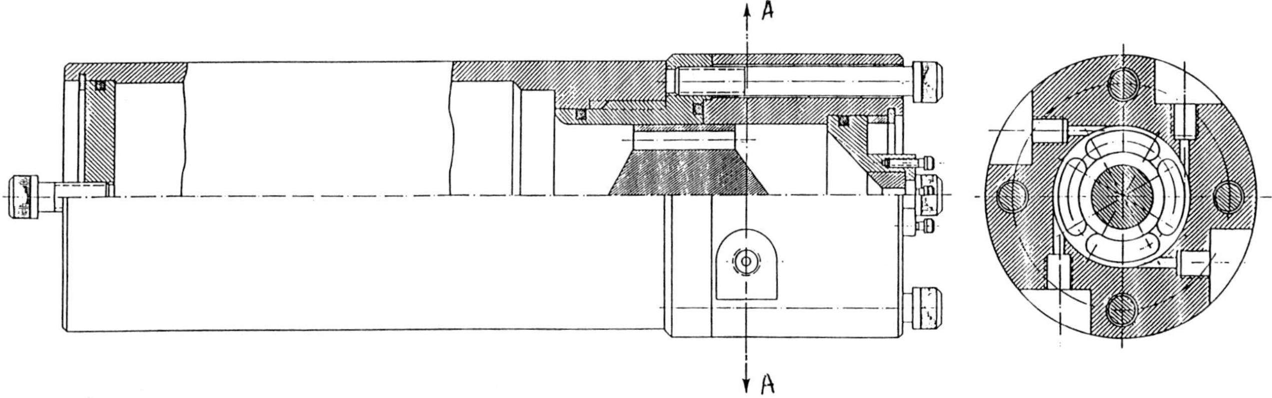

**Figure 2.31**   Schematic engine drawing (Goldman 1991).

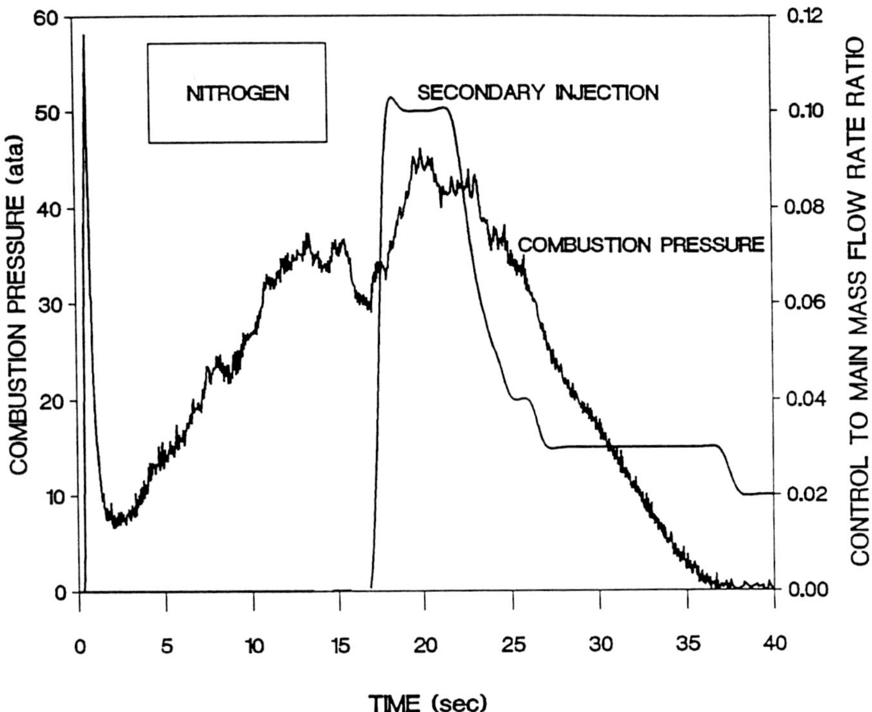


**Figure 2.32**   Representative experimental results with nitrogen as the secondary gas
(Goldman 1991).

consisting of 60% polybutadiene-acrylo-nitrile (PBAN), 20% ammonium perchlorate, and 20% potassium perchlorate was used. This is a propellant with a low specific impulse (160 sec), a low burning temperature (1300 K), having a burning rate law $r = 0.48\ p_c^{0.28}$ (MPa). A schematic of the engine is shown in Figure 2.31, while a representative experimental result is displayed in Figure 2.32. One can clearly see the influence of the nitrogen secondary injection on the pressure, which increases upon injection by about 40%. Figure 2.33 shows the influence of various gases (the lines are computed results). The amount of nitrogen required is somewhat larger than that of methane and much larger than for hydrogen. On the other hand, nitrogen gives

the best agreement with theory as shown for both fuel flow (Figure 2.34) and thrust modulation (Figure 2.35).

## 2.4 Ramjet Applications

The United States was among the first to use ramjets for operational application, but stopped employing them relatively early; the U.S. program on ramrockets started relatively late but little use has been made of them. The British were also among the early developers, and their Bloodhound II and Sea Dart were still in service in 1980. It appears that the Russians and the French have brought ramjets and ramrockets to their most extensive operational use,

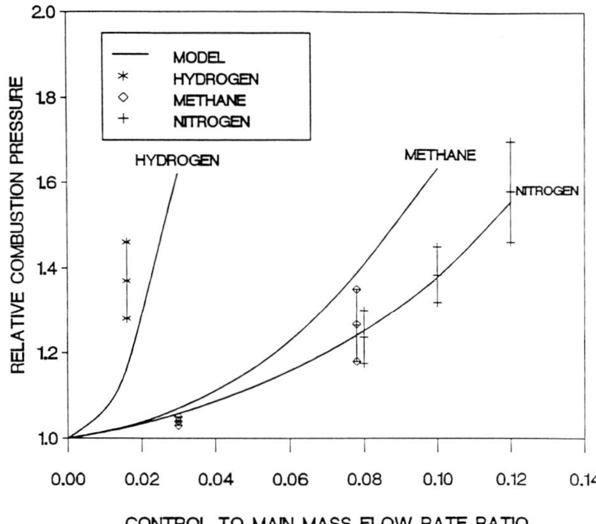



**Figure 2.33** Relative pressure as a function of control to main flow rate for different gases (Goldman 1991).

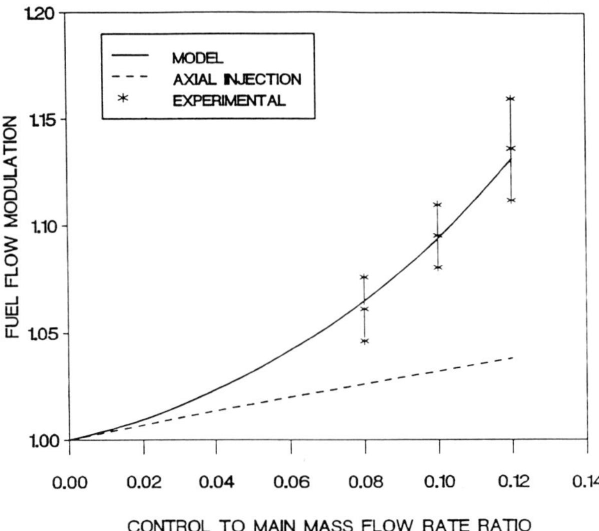



**Figure 2.34** Secondary flow influence on mass flow rate from gas generator—nitrogen injection (Goldman 1991).

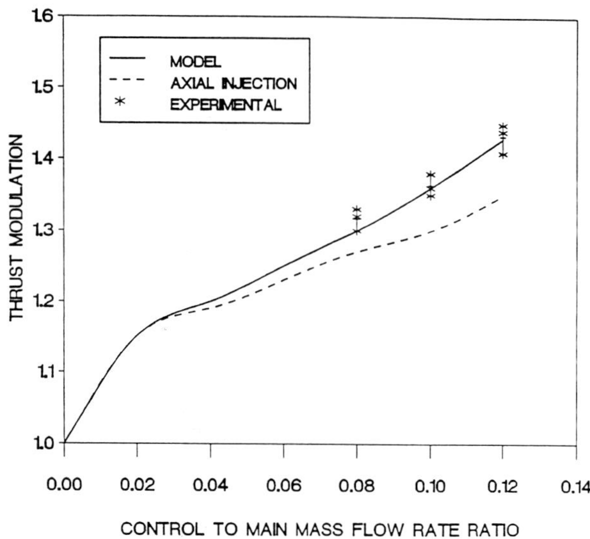



**Figure 2.35** Thrust modulation as function of control to main flow rate—nitrogen injection (Goldman 1991).

and they are continuing to develop and utilize them (especially ramrockets). The Chinese, who entered the field late, have two supersonic, sea-skimming anti-ship missiles—the C101 and C103, with ranges of 50 and 100 km respectively. Both are powered by liquid-fueled ramjet engines. Taiwan is also developing a medium/high altitude ground-to-air missile, using ramjet propulsion to achieve a range of more than 100 km. Like the C101 and C103, this is most likely also a ramrocket.

The USSR pioneered a number of operational applications of the ramjet. By the early 1960s the Soviets had introduced the surface to air SA-4-Ganef missile, propelled by four jettisonable solid-propellant booster rockets and one liquid fuel ramjet. The SA-4 has been produced in four versions. The last one, an 1800 kg missile designated SA-4B, is 900 mm in diameter and 8.3 m long. It has a range of 9.3 to 50 km and reaches Mach 2.5 at altitudes from 1.1 to 24 km. The configuration of the SA-4 is dictated by its propulsion system, consisting of a large diameter cylinder centered on the ramjet sustainer and accommodating the kerosene fuel in its inside walls. There is an inlet centerbody for the warhead and the electronics, and it is equipped with control fins. The four solid boosters are attached to the outside of the body and they detach after the missile has accelerated to ramjet speed, when the sustainer is ignited.

The SA-6-Gainful surface-to-air missile entered service in the mid 1960s and was the first operational ramrocket. With a 355 mm diameter, length of 5.8 m, and weight of 550 kg, it has a range of 4 to 30 km at low altitude or 4 to 60 km at high altitude, the limits being 0.1 to 12 km. The missile is shown in Figure 2.36. It has four inlets for the ramjet set at 45° to the wings; they are at a small distance from the body to keep them clear from the turbulent, slow-moving boundary layer of air close to the body. The solid-propellant booster accelerates the missile to Mach 1.5; after burnout its small diameter nozzle is jettisoned. The empty envelope becomes a ramjet, with air fed through the four inlets to mix with the hot gases produced by a solid-fuel generator. The resultant mixture is burned and exhausted through the now exposed larger nozzle, accelerating the missile to Mach 2.8. As already mentioned this missile was used very effectively by Egypt and Syria against Israeli attack and close support aircraft in 1973.

The French contribution to the field, which spans a period of 35 years, is described very lucidly in a paper by Marguet (1989). Since 1986 France has possessed an air-launched supersonic missile propelled by ramjet; it is capa-

**Figure 2.36**  The SA-6 missile (Chant 1989).

**Table 2.3**  History of ramjet efforts in France (Marguet 1989). *Copyright © AIAA 1989. Used with permission.*

| Period | National activity | Mission |
|---|---|---|
| 1955–1965 (separated booster) | Leduc (airplanes) Griffon | Experimental M < 2 Flights test, *but not operating* M < 5 |
| | Vega, CT 41, Matra 431 Statex, Stataltex (missiles) | |
| 1965–1974 (separated booster | (Hypersonic period) SCORPION (missiles) ESOPE | Experimental no flight tests M < 7 |
| 1974–1989 (separated booster) | – ASMP, ANS – Chocked ducted Rocket – Unchocked Ducted Rocket (Rustic) | Flight tests *Operating* and Experimental |

ble of a range of several hundred kilometers, is highly maneuverable, and can be operated at very low as well as high altitudes. This is the ASMP (Air-Sol Moyenne Portée-medium range air to ground missile), carried by the Mirage IV, the Mirage 2000, and the Super-Etendard military aircraft. Table 2.3 describes the history of the ramjet program in France.

The Scorpion surface to air missile designed to fly at an altitude of 30 km and a Mach number of 6 (Laruelle 1985 and 1987) was qualified on the ground. The study was ended by tests simulating Mach 6, reaching a maximum pressure

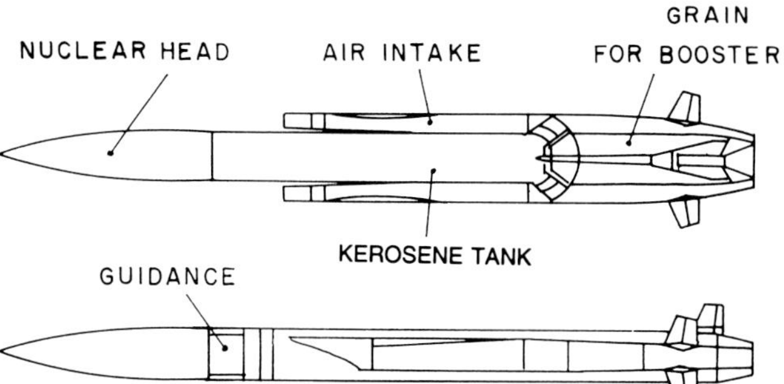


**Figure 2.37**  The ASMP missile (Marguet 1989). *Copyright © AIAA 1989. Used with permission.*

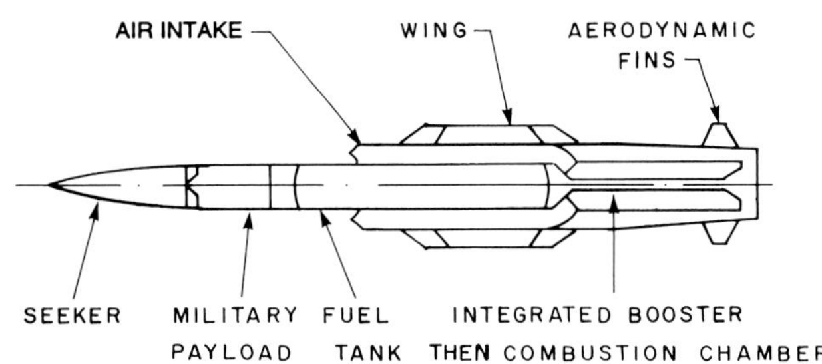


**Figure 2.38**  ANS cutaway (Marguet 1989). *Copyright © AIAA 1989. Used with permission.*

of 2.5 MPa and a temperature of 1650 K, in the hypersonic blow down wind tunnel S4 at Modane (see Figure 3.23).

The ASMP (see Figure 2.37) is a liquid fuel ramjet, burning kerosene, with an integrated solid fuel booster. The flight sequence starts with a boost phase, lasting about 4 seconds, which takes the missile up to speeds close to Mach 2; the cruise thrust can be modulated by a factor of approximately 15 through variation of the fuel feed rate. The weight is about one ton and the caliber 450 mm. At low altitudes the missile can fly at Mach 2 for over a hundred kilometers, while maximum range is achieved at altitudes above 20 km.

Another liquid fueled missile is the ANS (Anti Navire Supersonique or supersonic antiship), which, in cooperation with West Germany, is in the final development stage. In this case also the propulsion unit is a ramjet with integrated booster, but here a synthetic, high density hydrocarbon fuel is employed. The configuration, shown in Figure 2.38, has four circular air intakes. The ANS can accomplish a variety of missions, both sea and air launched; it has a minimum range of 6 km, reaching a maximum of 250 km (see Figure 2.39).

For shorter ranges the French developed a ramrocket, named "Rustic," whose advantage lies in simplicity of design, storage, and operational use. The fuel gas generator is of the unchoked type, its operation depending on the pressure inside the ram combustor. An appropriate solid fuel follows regression laws that are a function of the pressure in the combustor, which in turn depends on altitude and flight

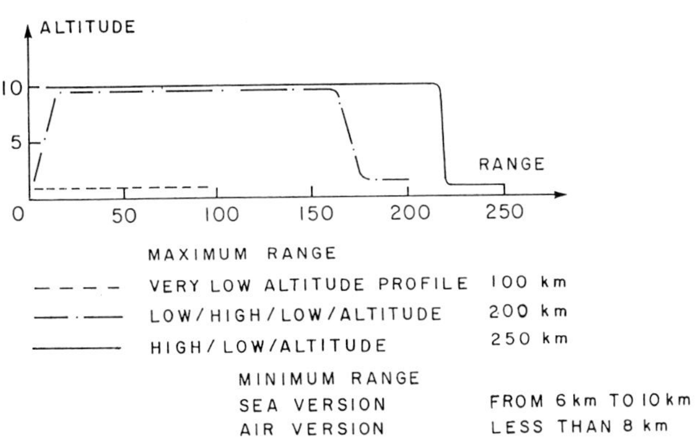



**Figure 2.39** The ANS missions (Marguet 1989). *Copyright © AIAA 1989. Used with permission.*

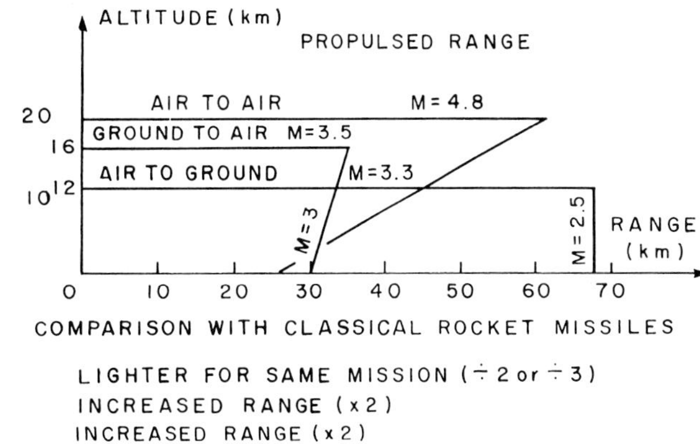



**Figure 2.40** "Rustic" ducted rocket applications (Marguet 1989). *Copyright © AIAA 1989. Used with permission.*

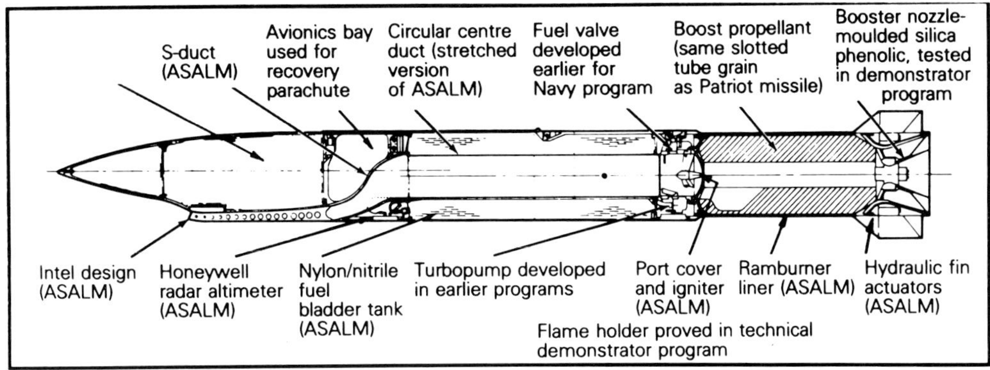



**Figure 2.41** Sectioned Martin Marietta YAQM-127A (Wanstall 1984). *Copyright © AIAA 1984. Used with permission.*

speed. This controls the fuel flow and regulates the ram thrust, according to flight conditions, as shown in Figure 2.40. A variety of missions are planned for the "Rustic": air to air, ground to air, and air to ground, covering a range up to 70 km and speeds up to Mach 4.8.

A further French development is the Aster 30 ramrocket which uses a liquid ramjet and has a range of over 25 km. It is planned in two versions: surface to air and ship to air.

In the late 1970s, the United States developed an advanced strategic air-launched missile (ASALM), which used liquid fuel and demonstrated high maneuverability during a successful seven-flight test program in 1979–80 when it reached Mach 4 at over 30 km height. The program was, however, terminated for fiscal reasons. Most of the components developed for this program were employed in a supersonic low altitude target, YAQM-127A, developed by Martin Marietta for the Navy (see Figure 2.41). This missile has a Mach 2.5 capability and an air launched range of over 100 km.

# Chapter 3

# *The Scramjet*

In order to avoid the effects of dissociation caused by the high temperature increase due to the strong deceleration in the subsonic combustor ramjet, attention was focused in the late fifties (Weber and McKay 1958) on the possibility of performing the combustion at supersonic speed.

The theoretical and experimental basis was laid mainly in the United States and England. The most important contributions came from Ferri (1964 and 1968), the Applied Physics Laboratory, NASA—Langley, the Marquardt company, and Swithenbank (1966).

The basic components of the scramjet (see Figure 3.1) are the air inlet, the combustor, and the nozzle. The air velocity is reduced in the inlet by a few percent only, while the static pressure increases by one to two orders of magnitude; after this, fuel is injected and burned with air in the combustor, thus creating supersonic combustion which accelerates the products to a speed slightly higher than the flight speed, while the exit pressure is just above the free stream static pressure. The net thrust of the engine comes from the increase in gas velocity and is transmitted as pressure on the engine surface. The flow through this type of engine is supersonic; although the absolute speed does not change much, the variation in Mach number is significant (up to an order of magnitude) because of the high rise in static temperature. Since the flow speed in the combustor is close to the flight speed, it is possible to approximate the dwell time by the quotient of the combustor length by the flight velocity. The thrust can be estimated by remembering that the exit momentum is equal to the inlet momentum with an increase due to the fuel enthalpy. As a consequence the efficiency of the diffuser and the nozzle will be better than in the ramjet, where the changes in velocity are high.

In order to determine the velocity range of interest for such an engine, one must consider the possible applications. These can be divided into boost and sustain missions; the upper limit for boost missions is the satellite velocity, while the upper limit for a cruise mission might be taken as the speed that would allow a range of about half the earth circumference (about 20,000 km). We shall, therefore, consider a speed range between Mach 5 and Mach 25. It is clear that a propulsion system using a scramjet must include means of reaching Mach 5, such as rockets, turbojets, ramjets or a combination of them.

There have been many theoretical investigations on this type of propulsion; a comprehensive review of computational fluid dynamic procedures for analyzing hypersonic propulsion was published recently by Barber and Cox (1989). The possibility of supersonic and hypersonic combustion was proven experimentally and the main problems treated today deal with the development of new materials and techniques that can withstand the high temperature and stresses caused by such velocities.

## 3.1. Theoretical Cycle Analysis

An upper limit for the performance available can be obtained from energy considerations. The potential and kinetic energy of one kg mass in orbit is one fifth of the chemical energy contained in one kg of hydrogen (the preferred fuel for scramjet applications). Thus the upper limit for the weight in orbit will be 80% of the launch weight, assuming complete release of energy from the hydrogen by combustion with the air and no drag.

In order to calculate the specific impulse as a function of flight velocity one must include also the kinetic energy of the fuel. Thus the power derived from the chemical energy

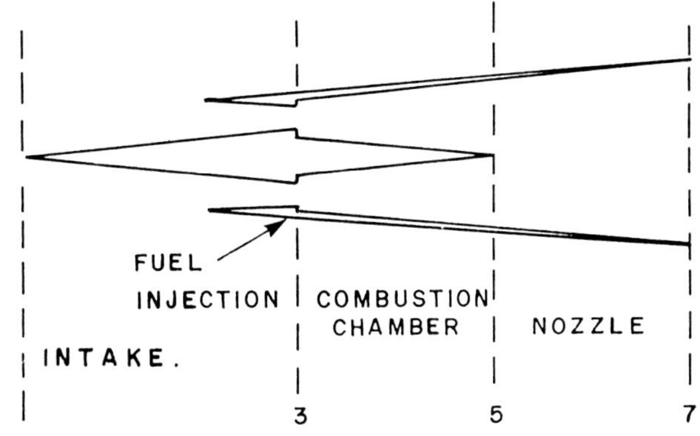


**Figure 3.1**  The basic components of a scramjet (Swithenbank 1966). *Reprinted with permission from Pergamon Press Ltd.*

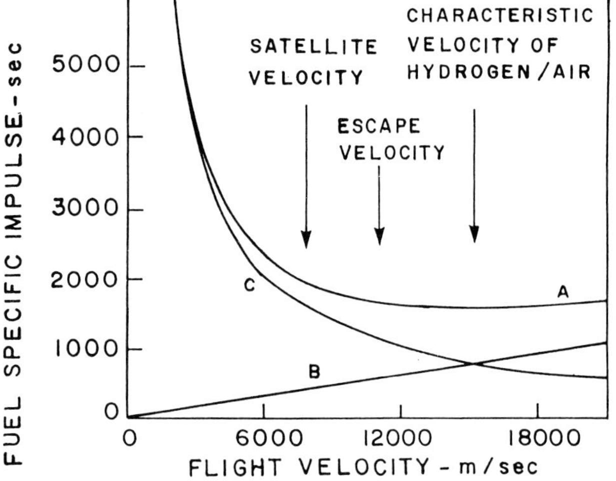



A MAXIMUM POSSIBLE  SPECIFIC IMPULSE
B IMPULSE DUE TO KINETIC ENERGY OF FUEL
C IMPULSE DUE TO CHEMICAL ENERGY
   OF HYDROGEN

**Figure 3.2** Maximum specific impluse of hydrogen vs. flight velocity (Swithenbank 1966). *Reprinted with permission from Pergamon Press Ltd.*

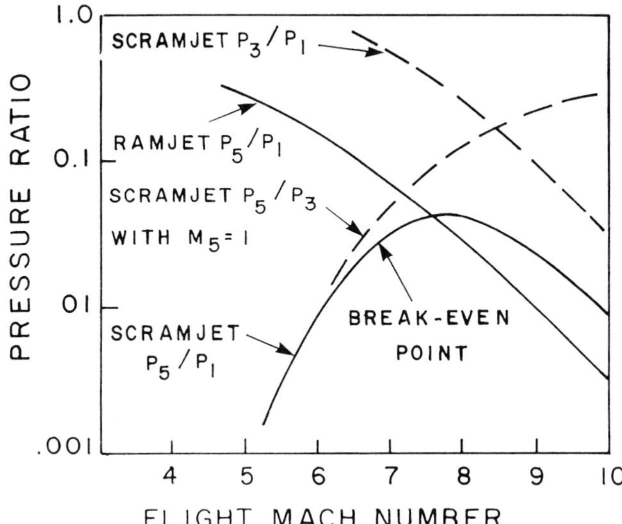



**Figure 3.3** Simplified treatment of ramjet-scramjet break-even point (Swithenbank 1966). *Reprinted with permission from Pergamon Press Ltd.*

is (see Swithenbank 1966)

$$T_C V_1 = g \, m_f \, h_f \tag{3.1a}$$

while that derived from the kinetic energy is

$$T_K V_1 = \frac{1}{2} m_f V_1^2 \tag{3.1b}$$

Therefore the maximum thrust is

$$T_{max} = T_C + T_K = m_f[(gh_f/V_1) + (V_1/2)] \tag{3.2}$$

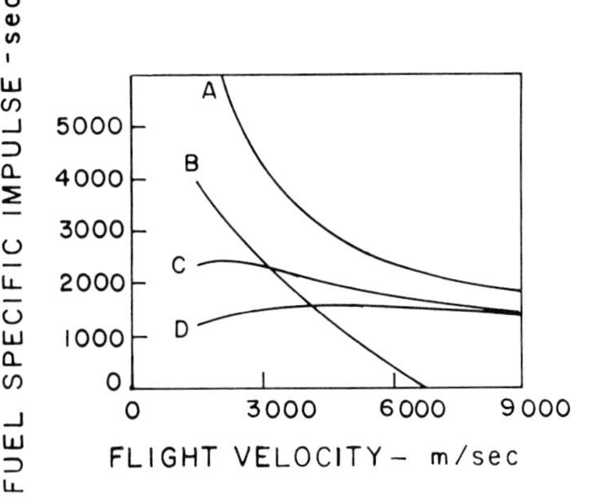



A MAXIMUM SPECIFIC IMPLUSE
B IMPULSE WITH SUBSONIC COMBUSTION
C IMPULSE WITH SUPERSONIC COMBUSTION ($V_3 / V_1 = .90$)
D IMPULSE WITH SUPERSONIC COMBUSTION ($V_3 / V_1 = .95$)

**Figure 3.4** Specific impulse from simplified equations with subsonic and supersonic combustion (Swithenbank 1966). *Reprinted with permission from Pergamon Press Ltd.*

and the maximum specific impulse is

$$I_{sp\,max} = T_{max}/gm_f = h_f/V_1 + V_1/2g . \tag{3.3}$$

Figure 3.2 shows that for hypersonic velocities the kinetic energy contribution is significant. In the speed range between 1700 and 7800 m/s the ideal specific impulse varies between 700 and 2000 seconds as compared to a maximum of 500 s for chemical rockets.

The main advantages of hydrogen as fuel are its low molecular weight (one should remember that $I_{sp} \sim M^{-1/2}$), the high cooling capability of liquid hydrogen, its thermal stability, and its high reactivity, which allows the use of a shorter combustor length. The main drawbacks are handling and storage problems, since cryogenic facilities are required (although suitable techniques developed for liquid rockets are available) and its low density, which demands larger fuel tanks and consequently larger flight vehicles.

In order to compare supersonic and subsonic combustion one can make a number of simplifying assumptions (Swithenbank 1966): constant area combustion chamber, nozzle entry at Mach 1, an inlet in which supersonic combustor entry conditions are achieved by terminating the diffusion at the appropriate point and subsonic conditions are obtained by continuing diffusion to subsonic velocities. Furthermore, one assumes flow with heat addition corresponding to hydrogen treated as an ideal gas and compares performance with the aid of the product of stagnation pressures $(p_{01}/p_{03})(p_{03}/p_{05})$. Figure 3.3 shows the results of such a comparison, giving a break-even point slightly above Mach 7; more accurate calculations show that this comes at a lower speed between Mach 5 and 6. Results of similar calculations for the specific impulse are presented in Figure 3.4. Here the inlet efficiency is assumed to be 0.92 for sub-

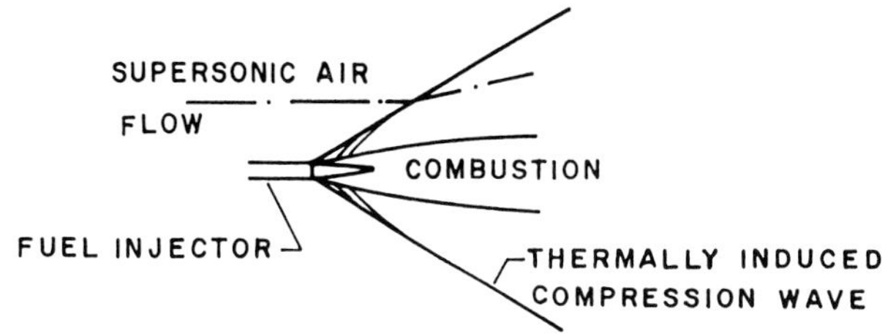


**Figure 3.5** Sketch of supersonic combustion flowfield (Ferri 1968). *Copyright © AIAA 1968. Reprinted with permission.*

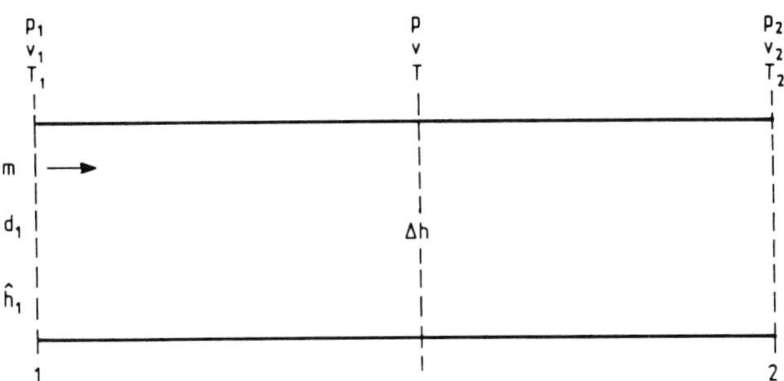

**Figure 3.6** Cylindrical combustion chamber (Barrère 1988). *Courtesy L'Aéronautique et L'Astronautique.*

sonic combustion, and 0.96 for the supersonic case at 1700 m/s, reaching 0.995 at 9000 m/s; the respective values of the nozzle efficiency $\eta_N$ are 0.95 and 0.99. Two cases are shown: $V_3/V_1 = 0.9$ and 0.95. The main conclusions are that below Mach 4 scramjet performance is worse, because of higher losses due to heat addition to the supersonic combustion; between Mach 4 and 8 one can use both subsonic and supersonic combustion according to the mission (as already mentioned, more accurate calculations show that above Mach 5 supersonic combustion is preferable). Between Mach 8 and 18 good performance is expected from the scramjet, while above Mach 18 friction losses become high and scramjet operation becomes doubtful.

Experimental work has shown that it is easier to maintain supersonic combustion at constant pressure, using a divergent combustor. The flow velocity then remains approximately constant. Calculations show that best results are obtained for $M_1/M_3 \approx 3$ with $T_3$ between 1000 and 2000 K. For these conditions, one has spontaneous ignition and the dissociation of the combustion products is minimal. For $M < 10$ a fuel-lean mixture gives the best results, while above Mach 10 a fuel-rich mixture is preferable.

## 3.2 Supersonic and Hypersonic Combustion

The combustor is probably the most complex component of a supersonic engine; therefore, great efforts were made as early as the 1960s in order to gain a basic understanding of the phenomena involved and considerable progress was achieved (Ferri 1968).

The fundamental problem of supersonic combustion is that it may cause a considerable pressure rise, even bringing about choking of the flow. As a consequence of the local injection of the flow, the combustion is controlled by mixing; therefore, it starts near the injection zone and transmits waves in the air outside of the combustion region (see Figure 3.5).

The principles of supersonic combustion are first analyzed following Barrère (1988), after which three different types of models are discussed, using the methods of Billig (1988), who treats the entire spectrum above Mach 1. The section is concluded with a brief review of what Stalker calls hypersonic combustion, which occurs when the flight Mach number is so high that the combustor entrance airflow is above Mach 5.

### 3.2.1 Principles of Supersonic Combustion

Barrère (1988) analyzed supersonic combustion by considering a cylindrical chamber, shown in Figure 3.6 with entrance section 1 and exit section 2; he wrote the three conservation equations together with the equation of state and utilizes von Karman's representation, defining the energy added, $\Delta h$, through the variable $x = a(\hat{h}_1 + \Delta h)$, the velocity $v$ by $y = bv$, the temperature $T$ by $z = cT$ and the pressure $p$ by $p = (p_1 + mv_1)\,[1 - (\gamma/\gamma + 1)y]$. Here $a$, $b$, $c$ are constants, depending on the entrance conditions (mass flow $m_1$, pressure $p_1$, velocity $v_1$ and specific heat ratio $\gamma$ assumed constant) given by

$$a = \frac{\gamma^2 - 1}{\gamma^2}\left(\frac{m}{p_1 + mv_1}\right)^2;$$

$$b = \frac{\gamma + 1}{\gamma}\,\frac{m}{p_1 + mv_1};$$

$$c = \frac{\gamma - 1}{\gamma}\left(\frac{m}{p_1 + mv_1}\right)^2 c_p T$$

Figure 3.7 presents the change of $y$, $z$, and $p$ as function of $x$; $y(x)$ is a parabola with the axis parallel to $Ox$. Point $C$ corresponds to $y = 1$, i.e. to a speed $v_c$ equal to the speed of sound $a_c$ and therefore $M_c = 1$. It defines two combustion zones: the subsonic one for $y < 1$, and the supersonic one for $y > 1$. Starting from the chamber entrance $B_1$ with stagnation enthalpy $\hat{h}_1$, the heat addition $\Delta h$ gives the exit conditions at $B_2$ ($h_2$). The velocity and the Mach number are lower due to the added energy; the maximum heat that can be added corresponds to point $C$, giving

$$(\Delta h)_{\max} = \frac{\gamma^2}{2(\gamma^2 - 1)}\left(\frac{p_1 + mv_1}{m}\right)^2 - \hat{h}_1 \qquad (3.4)$$

while point $E$ corresponds to Mach infinity and depends on the limiting velocity $v_L = (p_1 + mv_1)/m$ according to $v_E = (\gamma/\gamma + 1)v_L$.

The temperature, which is represented by the variable $z$,

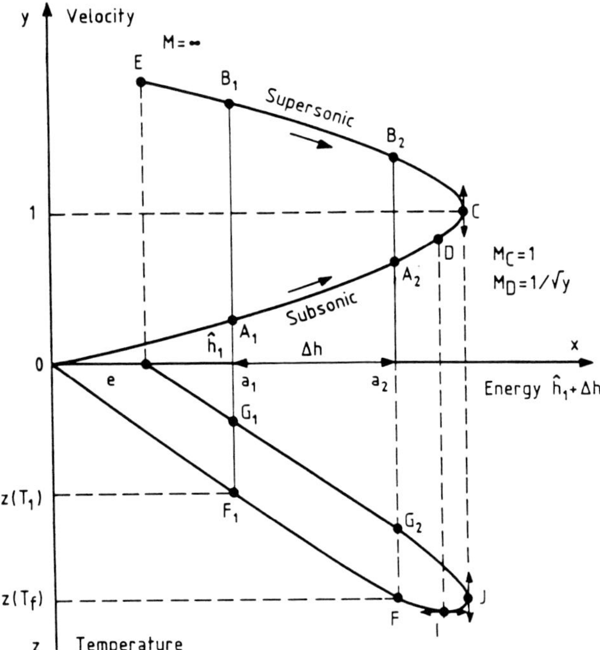

**Figure 3.7** Different types of aerobic propulsion (Barrère 1988). *Courtesy L'Aéronautique et L'Astronautique.*

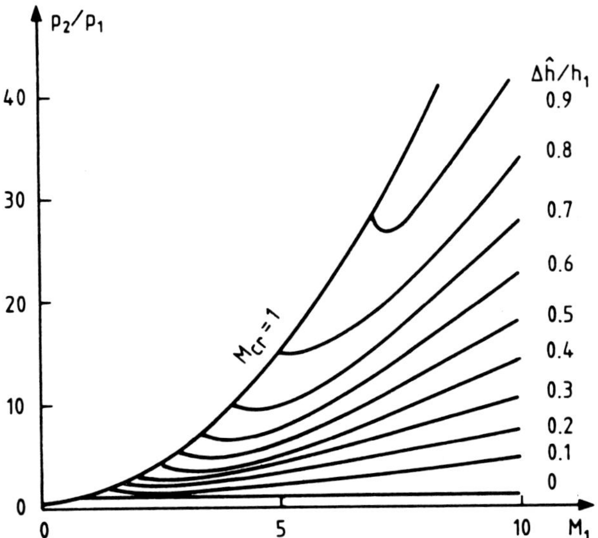

**Figure 3.8** Variation of the ratio of the exit pressure ratio $p_2$ to the entrance pressure $p_1$ as a function of the entry Mach number, $M$, for different values of the ratio of heat addition to the entry stagnation enthalpy, $\Delta h/h_1$ (Barrère 1988). *Courtesy L'Aéronautique et L'Astronautique.*

has two zeros on $Ox$. One corresponds to the subsonic branch, the other to the supersonic one $y = (\gamma + 1)/\gamma$; the Mach number is infinite ($z = T = 0$). The supersonic branch starts from point $G_1$ for the entrance value and increases to $G_2$ at the exit; the energy increase $\Delta h$ augments the static temperature. The highest possible temperature in the supersonic branch is at point $S$, corresponding to the velocity $v_c$ being equal to the sound velocity $a_c$.

The third parameter of interest is the static pressure $p$; as a function of $y$ it corresponds to a straight line going from $z = p_1 + mv_1$ for $y = 0$ to the value $z = 0$ for $y = (\gamma + 1)/\gamma$. For supersonic combustion, i.e. at point $B_1$ (static pressure $p_1$), the pressure goes up to the value $p_2$, corresponding to point $B_2$. Assuming a mean constant $\gamma$, the momentum equation gives the pressure ratio as

$$p_2/p_1 = (1 + \gamma M_1^2)/(1 + \gamma M_2^2) . \qquad (3.5)$$

It is important to express the change of the ratio $p_2/p_1$ as a function of the entrance conditions and the energy addition $\Delta h$ or the quantity $\Delta h/\hat{h}_1$:

$$
\begin{aligned}
\frac{p_2}{p_1} = {} & \frac{1 + \gamma M_1^2}{\gamma + 1} \\
& \pm \left\{ \left[ \frac{\gamma}{\gamma + 1} (1 - M_1^2) \right]^2 \right. \\
& \left. - \frac{2\gamma^2}{\gamma + 1} M_1^2 \left( 1 + \frac{\gamma - 1}{2} M_1^2 \right) \frac{\Delta h}{\hat{h}_1} \right\}^{1/2}
\end{aligned}
\qquad (3.6)
$$

The two roots correspond to the subsonic and supersonic solutions; the last one is shown in Figure 3.8, which presents $p_2/p_1$ as a function of the entrance Mach number $M_1$ for different values of $\Delta h/\hat{h}_1$. Increasing $\Delta h/\hat{h}_1$ for fixed $M_1$ increases the static pressure ratio $p_2/p_1$ up to a critical value, $M_2 = 1$.

This simple analysis brings out the main characteristics of supersonic combustion; namely, heat addition diminishes the velocity and the Mach number, while increasing the static temperature and pressure. In order to obtain a more realistic quantitative evaluation of supersonic combustion, one must go to numerical methods which are described in the next section.

### 3.2.2 Calculation Methods

A thorough discussion of the combustion process in supersonic flow is given by Billig (1988). He distinguishes the following three types of models for describing the processes occurring in diabatic flows in ducts having supersonic entry conditions: integral techniques, finite difference methods, and exact two-dimensional planar flame models based on instantaneous heat release.

Figure 3.9 presents the main features of the flow field, for developing the integral and finite difference approaches. The blockage caused by the injection and the heat release generates a "shock train" disturbance, starting upstream of the fuel injectors and extending into the combustor. At moderate combustion inlet Mach numbers, e.g. $M = 1.5$ to $4$, corresponding to flight Mach numbers $M = 6$ to $11$ for typical heat release and fuel equivalence ratios, the pressure rise associated with the shock train is strong enough to separate the incoming boundary layer. The length $S$ of the shock train is given by the S-shaped pressure rise, the distance that

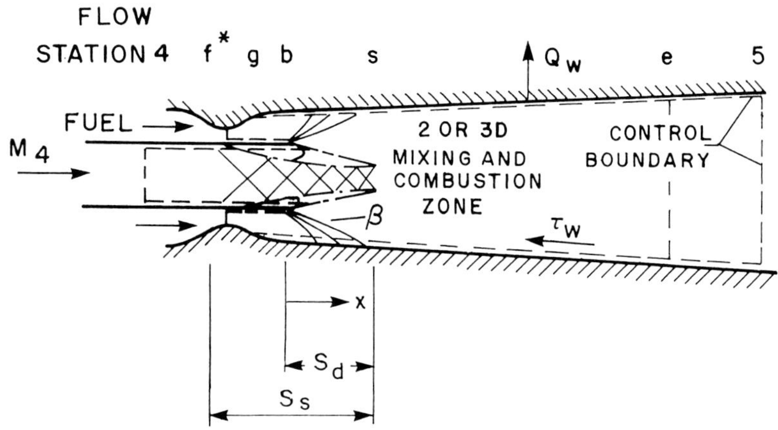



**Figure 3.9**  Flow processes in a combustor with supersonic diffusion flame (Billig 1988). *Copyright © AIAA 1988. Used with permission.*

the shock train extends into the combustor is $S_d$. Empirical relations for $S$ and $S_d$ have been obtained from experiments. As flight speed increases, $M_4$ increases, the pressure rise in the shock train decreases, and the separated zone may be absent.

Downstream of the shock train there is intense mixing and combustion, with large gradients in flow properties and chemical composition in the axial, radial, and possibly circumferential directions. This region, extending from station $s$ to station $e$ is labeled as a "2 or 3D mixing and combustion zone"; from station $e$ to station 5, mixing and combustion are less intensive and this is the "1-D approach zone." In the integral models the flow in this region is approximated by one dimensional mean flow properties at each axial station. In the finite difference models circumferential symmetry is assumed and, therefore, regions $s$ to $e$ and $e$ to 5 cannot be distinguished. As one approaches the combustor exit the gradients in flow properties and composition decrease. The

control boundaries for the integral models are the plane at station 4 upstream of the shock train, the throat of the injector, the injector and combustor walls, and the combustor exit plane at station 5. For simplicity, Billig assumes constant flow properties at the end planes, taking representative mean values of $\rho, u, p, h$, and $T$ at stations 4 and 5 with respective duct areas.

If the integral and finite-difference techniques are coupled, as shown in Figure 3.10, their deficiencies can be practically eliminated. The first step in this approach is to solve the integral equations for given initial conditions and combustor geometry, thus obtaining the axial pressure distribution $p(x)$ with modeled values for the wall shear $\tau_w$ and the heat transfer $Q_w$ and an assumed value for the combustion efficiency $\eta_c$. The next step is a simplified finite difference calculation that avoids calculating the flow in the separated regions so that $\tau_w$, $Q_w$, and $\eta_c$ are obtained analytically. In this step one must choose appropriate models for kinetic

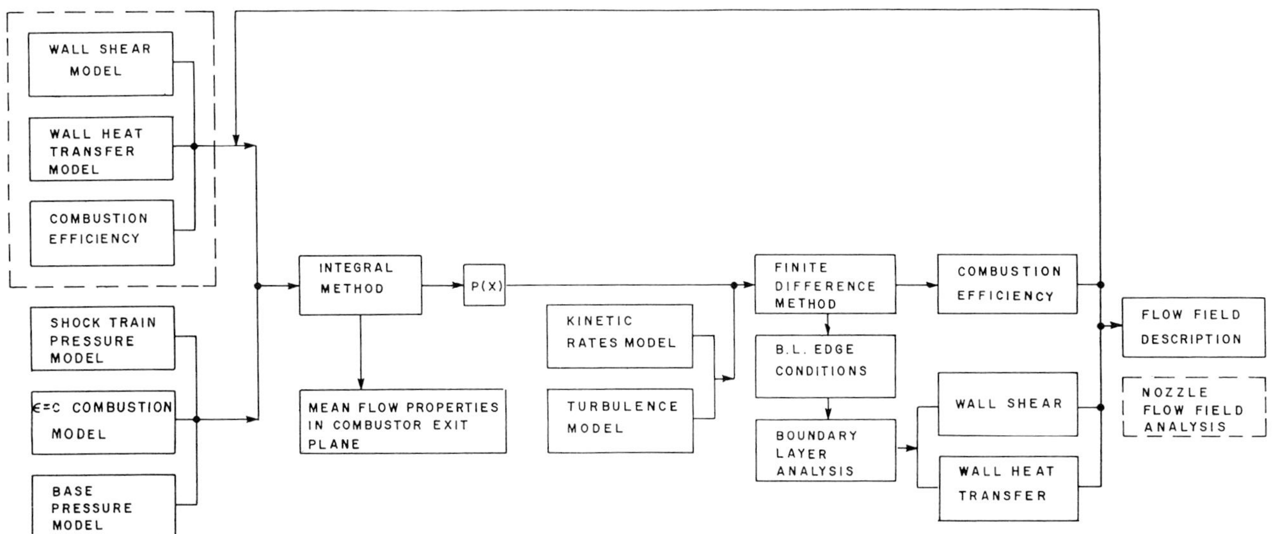



**Figure 3.10**  Method of computation for supersonic combustion analysis (Billig 1988). *Copyright © AIAA 1988. Used with permission.*

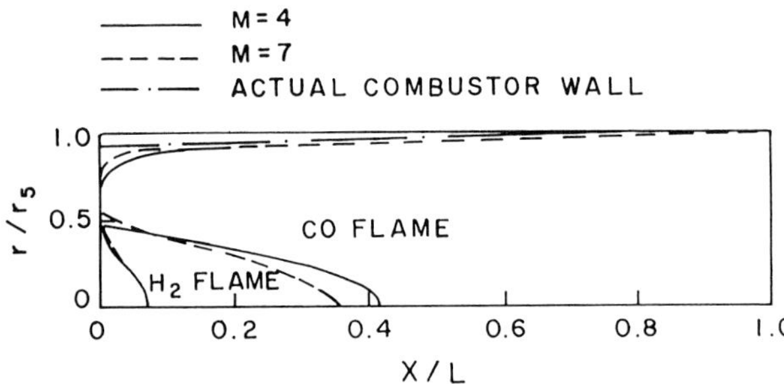


**Figure 3.11** Predicted duct contours and flame shapes for engine operating at Mach 4 and 7 with equivalence ratio 0.5 (Billig 1988). *Copyright © AIAA 1988. Used with permission.*

rates and turbulent mixing. The integral solution is then updated with these values of $\tau_w$, $Q_w$, and $\eta_c$, a new $p(x)$ is obtained and so on, till the flow area obtained from the finite difference solution agrees wit the geometric area downstream of $s$.

Typical results obtained for the combustion of Shelldyne H fuel at an equivalence ratio of 0.5 are shown in Figure 3.11. The predicted duct contours and the double flame sheet shapes correspond to flight Mach numbers 4 and 7 at 11.3 and 18.5 km height. The computations show that in both cases the separated zones extend about 20% into the combustor.

Billig (1988) discusses the use of the integral method alone for design and analysis, stating that this may be applicable if sufficient experimental data are available, for preliminary design information and for providing engine performance estimates, treating parametrically $\tau_w$, $Q_w$ and $\eta_c$. This approach requires much less computer time than the coupled technique described before. The planar flame models are useful for providing insight into the interplay of shocks, expansions and heat addition, but have not been used for actual predictions.

The terminology "hypersonic combustion" was probably first used with a special meaning by Stalker (1972 and 1987). By his definition, the term specifically refers to combustion in the flight speed regime of a scramjet engine where the combustor entrance airflow is at hypersonic Mach number, and the combustor bulk flow remains hypersonic throughout the fuel injection, mixing, and burning process. If hypersonic speeds are taken to occur at Mach numbers greater than five, then hypersonic combustion is relevant to scramjets at speeds approaching flight Mach numbers of 20 and above. In general terms, hypersonic combustion describes the airbreathing engine internal flow at speeds approaching orbital velocity.

At such speeds the kinetic energy of the free-stream air entering the scramjet propulsion cycle is large compared to the energy released by reaction of the oxygen content of air with fuel (say hydrogen). Thus, the effects of reaction at Mach 25 speeds, where heat release from combustion may be 10% of the total enthalpy of the working fluid, will be small compared to Mach 8 flight where the air kinetic energy and potential combustion heat release are roughly equal. Flow deflections due to heat release are small—a few degrees at most—and flow boundaries are conceived as contoured to control the pressure rise at the location of the flame and eliminate the possibility of strong shock formation.

To the contrary, at speeds of Mach 8 and below, combustion in ducted flows can generate large local pressure rise, flow deflection, and separation. This has been studied extensively by Billig and Dugger (1969) and is characteristic of supersonic combustion in constant area channels below flight speeds of about Mach 8. In such flows involving local separation and a bulk Mach number near $M = 1.0$, local wall static pressure is representative of the pressure across the entire flow at a given axial station. Therefore, a one-dimensional approximation to the flow can provide a reasonable description of the flow behavior. In hypersonic combustion, however, local Mach numbers remain high, Mach angles are quite shallow, and significant variations in static pressure are likely to occur across the combustor flow field at a given axial station. Typically, a fully three-dimensional representation of the flow field will be required, and as pointed out by Stalker (1989), special care must be taken to properly relate the implications of one-dimensional calculations to fully three-dimensional experimental data in a meaningful way.

So hypersonic combustion flows differ from supersonic combustion flows in that the flow remains hypersonic throughout in a bulk sense, the effects of heat release are smaller, and the pressure field is fully three-dimensional. Common features of hypersonic and supersonic combustion flows include: real gas effects, nonadiabatic wall boundaries, finite strength shock waves, dissimilar gas injection, turbulent mixing, finite rate chemical reaction, flow separation, etc. Also, because aerodynamic, fluid mixing and chemical rate process are all expected to be important, experimental simulation requires nearly full-sized hardware and close duplication of flight conditions.

Hypersonic combustion raises some additional uncertainty and concerns. First, the effects of extreme compressibility on turbulence generation and mixing are not well known or understood. Second, at about Mach 12 the velocity of the injected fuel stream equals the velocity of the combustor air stream, and at high flight speeds the air velocity exceeds the fuel velocity. The behavior of fuel-air mixing under these conditions is also not well known. In fact, compared to flight at Mach 8 and below, there is very little data on which to base confidence in our understanding of or ability to model hypersonic combustion flows.

The primitive variables available which describe the flow in a hypersonic combustor are: $P$ = pressure (or $\rho$ density),

$T$ = temperature, $u$ = velocity, $L$ = model length, and $v_i$ = gas composition. The important simulation parameters are: the Mach number M, the Reynolds number Re, the Stanton number St, Damkohler's first number $D_1$, Damkohler's second number $D_2$, and the wall enthalpy ratio GW. To these may be added certain second-order parameters such as the Prandtl number Pr and the Schmidt number Sc.

The following physical interpretation of the first three fluid dynamics and heat-transfer parameters is also well known: the Mach number represents the ratio of kinetic to thermal energy, the Reynolds number the ratio of inertial to viscous forces, and the Stanton number the ratio of heat flux to inviscid energy flux. However, the interpretation of the last three is less well known as they are more specific to hypersonic reacting flows. These three parameters are the ratios of flow transit time through the combustor to chemical reaction time ($D_1$), the ratio of heat added by reaction to the stagnation enthalpy of the inviscid flow ($D_2$), and the ratio of enthalpy at the wall temperature to stagnation enthalpy of the inviscid flow (GW). The second-order parameters represent gas properties, the ratio of viscosity to thermal conductivity, and the ratio of viscosity to diffusivity. However, to the extent that these parameters reflect turbulence characteristics, they are more dependent on the first-order simulation parameters than on molecular properties of the gas. The first-order parameters can, in turn, be related to the primitive variable as follows:

$$M \sim \frac{u}{\sqrt{T}} \tag{3.7}$$

$$Re \sim \frac{\rho u L}{\sqrt{T}} \sim \rho L M \tag{3.8}$$

$$St \sim \frac{q_w}{\rho u H} \tag{3.9}$$

$$D_1 \sim \frac{L}{u t_c} \tag{3.10}$$

$$D_2 \sim \frac{\eta_c \, \Delta h_c}{c_p T + \dfrac{u^2}{2}} \tag{3.11}$$

$$GW \sim \frac{c_p T_w}{c_p T + \dfrac{u^2}{2}} \tag{3.12}$$

where $t_c$ is a characteristic combustion time, $\eta_c$ is a combustion efficiency, $\Delta h_c$ is the heat of combustion, $c_p$ is a characteristic specific heat, and $T_w$ is the temperature of the model wall. The overall reaction time for combustion processes is generally proportional to pressure (or density) to an exponent around 1.75 and exponentially dependent on temperature. In certain restricted conditions wherein only binary (two-body) reactions occur, the combustion time is linear in density, leading to a direct relationship between Reynolds number and Damkohler's first number; namely,

$$D_1 \sim (\rho L/u) \, \exp(-T) \sim Re \, \frac{\sqrt{T}}{u^2} \exp(-T) . \tag{3.13}$$

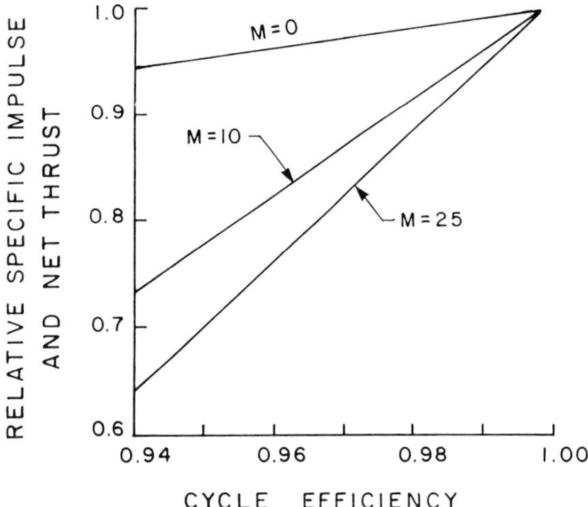


**Figure 3.12** Cycle efficiency impact on performance (Kors 1988)

Hence, if velocity and temperature were to be duplicated, then simulation of Reynold's number would also satisfy the requirements for simulation of binary reaction time, and vice versa. However, even in the simplest of situations, it is virtually impossible to manipulate the temperature, velocity, and model length in a fashion that will simultaneously preserve the values of Mach number, Reynolds number, and Damkohler's numbers.

Therefore, it is clear, in general, that in hypersonic combustion experiments it is necessary to duplicate the primitive variables, including model length and gas composition, to ensure a faithful representation of the coupled chemical and flow processes. This automatically satisfies all of the simulation parameter requirements, with the possible exception of wall temperature and wall reactivity simulation.

## 3.3 Aerodynamic Losses and Integration

At hypersonic speeds the location of the engine in the pressure field of the aircraft can profoundly affect the engine characteristics and lead to losses much higher than in the case of subsonic or low supersonic velocities. At such velocities shock waves are generated and they may adversely effect the overall efficiency of the vehicle. Design studies have shown that in order to provide sufficient volume, adequate lift and propulsion, the slender body aircraft configuration is unsuitable (Kuechemann 1965). The increasing effect of cycle efficiency with Mach number is shown in Figure 3.12. As a consequence one must design the propulsion system as part of the vehicle. This concept is known as integration. There are two main approaches—one is the use of a lifting body suitable for missiles (Waltrup et al. 1982), while in the other, which appears to be more appropriate for large vehicles, a number of modules are used. One of the early configurations proposed for the first approach is the Nonweiler or waverider wing (Nonweiler 1963) shown in Figure 3.13, which has an inverted V cross section with a plane

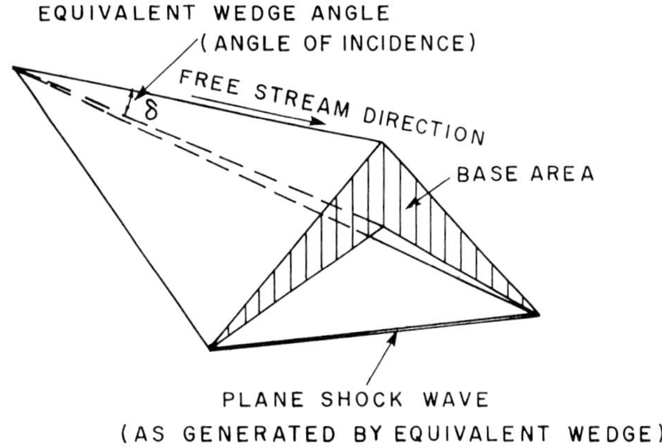



**Figure 3.13**  Nonweiler "caret" waverider wing, without engine (Swithenbank 1966). *Reprinted with permission from Pergamon Press Ltd.*

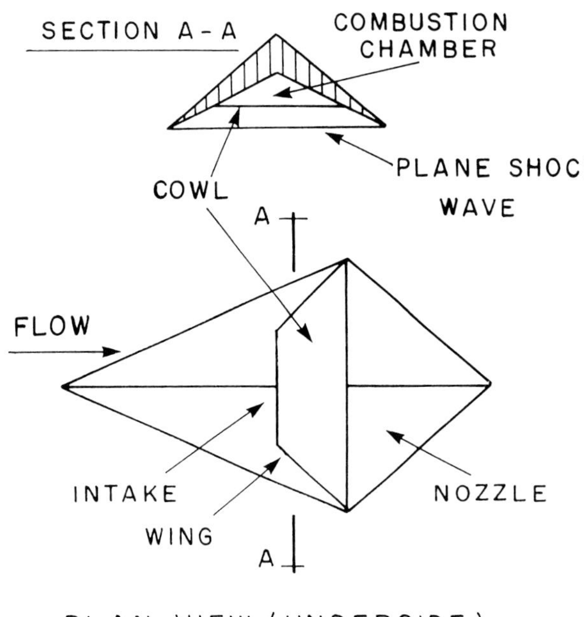



**Figure 3.14**  Diagrammatic integrated engine/airframe design (Swithenbank 1966). *Reprinted with permission from Pergamon Press Ltd.*

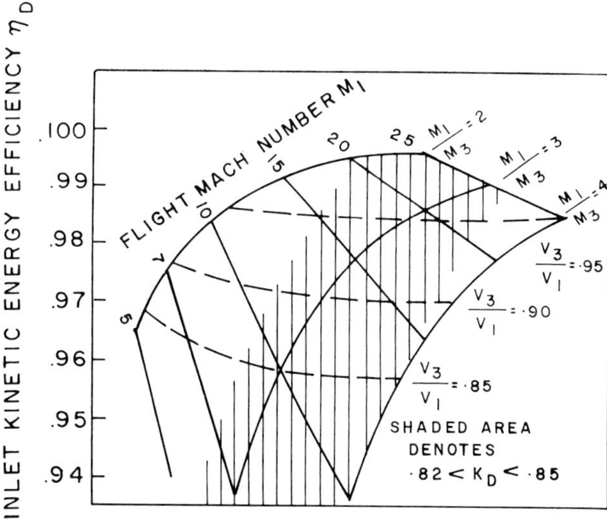



**Figure 3.15**  Efficiency of two-dimensional, two oblique shock wave inlets—inviscid flow (Swithenbank 1966). *Reprinted with permission from Pergamon Press Ltd.*

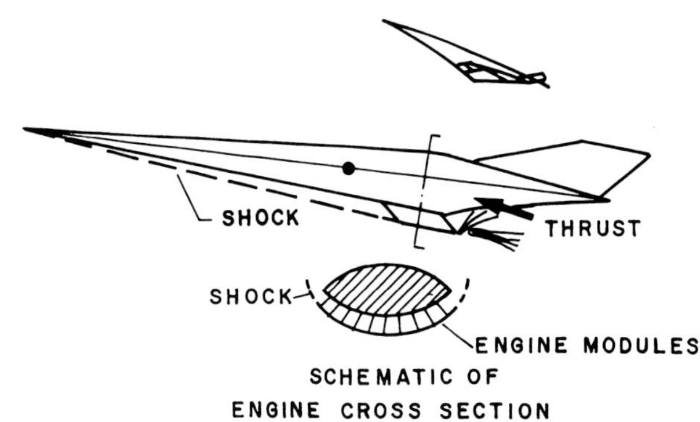



**Figure 3.16**  Scramjet vehicle integration (Northham and Anderson 1986). *Copyright © AIAA 1986. Used with permission.*

shock confined between the leading edges giving uniform pressure over the lower surface. Figure 3.14 shows a diagrammatic design for an integrated engine/airframe. The critical components are the inlet with its curve and the exhaust nozzle. Various parameters have been used to characterize the inlet efficiency, the more important ones being: (1) the kinetic energy efficiency $\eta_D$ defined by

$$\eta_D = (H_1 - h_1')/(H_1 - h_1) \tag{3.14}$$

where $H_1$ is the stagnation enthalpy and $h_1'$ is the hypothetical static enthalpy obtained by isentropic expansion from station 3 to station 1. (2) the process efficiency $K_D$ defined by

$$K_D = [\eta_D - (V_3/V_1)^2]/[1 - (V_3/V_1)^2], \tag{3.15}$$

and (3) the amount of diffusion specified by $V_3/V_1$, $h_3/h_1$, or $M_3/M_1$. Figure 3.15 presents the efficiency of two-dimensional oblique shock wave inlets, neglecting viscous effects. One sees that an increase in flight Mach number has a favorable influence on the efficiency, as well as a low diffusion ratio $M_1/M_3$. The required velocity rate, $V_3/V_1$, increases as flight speed increases, while the process efficiency $K_D$ remains nearly constant at $0.82 \leq K_D \leq 0.85$ for the range of inlet operation that is of interest, thus making $K_D$ a valuable parameter for the evaluation of inlet performance. Other parameters which must be considered are the skin friction, the heat transfer, and the boundary layer thickness.

In the modular approach, the whole underside of the vehicle is considered as part of the propulsion system. This concept, illustrated in Figure 3.16, utilizes the vehicle forebody for part of the inlet compression and the aftbody as part of the nozzle expansion. The inlet must capture most of the air processed by the vehicle undersurface bow shock. An annular inlet, divided into smaller rectangular units re-

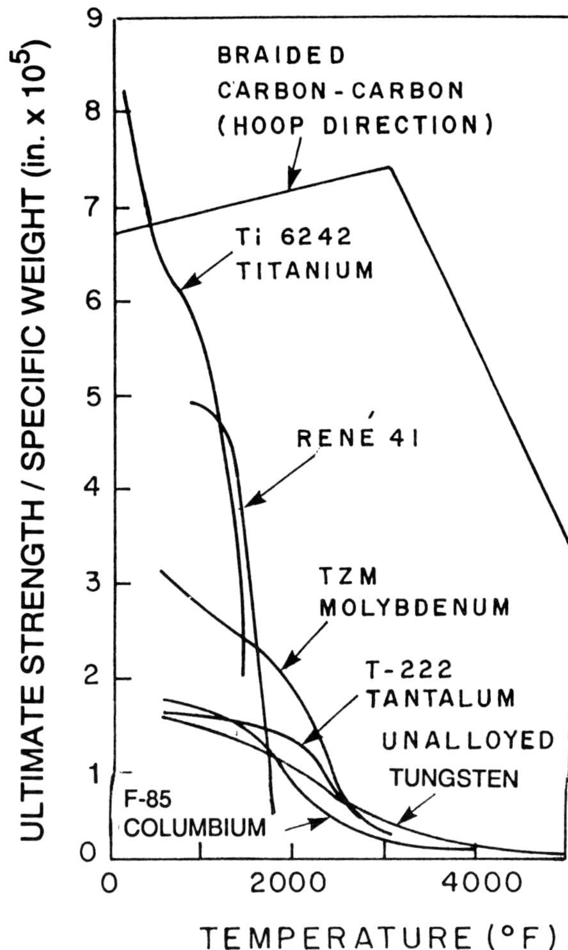

**Figure 3.17** Strength/weight of candidate materials as a function of temperature (Kors 1988). *Copyright © Acta Astronaut 1988. Used with permission.*

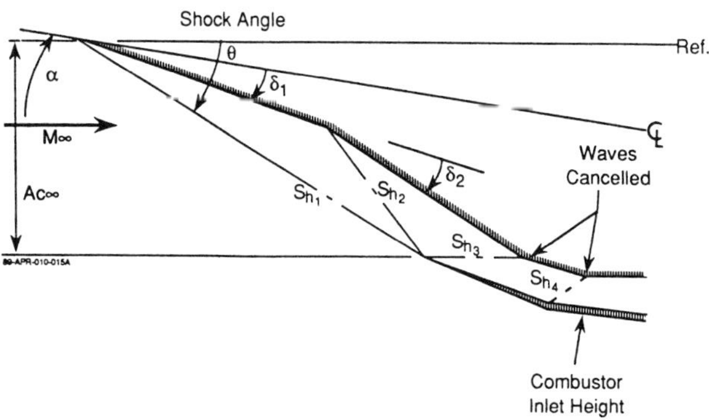



**Figure 3.18** Schematic of integrated forebody inlet (Ikawa 1991). *Copyright © AIAA 1991. Used with permission.*

sults in a number of identical modules suitable for ground testing.

Some of the demanding requirements of such a combined cycle engine involve the control system, which must assure stable operation over a wide operating range during which there is transition from one cycle to another; at hypersonic speed, where time constants are very short, a high degree of integration between vehicle and engine control is mandatory, so that engine forces (both thrust and drag components) are dynamically controlled by the vehicle control system.

Another area that demands further study is the development of light weight, high temperature materials; Figure 3.17 shows some of the candidate materials. It appears that braided carbon-carbon is particularly well suited, although other possible materials include also ceramic matrix and metal matrix composites.

With respect to the envelope, great progress is taking place in both thermoplastic and thermoset composites (Brown 1989). As material problems are common to ram and turbo-propulsion, they will be discussed in greater detail in Chapter 15.

## 3.4 Preliminary Design and Performance Calculations

For preliminary design purposes, as requested by the NASP program, a simpler engineering procedure for performance predictions was developed by Ikawa (1991). This method allows one to design and integrate scramjet vehicles for a prescribed geometry and given flight conditions, obtaining also off-design performance data. It also permits one to execute many design iterations quickly and economically in order to study trade-offs. For simplicity, a two-dimensional inviscid flow analysis is carried out, estimating subsequently the effect of viscous interactions; the calculation can be performed on a PC.

The program first computes the integral forebody-inlet surface, then the divergent combustor, and finally performs an afterbody nozzle analysis. Figure 3.18 is a schematic of the integral forebody-inlet. The lower surface of the vehicle forebody is designed to form part of the external compression device. Its aim is to capture all the freestream airflow at the relevant flight conditions which are defined by the Mach number, the angle of attack, and the altitude. One can observe the different shock waves appearing at the inlet and at the combustor entrance. The conditions there are used as initial conditions for the analysis of the divergent combustor; the physical effects of turbulent mixing, dwell-time, burning characteristics at high speed (3 km/sec) and low pressure (0.01 MPa), real gas effects and chemical reactions are accounted for through an energy balance, equating the work performed by the vehicle on the surrounding atmosphere to the heat content of the fuel. An efficiency factor, based on a reasonable engineering estimate, is also employed in the computation.

The next step is to analyze the combustor-afterbody nozzle process, as shown in Figure 3.19. This allows one to calculate the thrust coefficient, the drag, lift and pitching moment coefficients and to finally obtain the specific impulse.

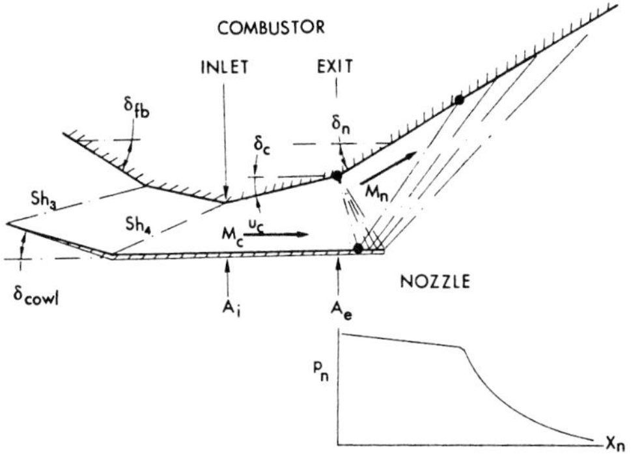



**Figure 3.19**  Combustor/afterbody nozzle process (Ikawa 1991). *Copyright © AIAA 1991. Used with permission.*

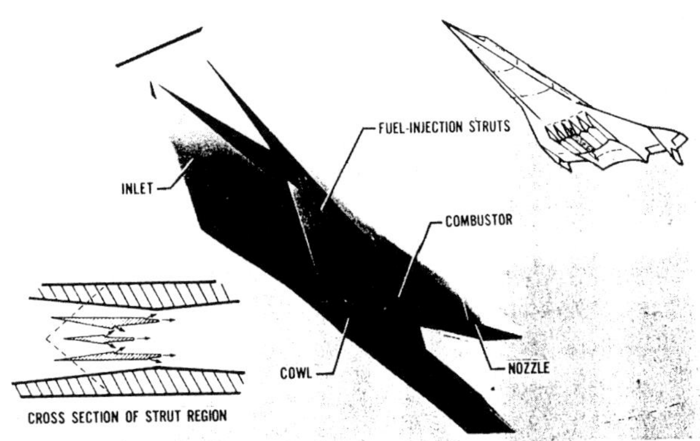



**Figure 3.20**  Airframe integrated supersonic combustion ramjet (Beach 1979). *Courtesy of NASA.*

As mentioned earlier, one can then estimate the viscous flow interaction in the forebody inlet and also take into account real gas effects.

## 3.5 Scramjet Research and Applications

The main center of research has been in the United States, where the activity was sponsored by NASA, the Air Force, the Navy, and industry (General Electric, Marquardt, Garrett). Significant work has been conducted at a number of industrial and university laboratories including United Aircraft Research Laboratories (UARL), General Applied Science Laboratories (GASL), Applied Physics Laboratory of John Hopkins University (APL), and at the NASA Lewis and NASA Langley Centers. Up to 1984 various programs using both hydrogen and hydrocarbons were carried out that included ground tests of components and complete experimental engines; in 1984 the ambitious national aerospace plane (NASP) program was launched, which is aimed at designing and building the experimental X-30 vehicle. In England an active center, headed by Swithenbank, did important work in the 1960s, but government support was low and Rolls Royce's Hotol program, started in the 1980s for a one-stage to orbit hypersonic vehicle, is looking for inves-

tors. The French have important programs at ONERA, but their interest is limited to Mach numbers up to 12. In Germany there is active research on the Sänger orbital launcher, whose first stage is powered by a combined cycle, including scramjet operation. Finally, the work being done in Australia should be mentioned; the unique free piston shock tunnel at the National University allows engineers to perform tests at very high enthalpies. Japan has started a promising aerospace plane program, based on cooperation of the industry (IHH) and the National Aerospace Laboratory. Russia conducted some early flight tests in 1991.

### 3.5.1 Early Efforts with Emphasis on the Ramjet-Scramjet Transition

During the 1960s four research engines were built and tested on the ground for the U.S. Air Force. They were fueled with hydrogen and tested various aspects of internal performance, such as a variable-geometry inlet, staged fuel injection for operating over a wide Mach number range, rearward-leaning steps for combustor-inlet isolation, and thermal compression. Navy sponsored research in the same period also took into account volume limitation and the use of storable fuels. The combustion configuration that evolved, using auto-ignitable hydrocarbon fuel, employed a rearward leaning step, a constant-area isolation section, and an expanding conical section with a final area ratio of 2. Free jet tests of such models, conducted at Mach numbers of 5, 5.8 and 7 have shown that net thrust is produced. Since 1979 the APL researchers introduced a new concept—the integral-rocket dual-combustion ramjet (Billig et al. 1979) for volume limited engines. It is an embedded dump combustor ramjet system serving as a fuel rich hot gas generator for a primary ramjet system. This has been further developed (Waltrup et al. 1982, Waltrup 1987) and is apparently under consideration for military applications. NASA also began its scramjet research in the 1960s, with the goal of advancing the technology for manned vehicles. This effort was centered on a hypersonic research engine program, which should have been tested on the X-15 research airplane. Although that goal was not reached, due to the X-15 cancellation, two engines were built and tested by Garrett for this purpose. These engines were axisymmetric and used inlet airflow in a range of Mach 4 to 8, with combustor operation controlled by staging fuel injection at different points. A rectangular module concept that takes advantage of this approach is shown in the center of Figure 3.20. Since the vehicle compresses the flow in the vertical direction, the module inlet side walls have wedge shapes to compress the flow horizontally. This tends to minimize flow distortion over a flight Mach number range. Sweep of these wedges, in combination with a recess in the cowl, allows spillage to occur efficiently with fixed geometry. Inlet compression is completed by three wedge-shaped struts located at the minimum-area section of the module. These struts also provide multiple planes for fuel injection and therefore

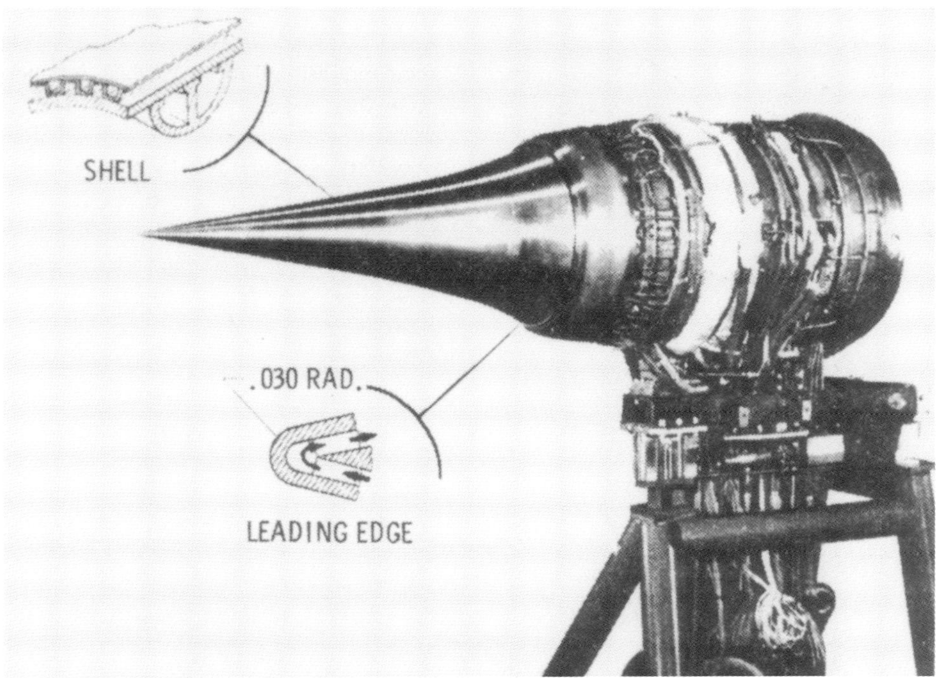



**Figure 3.21**   Structural assembly model of hypersonic research engine (Anderson 1975). *Courtesy of NASA.*

shorten the required combustor length. A cross-sectional view of the strut region, shown to the left in Figure 3.20, illustrates the technique used to control the distribution of heat release in the combustor. At Mach numbers of 7 and higher, almost all the fuel is injected perpendicular to the airflow over the struts. This is an inherently fast mixing process that results in rapid heat release close to the struts for maximum performance. At Mach numbers below 6, too much perpendicular injection produces too rapid mixing and reaction, and the flow is thermally choked. To prevent this, most of the fuel is injected from the base of the struts parallel to the local flow direction. Mixing thus occurs much more slowly, and the heat release is stretched out over a greater length. Structural tests of a complete flight-weight regenerative cooled engine (see Figure 3.21) were carried out in the Langley 8-foot temperature structures tunnel (Anderson 1975). Thrust performance was determined in separate tests at Lewis Plum Brook Station on a water-cooled, heavy-walled version of the engine. The measured internal performance in the range of Mach 5 to 7 was close to the design goal (Beach, 1979). These two sets of tests verified the structural and cooling design and manufacturing techniques and demonstrated the feasibility of good internal thrust over a range of flight speeds.

In parallel with this work, a great deal of research went on, in particular at NASA Langley, on airframe-engine integration (Anderson 1975, Beach 1979). Figure 3.22 presents results of tests conducted at Mach numbers between 2.3 and 6 ahead of the inlet, which correspond to flight Mach numbers from 3 to 8. They indicate variations in throat Mach number, contraction ratio, mass capture, and pressure

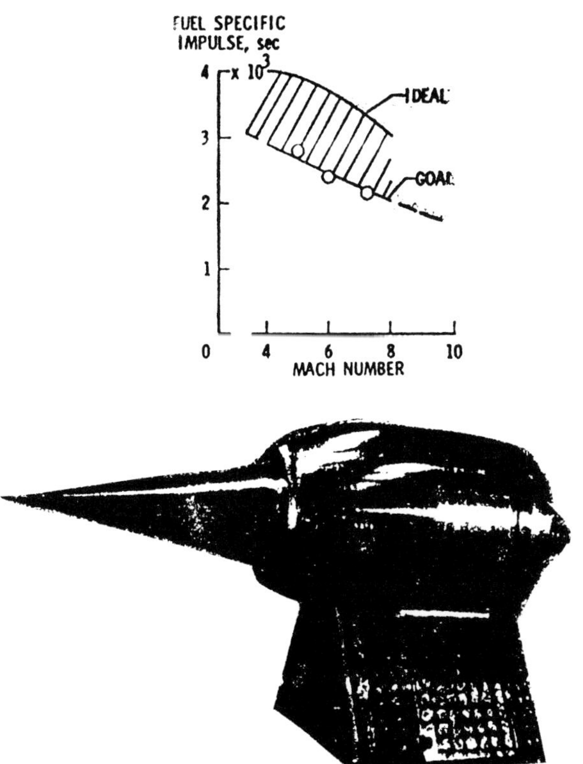



**Figure 3.22**   Aerothermodynamic integration model of hypersonic research engine and specific impulse based on internal thrust (Anderson 1975). *Courtesy of NASA.*

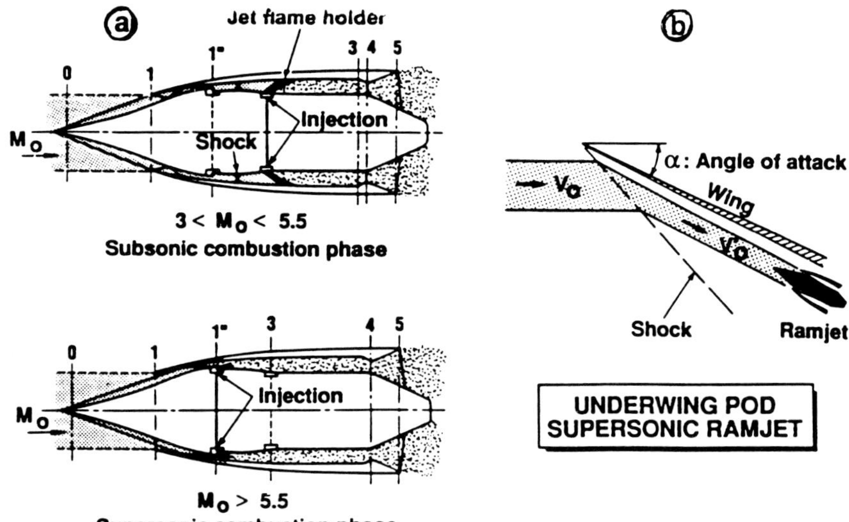



**Figure 3.23**  Esope concept (Contensou et al. 1973). *Copyright © ONERA 1972. Used with permission.*

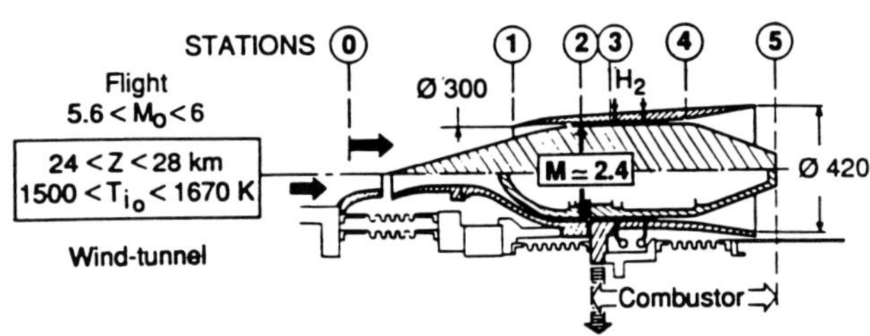



**Figure 3.24**  S-4 wind tunnel hypersonic testing (Contensou et al. 1973). *Copyright © ONERA 1972. Used with permission.*

recovery similar to those of a variable geometry inlet. It appears that the vehicle boundary layer can, be ingested without separation, at least as long as the effects of the combustor flow do not feed forward. Some effects warranting further study were observed, such as delayed ignition. The lower part of the figure shows a model built to study auto-ignition with the fuel injectors drilled near the rearward facing step; the experiments showed that the ignition location is critical. During the entire period theoretical work continued, with increasing reliance on computational fluid mechanics, as reported by Barber and Cox (1989).

In parallel with the work performed in the United States, the French ONERA conducted the Esope program between 1966 and 1974, Esope mainly treated the problem of transition from a vehicle with subsonic combustion to one with supersonic combustion. The possible applications envisaged were missiles operating between Mach 3.5 and 7 (or higher) as well as a hypersonic airplane. The concept of Esope, as shown in Figure 3.23, was based on a fixed geometry design, while changing the injection angle of the hydrogen propellant. The initial experiments showed that in the region of Mach 5 to 6 there were problems: ignition was not

assured and stratified combustion could appear. As a consequence, it was decided to investigate three combustion phases—subsonic for Mach 3.5 to 5, the transition region, between Mach 5 and 7, similar to the U.S. research mentioned above, and supersonic combustion for higher Mach numbers (Note: the last part of the program was not carried out at that time).

In wind tunnel experiments performed for the inlets at flight Mach numbers between 3.5 and 7, it was found that for $M_{\text{flight}} = 6$ the internal Mach number at the end of recompression was 2.4 and appropriate intakes were designed. As regards combustion, early experiments, conducted at Palaiseau with vitiated air, allowed a first evaluation of transition combustion. Experiments in the S4 high-enthalpy wind tunnel in Modane (Figure 3.24), where it is possible to simulate a Mach 6 flow in temperature and pressure with pure air, were conducted in an air-connected test arrangement (see section 5.2). Here the Mach number at the end of the recompression zone was again 2.4, the temperature was 1600 K, and the total pressure 1.5 MPa with an air flow of 1.5 kg/sec. The force operating on the combustor was measured with a dynamometer and from this the thrust was calculated after appropriate corrections. The experiments were run with gaseous hydrogen and showed that the flame can be stabilized aerodynamically. This was achieved by injecting the gas in two stages, using closer holes with a small diameter and more distant ones with a larger diameter, as shown in Figure 3.25. The experiments proved the feasibility of the transition combustion with gaseous hydrogen injected at Mach 6, a specific impulse of 2500 sec, and a combustion efficiency of 80% at stoichiometric conditions (see Figure 3.26).

Finally it is interesting to note that the Russians, who have been active in the hypersonic fields for many years as

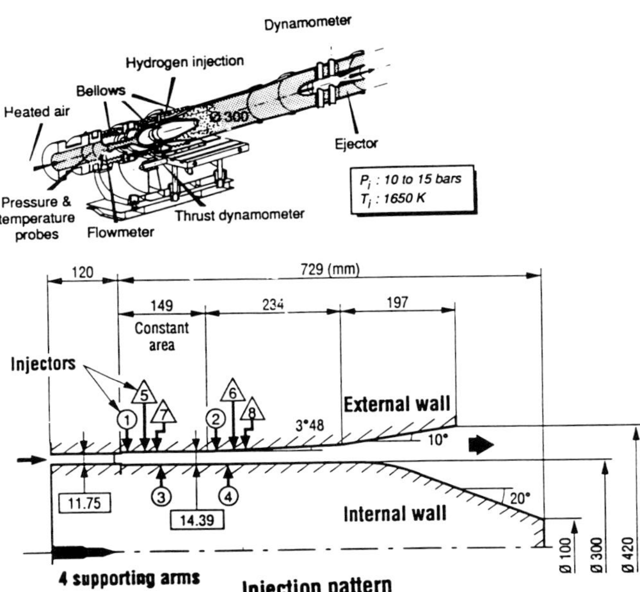



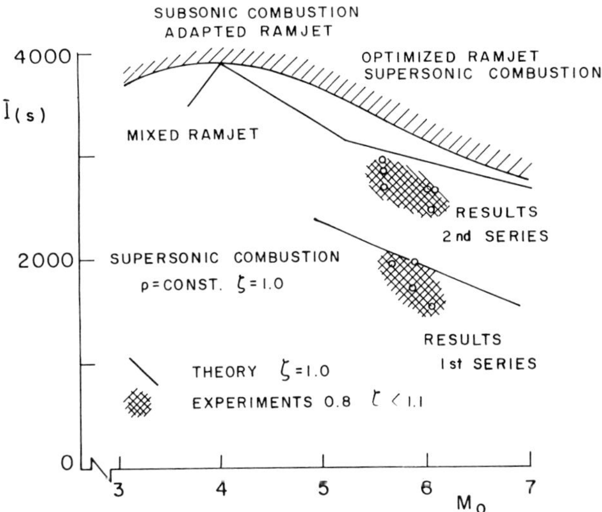



**Figure 3.25** Transition combustion test model (Contensou et al. 1973). *Copyright © ONERA 1972. Used with permission.*

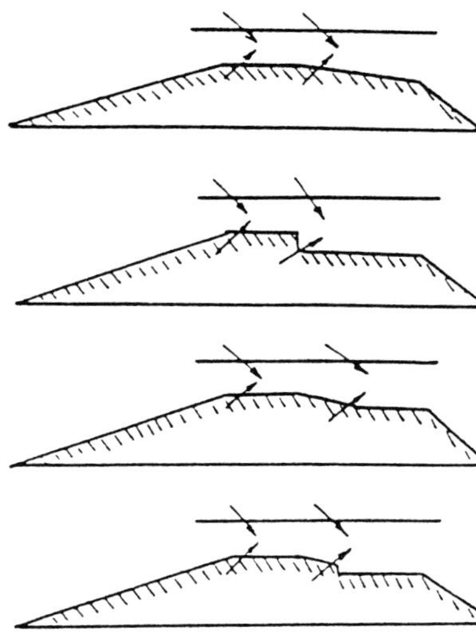

**Figure 3.27** Russian dual mode scramjet schemes (Vinogradov et al. 1990). *Copyright © AIAA 1990. Used with permission.*



**Figure 3.26** Comparison of results: specific impluse vs. flight Mach number. *Copyright © ONERA 1972. Used with permission.*

a consequence of their space program, came to very similar conclusions (Vinogradov et al. 1990); they propose a dual mode ramjet operating between Mach 3 and 10 and present a number of possible geometries, shown in Figure 3.27. Preliminary studies they performed showed that the use of a configuration in which the fuel is injected into a section with a larger cross-sectional area than the inlet throat (see Figure 3.28), gave satisfactory results for a subsonic combustion ramjet; however, injection of the fuel at the beginning of the duct in the inlet throat with combustion in the extending section of the combustor with supersonic burning caused large losses in total pressure, diminishing the ef-

ficiency. Consequently, a different configuration, the duel mode scramjet (somewhat similar to the one developed in the United States) was adopted.

Covault and coworkers (1992) report that the Russians succeeded in flying three to four tests with their dual model ramjet, evidently developed from the work described above, performed by their Central Institute of Aviation Motors. According to them the flight-tested scramjet is similar to the subscale unit shown in Figure 3.29 which is 1.7 m long with an inlet entry area of 0.2 × 0.2 m. It has four rows of hydrogen injectors on the upper and lower duct walls. Four cavity flameholders with electric igniters in each were used for combustion stabilization. In the first flight test, carried out in November 1991, the engine functioned for 20 sec at Mach 3.5 with subsonic combustion in the ramjet mode. The vehicle was then accelerated to Mach 5.6, where the transition to scramjet operation was effected. The maximum altitude reached was 30 km. The authors state that it is possible that a flight in December 1991 achieved full scramjet operation at velocities up to approximately Mach 7. They also conjecture that a test in February 1992 could expand the test envelope into a speed regime of about Mach 10. This should be compared with the status of the NASP program, which is not expected to fly before 1995 and may even be delayed beyond that date.

### 3.5.2 Aerospace Plane Projects of the 90's

Since the mid-eighties an increasing effort is taking place in the developed countries towards the achievement of an orbital capability which exploits atmospheric oxygen in an efficient manner. The leader in the endeavor is the United States, with its NASP program, but work is being performed in parallel in Europe (Russia, the German Sänger, the British Hotol, studies at ONERA, and a number of universities such as RWTH-Aachen, Imperial College, and the Univer-

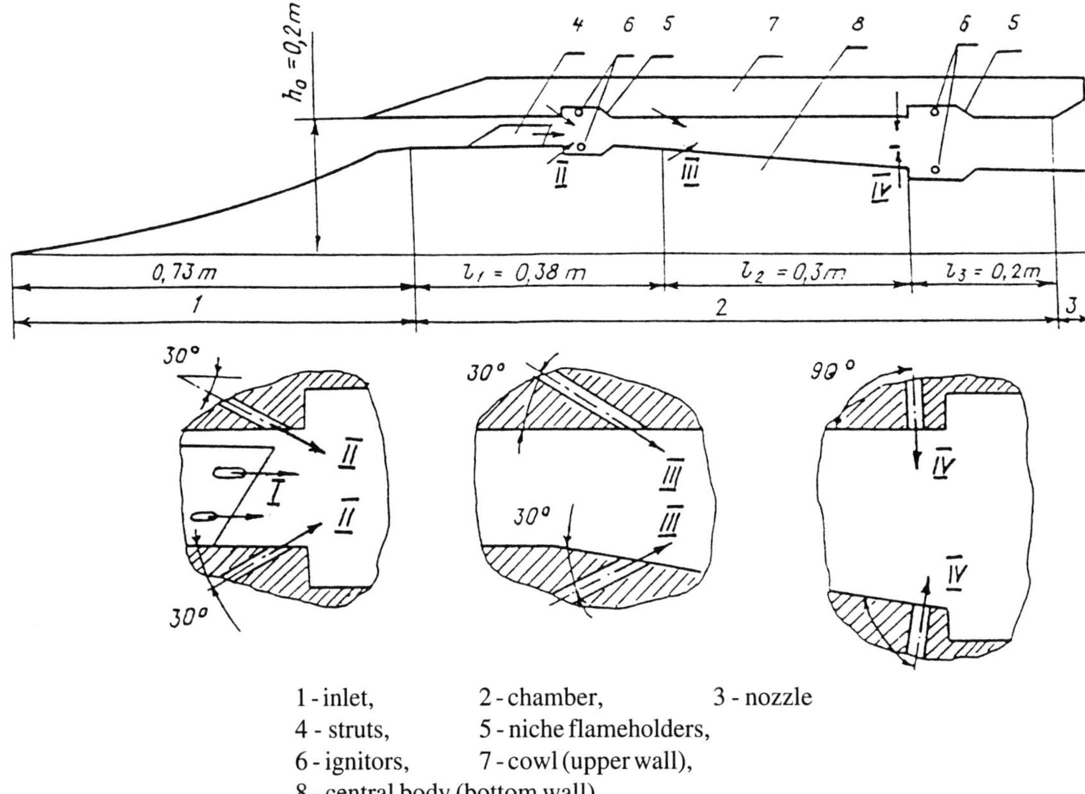


1 - inlet,          2 - chamber,          3 - nozzle
4 - struts,         5 - niche flameholders,
6 - ignitors,       7 - cowl (upper wall),
8 - central body (bottom wall)

**Figure 3.28**  Scheme of the 2D dual mode scramjet investigated by the Russians (Vinogradov et al. 1990) *Copyright © AIAA 1990. Used with permission.*

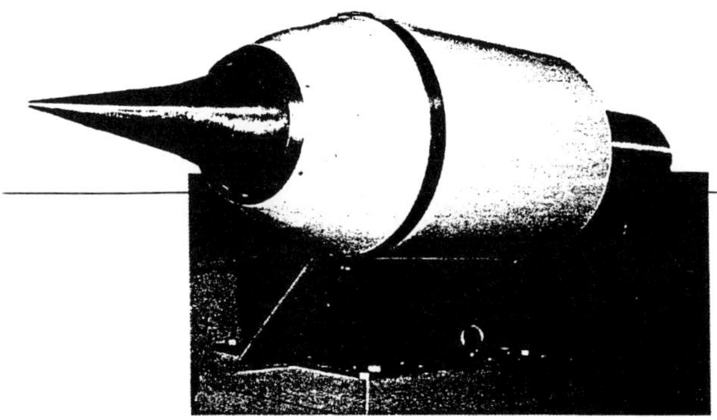

**Figure 3.29**  CIAM subscale scramjet model (Convault et al. 1992). *Photo courtesy of Aviation Week & Space Technology, copyright 1992, McGraw-Hill Inc. All rights reserved.*



**Figure 3.30**  NASP program goals (Waldman and Harsha 1990). *Copyright © AIAA 1990. Used with permission.*

sities of Southampton and Marseille), in Japan (industry and universities), and in Australia (academic work only).

### 3.5.2.1 The NASP Program

This program (Waldman and Harsha 1990) was started in 1984, mainly as a military venture, and transferred 5 years later to NASA with massive participation of industry. In the first phase the program goals, summarized in Figure 3.30 were defined. They include reaching orbit in a single manned stage with horizontal takeoff and landing from conventional runways. For this purpose the objectives shown in Figure 3.31 are the development of a technology base, then the design of the X-30 research vehicle, whose task is to demonstrate the capability of hypersonic cruise and of SSTO (single stage to orbit). This constitutes the second phase, lasting up to the beginning of 1993, when a decision

- DEVELOP TECHNOLOGY BASE
- CONDUCT RESEARCH USING THE X-30 VEHICLE
- DEMONSTRATE
  - AIRPLANE CHARACTERISTICS
  - HYPERSONIC CRUISE
  - SINGLE STAGE TO ORBIT
- DEFINE OPERATIONAL UTILITY

**Figure 3.31** NASP program objectives (Waldman and Harsha 1990). *Copyright © AIAA 1990. Used with permission.*

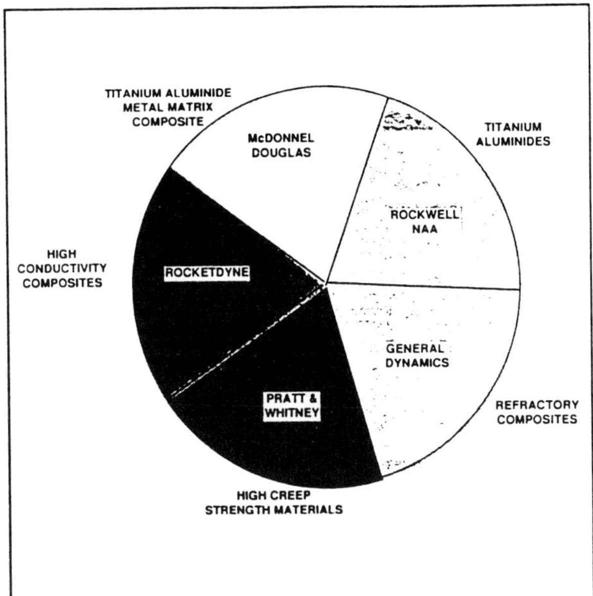

**Figure 3.32** Materials consortium work breakdown (Waldman and Harsha 1990). *Copyright © AIAA 1990. Used with permission.*

- SIX VEHICLE/ENGINE CONFIGURATIONS DEFINED
- TIP-TO-TAIL COMPUTATIONAL CAPABILITY DEVELOPED
- INTEGRATED SCRAMJET TESTED TO M = 8
- LARGE SCALE STRUCTURAL TEST ARTICLES FABRICATED
- CHARACTERIZATION BEGUN ON ADVANCED MATERIALS
- SUB-SCALE WIND TUNNEL TESTING TO FREE STREAM M = 14

**Figure 3.33** Program technical status prior to teaming (Waldman and Harsha 1990). *Copyright © AIAA 1990. Used with permission.*

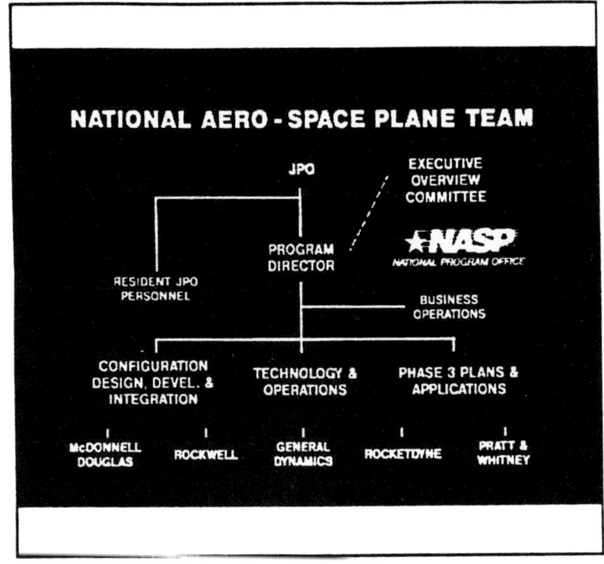

**Figure 3.34** National program office organization (Waldman and Harsha 1990). *Copyright © AIAA 1990. Used with permission.*

to build the X-30, which could start flying four years later, should be taken.

In the initial effort six vehicle/engine combinations were defined, representing General Dynamics, MDD, and Rockwell International aircraft designs with each of the P&W and Rocketdyne engine concepts. In March 1989 the five NASP contractors formed a consortium for materials development (see Figure 3.32 for the work breakdown) and later in that year a NASP national team was formed, on recommendation of the National Space Council. The technical state of the program before this teaming occurred is shown in Figure 3.33. The main advantage of having one team is that the best features of the different proposals could be integrated into one vehicle. Figure 3.34 depicts the organization of the National Program Office, while Figure 3.35 gives an overview of the whole program, showing clearly the three phases and what should be achieved in each one.

From the technical point of view, calculations showed that it would be advantageous to increase the density of the fuel by using a mixture of solid and liquid hydrogen, called slush hydrogen. By employing a 50% slush at 14 K and 0.069 bar one can diminish the volume by 13% (Dreher et al. 1990, Horsley et al. 1990); even more important is the fact that by rising the pressure to 0.136 MPa (a reasonable value

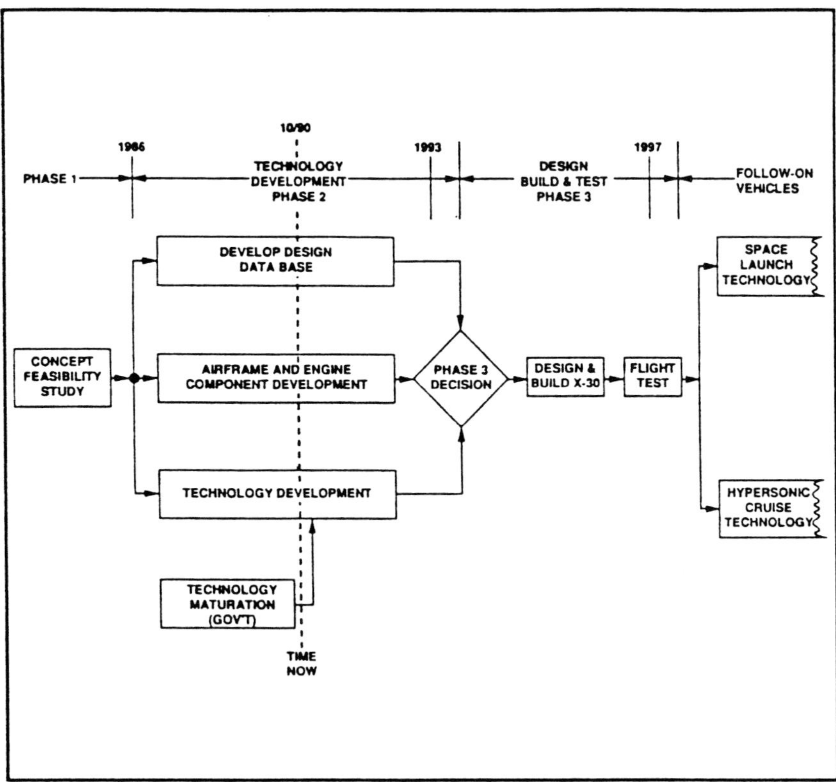


**Figure 3.35**   NASP program overview (Waldman and Harsha 1990). *Copyright © AIAA 1990. Used with permission.*

for a tank) this slush could absorb eight times more heat than liquid hydrogen. Later calculations have shown that the solid content of the slush should be at least 55% (Kandebo 1991). Both NASA-Lewis and Martin-Marietta have production facilities for hydrogen slush and have performed transport and flow tests.

Important work is being performed in the development of heat resistant materials, such as the silicon carbide reinforced 15-3 Ti metal matrix composite, from which fuselage type structures that can withstand 1000 K have been manufactured (see also Chapter 15). Another problem is that at low speeds the exhaust gases do not fill the physical volume of the nozzle, which includes the aft portion of the airframe. In order to decrease drag and improve combustion efficiency, gaseous hydrogen will be burned outside the nozzle and then used to fill that volume. Initial tests were conducted in a wind tunnel, but these data include interference effects, that arise because the combustion is performed in an enclosed area. To eliminate these an F/A18 aircraft was equipped with an external tank of gaseous hydrogen and an external burning model, which simulates part of the scramjet nozzle, was fitted to the wing tip. During 90 seconds external burning was achieved at speeds between Mach 1.1 and 1.4.

Recently, Nelson (1992) pointed out that the importance of hypersonic research, as is performed in the NASP pro-

gram, lies in the long-term impact it can have on the social, technological, defense and economic status of the United States vis-a-vis the rest of the world. This stems from the fact that hypersonic airbreathing propulsion offers the potential to cover the entire planet at high speed from horizontal takeoff for both civil and military vehicles, allowing great operational flexibility.

### 3.5.2.2 Scramjet Cycles for the Aerospace Plane

One of the important decisions to be reached on the propulsion of the aerospace plane is the engine cycle selection. Kanda et al. (1989) present a comparison of four cycles commonly used in rocket engine fuel fed system analysis: expander, staged combustion, coolant bleed, and gas generator cycles. The factors influencing cycle performance are the flight Mach number, the dynamic pressure, and the ratio of the fuel injection to air dynamic pressure. They examined the first for a range of Mach 6 to 12, taking Mach 10 as reference; the dynamic pressure range was 25 to 150 kPa, with 100 kPa as reference and the pressure ratio varied between 0.5 and 5 with 3 as reference. Figure 3.36 gives schematics of the four cycles—the first two are closed loop cycles, in which the gas driving the turbine is injected into the combustor, while the other two are bleed cycles, in which it is dumped outside the combustor. The energy required to drive the turbine is obtained in the expander and coolant bleed cycles from the regeneratively cooled jacket, while in the hot gas

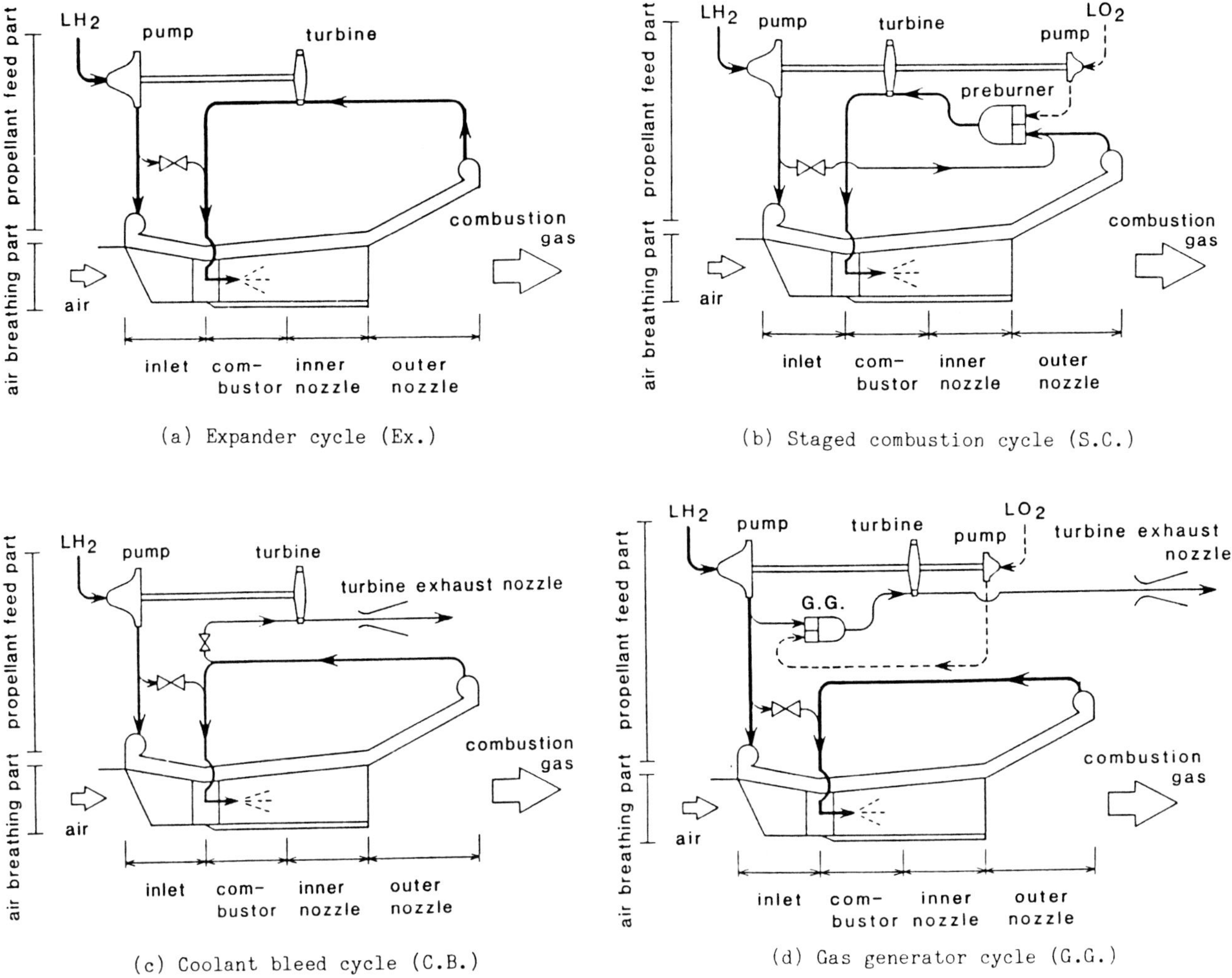



**Figure 3.36** Engine cycle schematics (Kanda et al. 1989). *Copyright © AIAA 1989. Used with permission.*

cycles (staged combustion and gas generator) it comes from precombustion in a preburner or in a gas generator.

Figure 3.37 shows the engine configuration studied, which comprises six modules, each with two fuel injector struts, fed from a single turbopump. At reference conditions the engine thrust is $2.5 \times 10^6$ newtons. After stating the properties of the propellants and the assumptions made in the quasi-one-dimensional model used to calculate temperature, pressure, velocity, and specific impulse in the air-breathing part, the power of the pump ($W_p$) and that of the turbine ($W_t$) are calculated as follows:

$$W_p = \frac{m_p(P_{p_2} - P_{p_1})}{\eta_p \rho_{p_1}} \tag{3.16}$$

and

$$W_t = \eta_t m_t c_p T_{t_1}[1 - (p_{t_2}/p_{t_1})^{\gamma-1/\gamma}]. \tag{3.17}$$

The tank storage pressure is 0.2 MPa for the hydrogen and 0.3 MPa for the oxygen; the efficiencies of the turbopumps are taken as 0.6 and 0.3 for the closed and open loop cycle

turbines, 0.6 for the hydrogen pump and 0.4 for the oxygen pump. For the hot gas cycles they assume a turbine entry temperature (TET) of 850 K.

In the calculations only the regenerative cooling of the engine modules is considered, treating the various components (inlet, combustor, injection struts, inner and outer nozzles) individually. For simplicity the wall temperature on the hot gas side is fixed at 1000 K; the heat flux is lower than in a rocket engine, due to the lower pressure. Therefore, the wall is assumed to be made of nickel alloy. The hydrogen temperature at the cooling jacket exit is set at 700 K; circular regenerative cooling channels are assumed. Pressure losses due to friction and expansion of the heated coolant in the tubes are considered, as well as those at the entrance and exit manifolds of the cooling jacket and the turbine and at the injector manifold.

Thrust is produced by the engine modules and the turbine exhaust nozzle, where the flow is taken as quasi-steady and 100% efficiency is assumed. The thrust and the specific impulse produced at stoichiometric conditions are the refer-

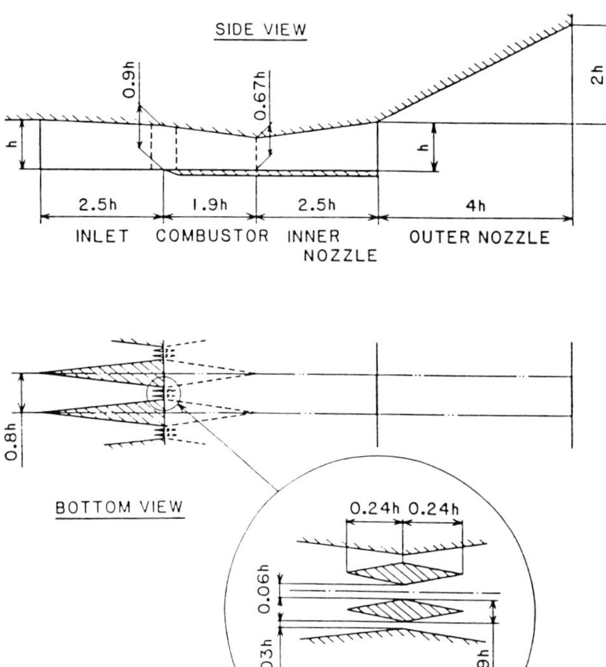



**Figure 3.37**   Engine configurations (Kanda et al. 1989). *Copyright © AIAA 1989. Used with permission.*

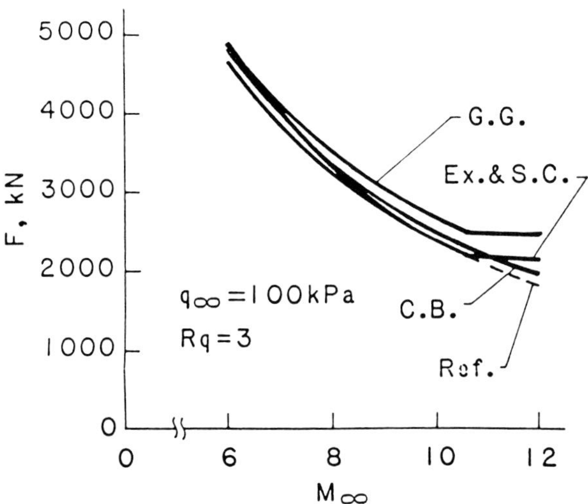



**Figure 3.38**   Engine overall thrust vs. Mach number (Kanda et al. 1989). *Copyright © AIAA 1989. Used with permission.*

ence values. The authors present the results in a series of graphs, showing the pump exit pressure, the hydrogen flow rate, and the specific impulse as functions of the following three variables: flight Mach number, flight dynamic pressure, and fuel injection to air dynamic pressure ratio. Finally, they present the overall thrust of the engine as a function of the flight Mach number (see Figure 3.38).

Their main conclusions are that the system pressure is higher in the closed loop cycles, but the specific impulse is also higher. The coolant bleed cycle shows a well balanced performance, while the expander cycle has the best $I_{sp}$, but may require high pump exit pressure. A way of reducing it could be to combine it with a staged combustion cycle. The authors recommend more detailed studies be conducted.

### 3.5.2.3 Other Single-Stage to Orbit Programs

There are two more proposals for single-stage to orbit (SSTO) vehicles—one is the British Hotol[1], the other is pursued in Japan. Hotol, studied by British Aerospace (Parkinson and Conchie 1990), is based on a radically new hybrid rocket engine, the RB 545 developed by Rolls-Royce. This engine breathes air while in the atmosphere and uses onboard oxygen in space; the fuel is liquid hydrogen. It is designed (Figure 3.39) with a payload bay diameter of 4–6 m,

[1]Acronym for Horizontal Takeoff and Landing

providing compatibility and interchangeability of payloads with the Shuttle Orbiter. At takeoff it would be eight times higher than the Orbiter assembly with double the payload-takeoff mass fraction. Its thrust to weight ratio at takeoff is less than 0.6, compared with a value of 1.6 for the shuttle. The horizontal takeoff and landing capability should allow up to 80% reduction in launch costs. The vehicle design is based on an optimization between mass, shape, and dry engine parameters. The configuration is dominated by the large liquid hydrogen ($LH_2$) tank, which constitutes an integral part of the load carrying structure; an aeroshell, consisting of metal panels backed by high temperature insulation and supported by the tank structure, is used to avoid $LH_2$ boiloff at 20 K. The thermal protection panels should be manufactured from a carbon-carbon metal matrix. The liquid oxygen (LOX) tank is only 1/8 in volume of the $LH_2$ tank, but, when fully loaded, accounts for over 50% of the vehicle's takeoff mass. 55% of the 240 tons total weight is LOX, 25% $LH_2$, 10% airframe, 5% engine and 2% systems; the remaining 3% are for the payload. The dimensions of the payload bay are 7.5 by 4.6 m and can accommodate large satellites. The missions envisaged for Hotol include satellite launch and recovery, servicing of manned space stations and unmanned platforms, microgravity experiments and military operations.

The feasibility of the engine has been proven. Detailed operational plans include takeoff at 70 km/h form a trolley and a conventional 3500 m runway; rotation is initiated by means of a mechanical pusher on the trolley and then the aerodynamic controls take over. After takeoff the vehicle will accelerate to between 900 and 1100 km/h, then climb at constant speed during the air breathing phase to Mach 5 at 26 km, when the air intake closes. The engine then converts to pure rocket power using LOX and $LH_2$. The devel-

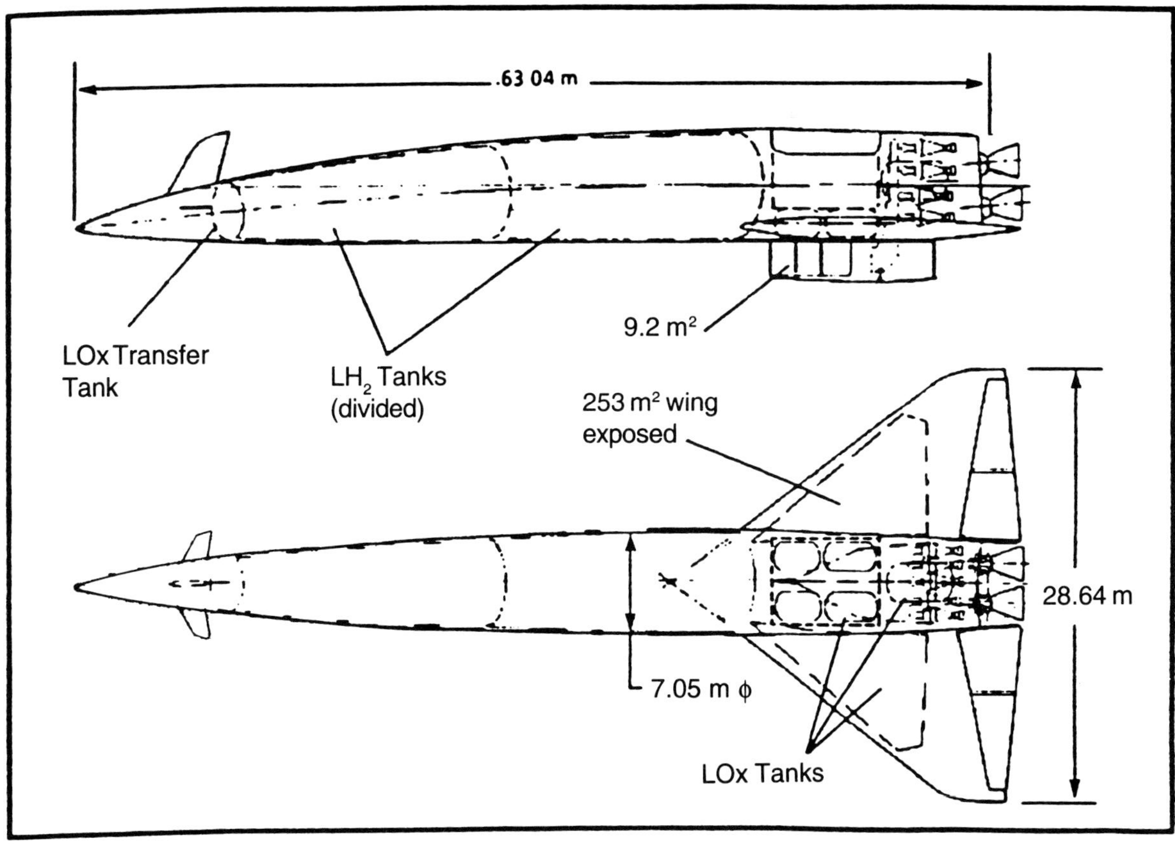



**Figure 3.39**  Hotol vehicle concept—Configuration K (Parkinson and Conchie 1990). *Copyright © AIAA 1990. Used with permission.*

opment of the vehicle could be completed in ten years, but the big question is financing, since the British government has shown little interest in the venture.

A six month study between British Aerospace and the Soviet Ministry of Aviation Industry, performed in 1991, has blended the attributes of a slightly modified Soviet An-225 heavy cargo aircraft and the British Hotol spacecraft to create a fully reusable, low recovering cost launch vehicle. The An-225 could haul the 250 ton Hotol on its back to an altitude of 9 km, where it would be released and then it would power itself into orbit, using its own oxygen-hydrogen rocket propulsion. Future work on this concept will depend on ESA interest, making it a Russian-European project. According to British and Russian officials, this air-launched interim Hotol could be operational ten years after a go-ahead is given.

The Japanese, on the other hand, have well defined, government encouraged plans to develop their space activities which also includes the development of air breathing propulsion. In this field they perform feasibility studies on SSTOs and two-stage to orbit vehicles in parallel with research on specific topics, like ignition problems and fuel in-

jection geometry (Kanda et al. 1991, Chinzei et al. 1991). The work is coordinated by the National Aerospace Laboratory (NAL), which has a ramjet performance section and a ramjet structure section and is carried out in cooperation with industries, like IHI and Nissan Motor Co.

The space plane program scenario, shown in Figure 3.40 (Maita 1991) depicts three phases. The planning phase includes technology development and system studies. These should be followed by the verification of technology readiness and, after the year 2000, by the development of an experimental space plane. The key requirements in achieving this objective are shown in Table 3.1; their aim is a reusable SSTO vehicle with horizontal takeoff and landing capabilities. The projected takeoff gross weight is 350 tons. The system layout of the space plane is shown in Figure 3.41, while the propulsion system is presented in more detail in Figure 3.42. For low speed propulsion there are three options—a liquid air cycle engine (LACE), a turboramjet, or a ducted rocket engine. In Figures 3.41 and 3.42 a LACE system is illustrated. The LACE concept is based on the Japanese experience in the handling of cryogenic engines, acquired in their H-II program (LE5 and LE7 LOX-LH₂ rocket engines). This engine would use slush hydrogen as in the U.S.

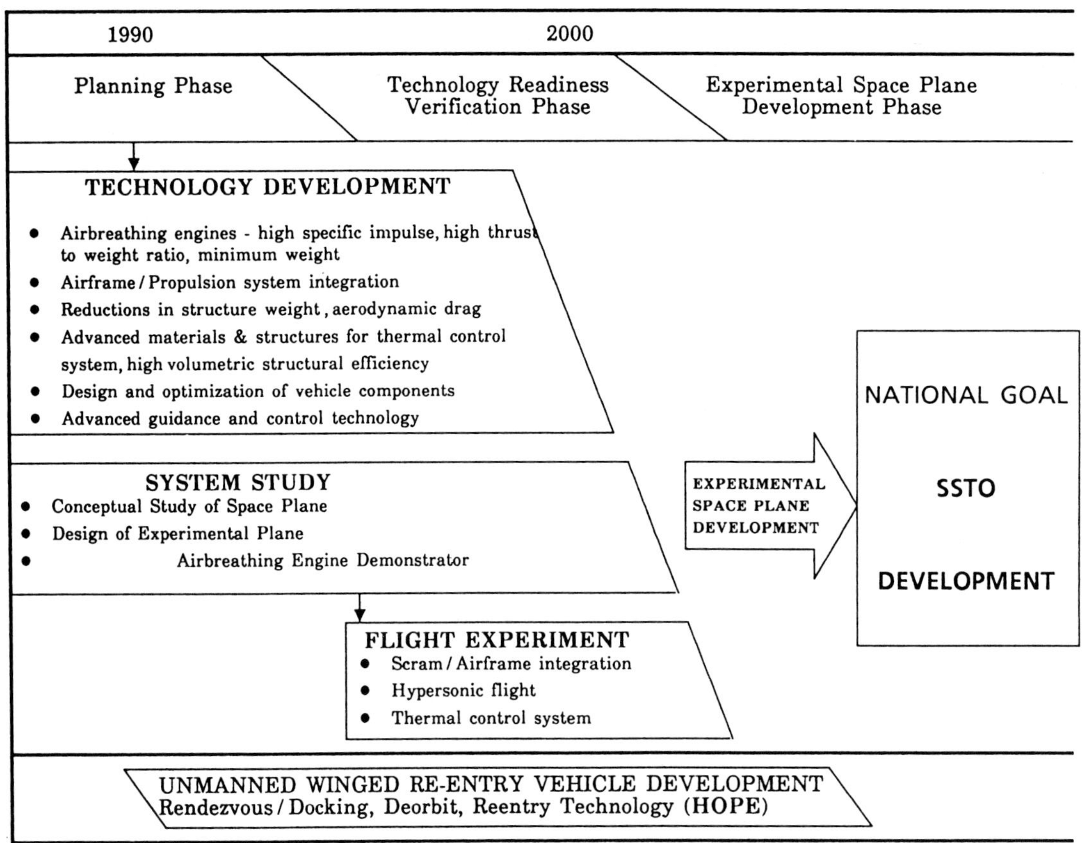


**Figure 3.40**    Japanese space plane R&D program (Maita 1991). *Copyright © AIAA 1991. Used with permission.*

**Table 3.1**    Key requirements for Japanese space plane (Maitia 1991). *Copyright © AIAA 1991. Used with permission.*

---

OBJECT:
- Enhance Operational Flexibility, Safety/Reliability
- Reduce Operational Cost

---

REQUIREMENTS
- Totally Reusable
- Single Stage To Orbit (SSTO)
- Horizontal Take-Off and Landing
- Ability to Abort Safely at All Times following an Initial Failure
- Manned Operation (Largely Autonomous)
- Hypersonic Propulsion System
  Accelerates to $M \simeq 20$

---

**Mission Requirements**
- Destination Orbit, 500 km (28.5°)
- Inclination Angles of the Orbit, 28.5° – 100°
- Number of Crew, 10 (Pilot, Co-Pilot and 8 Astronauts)
- Orbital Stay Time, 4 days

---

X-30 space plane. According to the present conception a LACE engine in the 100 ton thrust class would be used up to Mach 5, operating on an air compressor cycle, going over to the scramjet engine for acceleration to Mach 20, with the final acceleration to orbit given by the LACE engine, re-started and operated now in a rocket mode. The LACE is best described as an airbreathing rocket. It is based on a hydrogen-induced air liquefaction technology. Pump-supplied $LH_2$ is used to liquefy inlet air in a compact cryogenic heat exchanger. From there it goes to a turbopump, after which it is injected into the combustor with the fuel. The utilization of the atmospheric air as the oxidizer raises the specific impulse considerably. The Japanese studied a number of LACE cycles for space planes (Yamanaka 1990) before proposing the air compression cycle with tank circulation for the booster engine.

### 3.5.2.4 Two-Stage to Orbit Programs

Germany began working actively in the mid 1980s on the two-stage piggyback aerospace plane concept known as Sänger (Birch, 1989). This is a reusable horizontally launched system with airbreathing hydrogen propulsion, for which MBB is responsible. An initial phase, up to 1992, was funded by the government as a technology program. An airbreathing, hydrogen-powered engine is under development for the first stage which would use five turbo-ramjet engines, each with a 30 kN take-off thrust. The turbojet engines would accelerate the vehicle to Mach 3.5 and a 25 km height; the ramjet would take Sänger to 31 km and $M = 6.8$, where the second rocket propelled stage would separate.

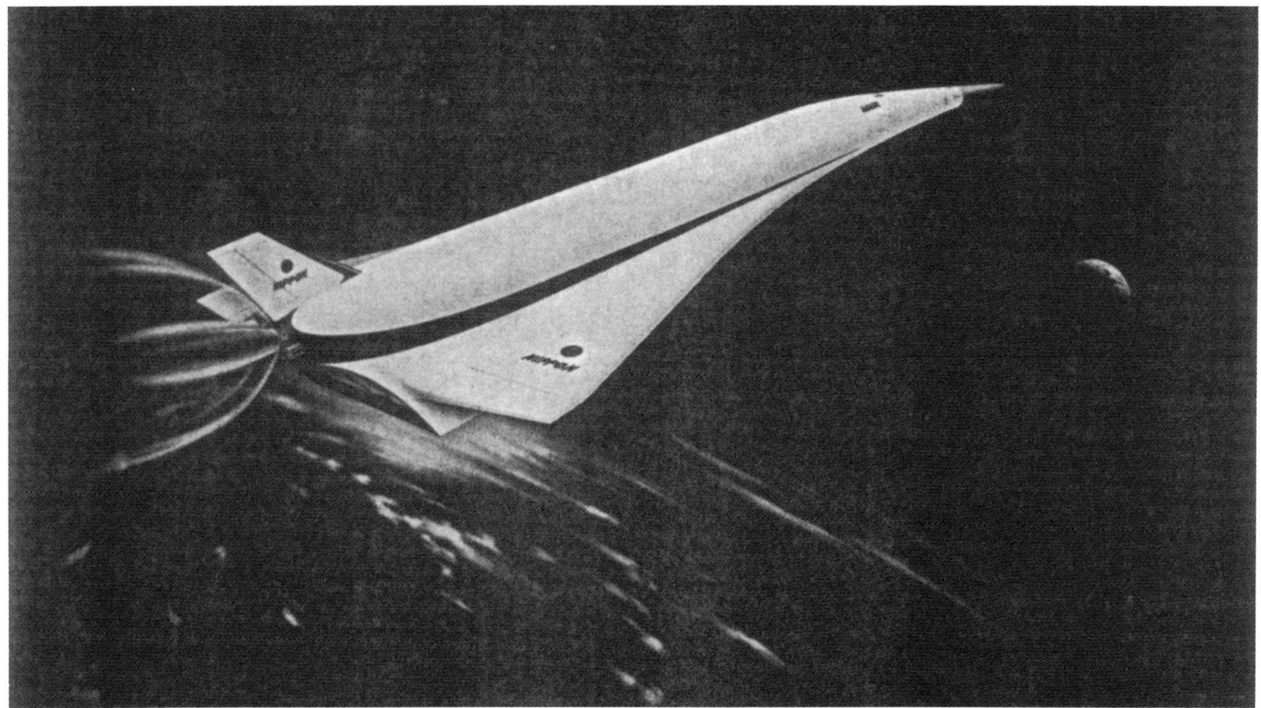

**Figure 3.41** Perspective sketch of SSTO space plane system (Maita 1991). *Copyright © AIAA 1991. Used with permission.*

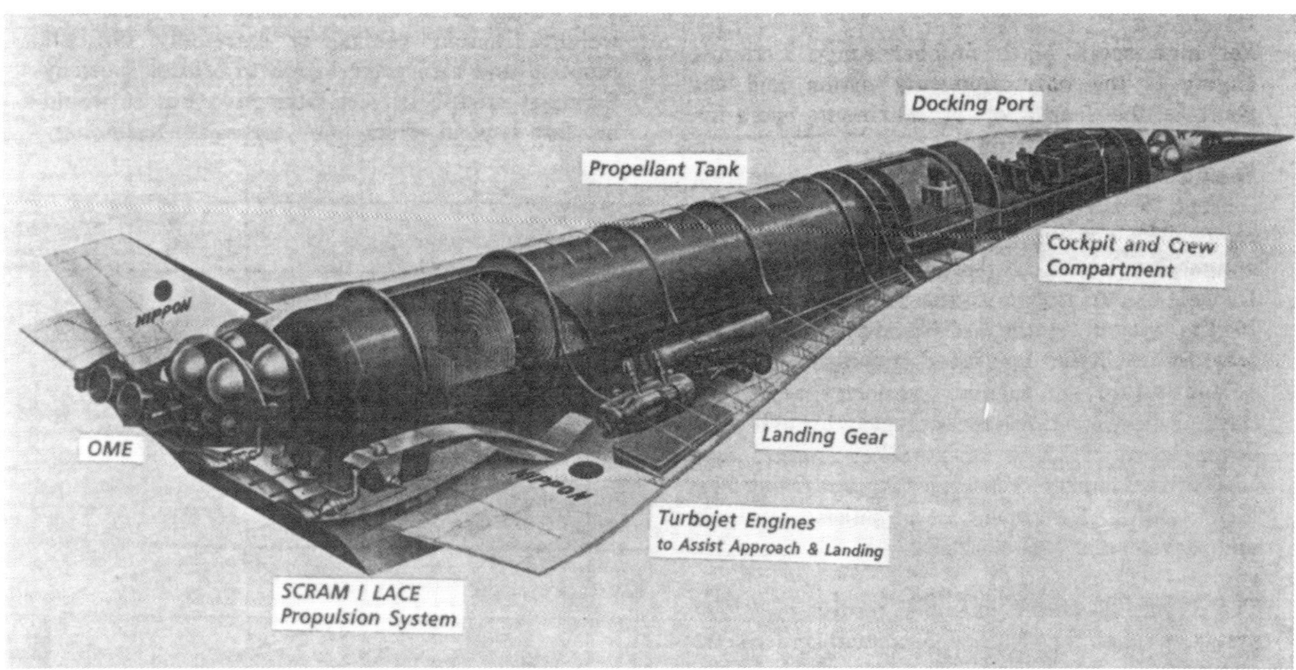


**Figure 3.42** Layout of SCRAM/LACE propulsion (Maita 1991). *Copyright © AIAA 1991. Used with permission.*

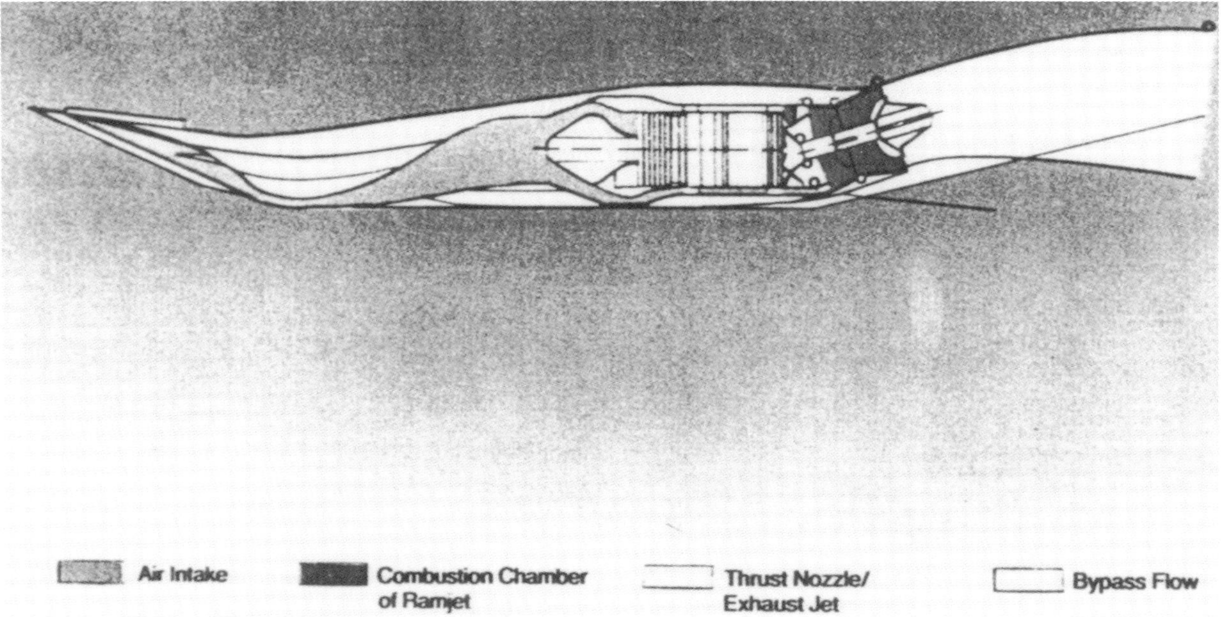



**Figure 3.43**   Sectional view of Sänger's first stage integrated turboramjet engine (Birch 1989). *Courtesy MBB.*

**SÄNGER with HORUS**

**SÄNGER with CARGUS**

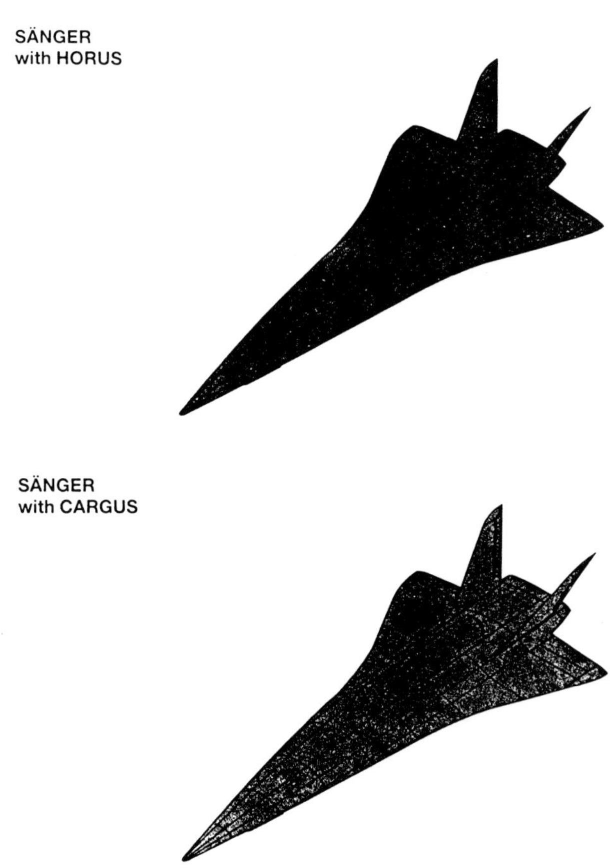

**Figure 3.44**   Sänger with Horus and Cargus (Birch 1989). *Courtesy MBB.*

The following four airbreathing propulsion concepts are being examined and tested for the first stage: a turbojet-ramjet engine combination operating in parallel, another in a coaxial arrangement, an integrated dual-circuit turbofan and ramjet system, and an integrated turbo-expander-ramjet with a hydrogen-powered turbine (see Figure 3.43). Initially Sänger was planned with two different upper stages —a reusable one, called Horus for manned missions and an expendable one, Cargus, for payloads up to 15 tons (see Figure 3.44). A second design cycle was completed in mid 1990. While it confirmed the Sänger concept and the configuration of the Mach 6.8 aircraft, an important change was made for the upper stage; namely, instead of the expendable Cargus, an unmanned version of Horus with a cargo bay was found to be a better solution, mainly on economic grounds. The payload capability will be somewhat smaller, but since Ariane 5 will be available at that time for heavy cargo launches, this is not seen as a limitation. Figure 3.45 shows the new pair, Horus M and Horus C. The mission requirements and the operational capability were restated as follows: for manned space station supply, a useful mass of 3000 kg including a crew of three plus one pilot. The unmanned payload capability was fixed at 7.5 tons for a 200 km orbit and at 6 tons for a 28.5 deg/460 km orbit. The required GTO launch capability is 2300 kg for a 200–36,000 km transfer orbit with an expendable propulsion module; this is equivalent to 1300 kg in GEO. The maximum period of the upper stage mission is 50 hours. The cross-range capability of the second stage must enable it to reach landing sites in Europe up to 52 deg North latitude (this is equivalent to 2700 km cross range from a 28.5 deg orbit). The first stage is required

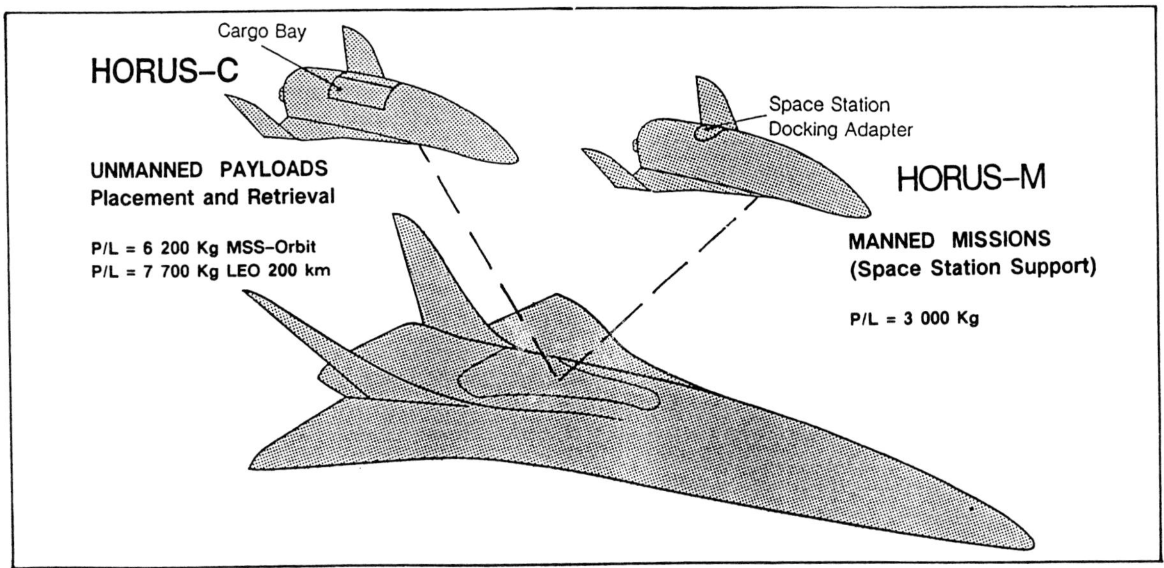



**Figure 3.45** Sänger system configuration for the twin upper stages, Horus C and Horus M (Koelle 1990). *Copyright © AIAA 1990. Used with permission.*

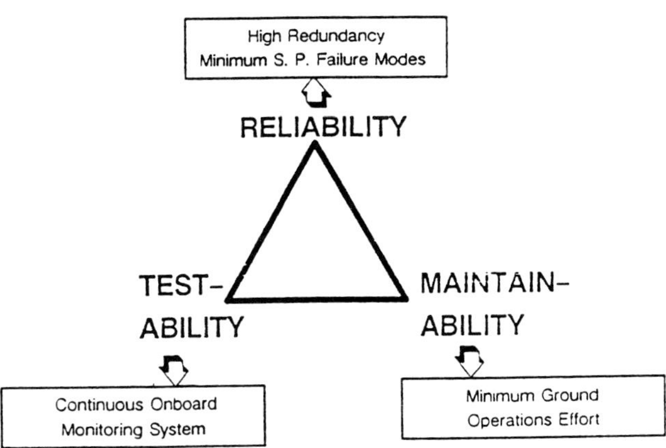



**Figure 3.46** Design requirements for fully reusable vehicles (Koelle 1990). *Copyright © AIAA 1990. Used with permission.*

to have a 2700 km flight range capability to the stage separation point. An operational capability of up to forty launches per year is envisaged, using two vehicle sets.

The main vehicle system design requirements are shown in Figure 3.46 and include very high reliability, achieved by a continuous diagnostic system (at least 100 reuses of the upper sage and 500 for the lower one), horizontal takeoff and landing from conventional airport runways with minimum ground support equipment and maintenance effort between launches, and a reduction of atmospheric pollution. The resulting mass values are shown in Table 3.2. The net mass of the upper stage is the critical performance parameter and it has been fixed at 112 tons. By optimizing this stage (see Figure 3.47), the propellant tank volume was increased by less than 4% over the previous version, leaving the external vehicle dimensions almost unchanged. Figure 3.48 presents a typical flight profile, while Figure 3.49 shows possible missions.

In order to get confidence in the propulsion system, after extensive testing of the regeneratively cooled combustor by MBB, it was proposed (Schaber et al. 1991) to build a hypersonic technology experimental demonstrator (HYTEX), powered by two $LH_2$-kerosene turboramjets derived from the Eurojet EJ200 turbofan which was being developed for the European fighter. The airframe would be built of Titanium, limiting it to a speed of Mach 5.6 for about one minute within the atmosphere. This speed is considered adequate to verify propulsion technology, aero-thermal behavior, and flight-control systems. The Hytex, shown in Figure 3.50, would be 27 m long with a span of 10 m and a takeoff weight of 16 tons, just 4.6% of the final vehicle weight.

In both single and two stage to orbit vehicles the problem of liftoff is solved more economically using airbreathing engines. Among these we shall discuss the turboramjet and various types of air-turboramjets (ATR) following Minoda et al. (1991). Figure 3.51 shows a number of alternatives. The tandem type turboramjet is a direct combination of the turbojet engine (operating up to Mach 3) and the ramjet which extends the flight speed to approximately Mach 7. It maintains a high $I_{sp}$ over a wide Mach range, but needs a complex operation control system, to take care of the transition between the two nozzles. The over/under type, which has two separate units, while simpler to operate, is heavier.

The ATR cycle is characterized by the use of hydrogen as a working fluid in the turbine that provides power to the fan. We consider here the following three types of ATR: (1) The first is equipped with a gas-generator that supplies fuel

**Table 3.2** Preliminary Sänger mass summary (Koelle 1990). *Copyright © 1990. Used with permission.*

| | |
|---|---:|
| TOTAL LAUNCH MASS (kg) | 366,000 |
| First Stage (EHTV): Total Mass | 254,000 |
| Nominal Net Mass | 156,000 |
| Maximum Propellant Mass (LH$_2$) | 98,000 |
| | |
| Second Stage HORUS-M: Total Mass | 112,000 |
| Nominal Net Mass | 28,100 |
| Propellant Mass (LOX/LH$_2$) | 80,900 |
| PAYLOAD 28 deg-Space Station Orbit | 3,000 |
| | |
| Second Stage HORUS-C: Total Mass | 112,000 |
| Nominal Net Mass | 24,000 |
| Propellant Mass (LOX/LH$_2$) | 79,400 |
| PAYLOAD LEO, 200 km-Orbit | 7,700 |

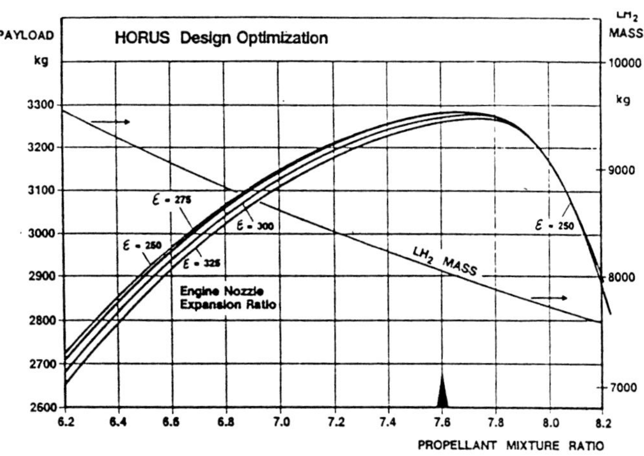


**Figure 3.47** Optimization of propellant mixture ratio and nozzle area ratio for Horus (Koelle 1990). *Copyright © AIAA 1990. Used with permission.*

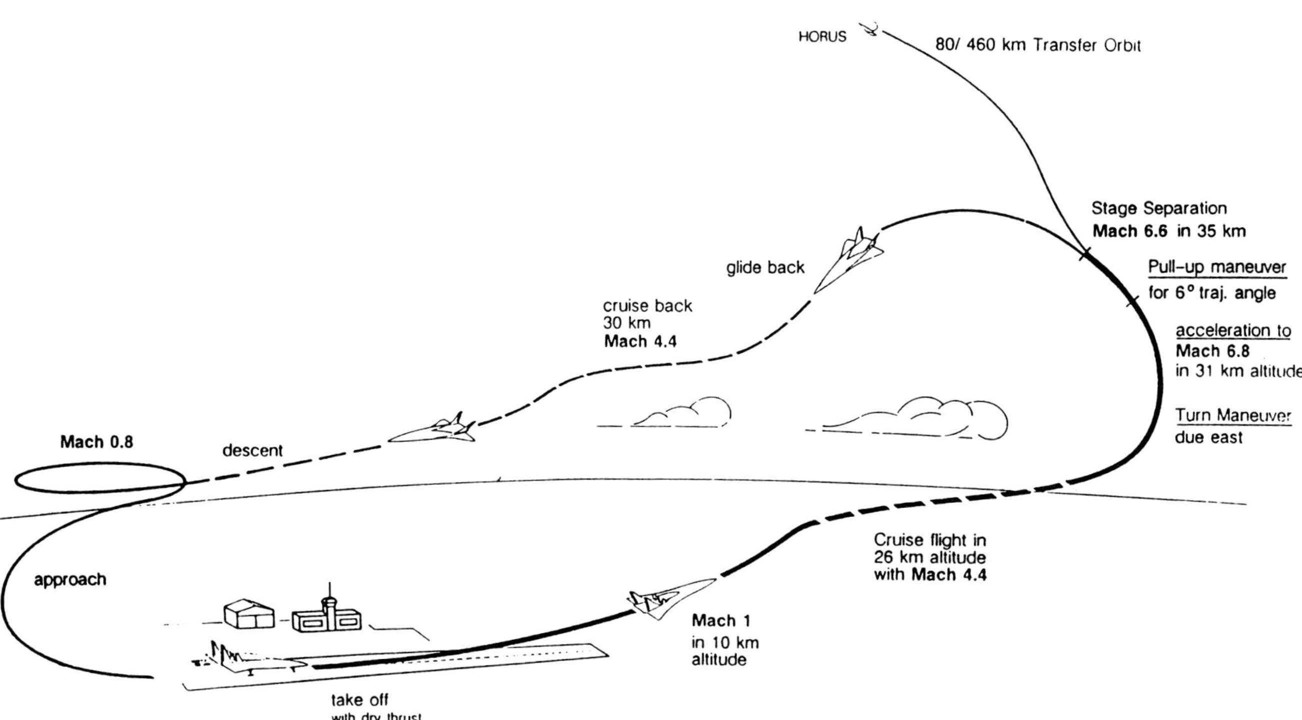


**Figure 3.48** Sänger ascent trajectory with Horus stage separation (Koelle 1990) *Copyright © AIAA 1990. Used with permission.*

rich combustion gas using fuel and oxygen; it is the simplest and lightest of these engines, however its specific impulse is somewhat lower (around 3000 sec). (2) In the expander ATR system the turbine is driven by fuel heated in a heat exchanger located in the combustion gas flow; therefore no oxidizer is required and the $I_{sp}$ is higher (up to 4000 sec). (3) The partial expansion ATR, which has a heat exchanger only around the exhaust system is intermediate between the other two types; here, hydrogen is heated to 500–900 K by the outer heat exchanger before it is burned in the gas generator. Less oxidizer is needed than in the gas-generator

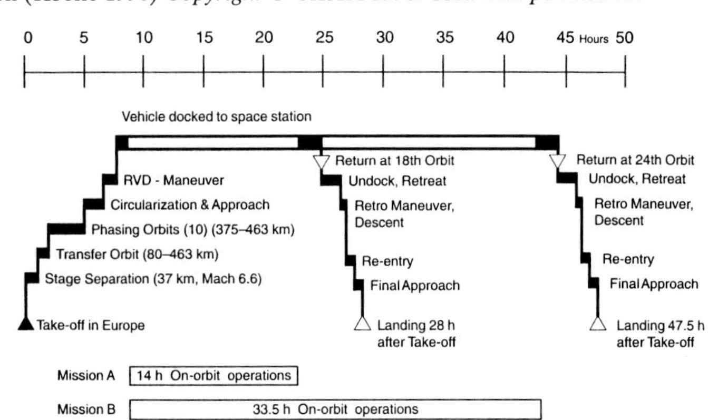


**Figure 3.49** Horus mission operations (Koelle 1990). *Copyright © AIAA 1990. Used with permission.*

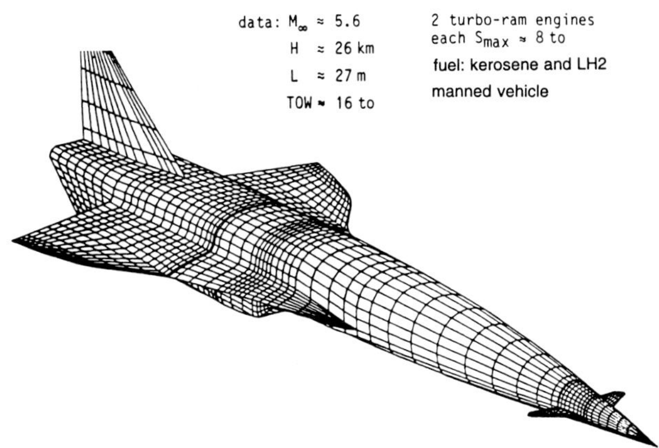



**Figure 3.50**  Sketch of geometry of HYTEX (Bullock 1991). *Courtesy DASA.*

ATR. The specific impulse is between that of the two other variants.

Minoda and coworkers (1991) then analyze the performance of the first two types. In the design of the Sänger different types of ATR's were considered and as we have seen, the expander ATR was chosen.

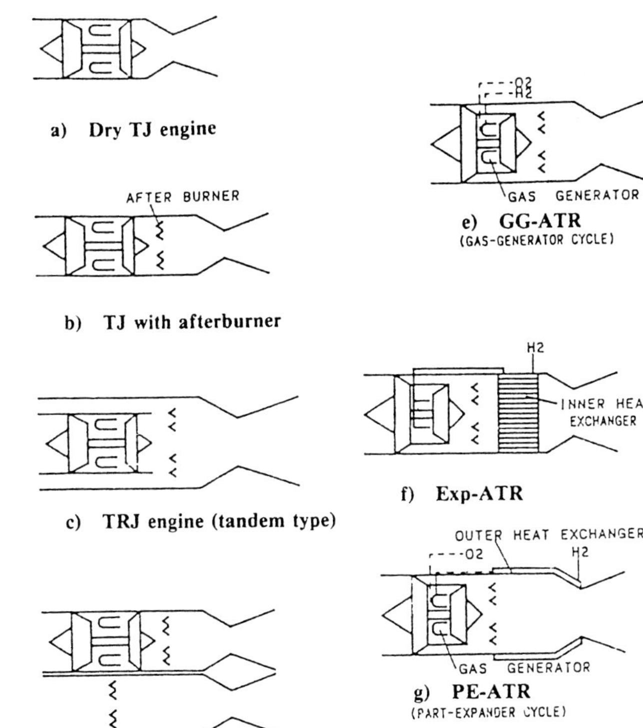



**Figure 3.51**  Different types of ATR (Minoda et al. 1991). *Copyright © AIAA 1991. Used with permission.*

# Chapter 4

# *The Pulse Jet*

In pulsating combustors a proper feedback between the flow and combustion processes results in periodic variations of the flow properties inside the combustor. The German Second World War V-1 vehicle (better known as the Buzz bomb—see Gosslau 1957), that caused such damage and fear in London is probably the best known example of a pulsating combustor. While the V-1 flying bomb was based upon the Schmidt tube principle, other well-known pulsating combustors operate on the Helmholtz resonator and Rijke tube principles. As an example, the operating principles of a Schmidt type combustor will be briefly outlined in the following paragraphs (Thring 1961).

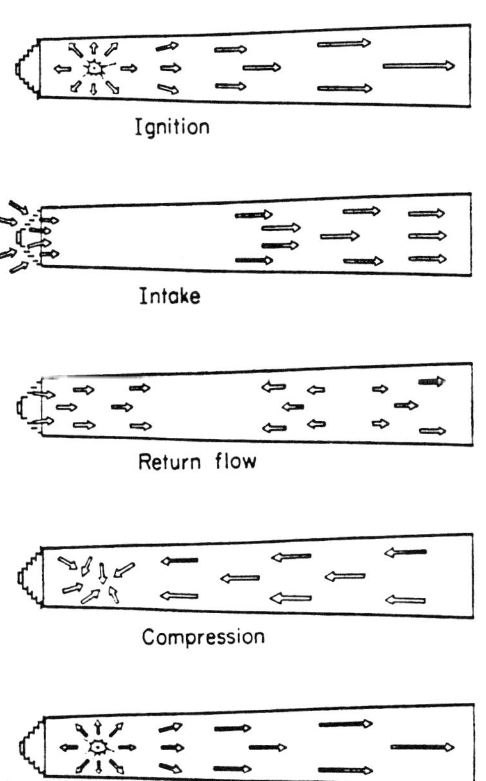



**Figure 4.1**  Processes in the Schmidt tube in one cycle (Thring 1961). *Copyright © Pergamon Press Ltd. 1961. Used with permission.*

Figure 4.1 is a diagram of the tube, which is closed at one end by flap valves. A stoichiometric fuel/air mixture is let into the tube and ignited once only by means of a spark plug. The result is combustion going over to an explosion producing a steep pressure rise and hence the flow of the burning mixture towards the open end of the tube, the inlet valves being closed by the pressure rise and thus preventing movement in the opposite direction. The inertia of the flowing gas column eventually produces negative pressure in the tube, so that the valves are reopened and air flows into the tube, mixing with the fuel also entering the tube from a special volume capable of responding to oscillation, to form a new explosive mixture. In the meantime, the column of exhaust gas flows back into the tube from the open end and overcomes the suction or is followed by air which causes a pressure rise. The air entering from the open end is of the same order as the quantity drawn in through the flaps. Through the combined action of pressure, temperature, and radical content of the exhaust gases, the fresh mixture is ignited, and ignition completed by explosion, thus starting a fresh cycle. In- and outflow are superimposed on the oscillation of the gas column, which must follow acoustic laws. The frequency of the tube is that of a pipe open at one end, that is to say that the wave length of the oscillation is four times that of the acoustic tube length. Of particular interest are the temperature curves and the effects on the oscillation intensity, during this reaction.

Apart from use during World War II as propulsion units for the German V-1 "flying bomb," pulsating combustors first came on to the market in 1948 and were developed to heat the passenger area of public service vehicles. These early pulsating combustors, with a heat output of 4.81 kW (5,000 kcal/h), were used in cold weather to heat a mixture of water and antifreeze solution which flowed into hot water radiators in the passenger area where the warm air was circulated by an electric fan.

## 4.1 Types of Pulsating Combustors and Their Uses

There appear to be at least four basic types of pulsed combustors for practical applications. In the first type, with powered rotary inlet valves, air is supplied to the combustion chamber through a rotary valve running at a prescribed speed, hence controlling the operational frequency; however, the frequency may be adjusted to optimize performance. In the second type, with mechanical inlet valves, an array of automatic flapper valves is used in the entrance region of the air and/or fuel lines to prevent back flow through the inlet. In the third type, aerodynamically-valved units, which are often termed valveless pulsed combustors, an inlet is so shaped that the pressure drops in the two flow directions through the inlet are greatly different being much higher in the back flow direction, for equal flow rates (Bertin et al. 1954). In the fourth type, the valveless pulsed combustor is equipped with a heavy-current fluidic device; this arrangement may be referred to as the flow rectifier system. The device serves to maintain the pressure drop of the back flow, as well as that of the inflow, as low as possible. In addition, the backflow is redirected rearwards with a minimum of pressure drop. In all four types of pulsed combustors, the mechanism of operation is basically similar.

A wide variety of uses for pulsating combustion has been suggested. Some have been investigated experimentally with various degrees of success. Some of the advantages claimed for pulsating combustion are: (1) improved combustion intensity; (2) improved heat-transfer characteristics (Putnam and Brown 1974); (3) reduced $NO_x$, smoke, and slag formation (Vedikhin et al. 1979); (4) effective cleaning of heat-transfer surfaces by acoustic motion (Vedikhin et al. 1979); (5) self-aspiration of the oxidizer flow, which eliminates the need for fans or blowers for forcing the oxidizer flow through the combustor; (6) a potential for attaining a pressure rise in the combustor leading to improved cycle thermodynamical efficiency (Kentfield 1980); (7) a potential for utilizing higher maximum temperatures in the combustor for improved cycle efficiencies (Foa 1960). Countering these advantages are the noise and possible vibration of various system components that normally accompany pulsating combustion, disadvantages that could be overcome by modern technology in applications involving stationary power sources.

The best known application of pulsating combustion is the Schmidt-Argus pulse jet of the German V-1 "Buzz Bomb" used in the Second World War. A mechanically valved pulsed combustor was used to propel a small unmanned airplane with an explosive warhead (Gosslau 1957).

## *4.2 Research and Applications*

After World War II the valveless SNECMA "Escopette" (Blanderbuss) pulse jet (Bertin et al. 1954; Foa 1960) was developed for the propulsion of target aircraft. A later development was the Ecrevisse (Crayfish) (Tharrat 1965 and 1966) in which the inlet is bent to face the same way as the exhaust in order to use not only the main exhaust thrust but

also that developed by reverse flow through the air intake to counter the difficulty of the negative momentum transfer. The use of a separate short tube, a thrust augmenter, was found to improve aspiration. The specific fuel consumption of the "Escopette" was 1.8 kg/h-kg thrust.

Later the Thermo-Jet company developed a valveless unit in which three (sometimes four) air inlet tubes with bell-mouth entrances were used. Vaporized fuel (propane or butane) was injected axially in these tubes.

Lockwood (1964) applied two augmenters to the Ecrevisse pulse jet in developing a high thrust-to-weight ratio vertical lift device. This would be particularly useful in situations where ingestion of dirt could occur with consequent risk (for conventional engines) of foreign body ingestion.

Reynst, Thring (1961) and Lockwood (1964) also discussed various applications of pulsed combustors to aircraft wing surfaces to assist propulsion and achieve a drag decrease.

It would seem that unless the noise problem is solved, the application of pulsed combustors to propulsion will be largely confined to target drone type vehicles. Fundamentally, the problem of how much noise is likely to be emitted by a pulsed combustor, is a complex one. A study of the noise problem was reported (Hanby and Brown 1971) and the following techniques were investigated with the objective of minimizing the noise:

i) Coupling two combustors together so that pressure and velocity fluctuations are canceled out at the junction.

ii) Reducing the exit diameter of the combustor; this reflects a portion of the incident waves back without a change of phase.

iii) Using quarter wave damping tubes connected to the resonance tube.

In all configurations tried in this investigation a reduction in radiated noise was accompanied by a drop in combustor performance.

Putnam (1980) gives a short review of component options in pulse combustors, treating both valved and valveless systems. He gives a table listing the major components:

**Table 4.1**   Major pulse combustor components (Putman 1980).

1. Air-supply systems
2. Fuel-inlet valves
3. Back-flow systems
4. Resonance tubes
5. Combustion chambers
6. Inlet mufflers and terminations
7. Exhaust mufflers and terminations
8. Starting-air systems

Kentfield (1980) gives a review of recent progress in pressure gain combustion, which covers, apart from his own contributions, work done in Australia (Catchpole and Runacres 1974), France (Bertin 1954; Tharrat 1965 and 1966), South Africa, and the United States (Porter 1958).

Kentfield's own work is divided into three categories: performance prediction, fundamental experimental work, and applications-oriented experiments. The description of the performance prediction work traces the development of a method-of-characteristics approach through three phases starting with hand-drawn wave diagrams and advancing, via a computerized "cold-flow" simulation of the pulsed pressure-gain combustor cycle, to a more recently developed "hot-flow" computerized numerical model incorporating experimentally obtained heat-release rate data to assist in the representation of the combustion process.

The description of the fundamental experimental work carried out in connection with pulsed, pressure-gain combustion reviews briefly: experiments designed to reveal the sensitivity of valveless pulsed combustors to small changes of geometry, the use of multiple inlets as an aid to reducing combustor length, and increasing cyclic frequency, the rectification (catching and redirecting downstream) of the back flow from the inlets of valveless pulsed combustors, noise spectra, exhaust emissions, and the influence on specific performance of scaling-up from small size laboratory units. The description of the applications-oriented experiments with pressure-gain combustors refers to a forced-convection space heater for use, e.g., on construction sites, a compact high-intensity heater for arctic applications, and a pressure-gain combustor for use in gas turbines and large-scale heating systems.

Chiu and Croke (1980) deal analytically with the noise emission characteristic of pulse combustion and treat also the problem of combustion performance; their model gives good qualitative agreement with experiments performed at the Argonne Laboratories.

Dec and Keller (1989) performed an experimental heat transfer study in the tailpipe of a pulse combustor, varying pulsation frequency and amplitude and the mean flow rate. The Nusselt number, derived from thermocouple measurements, showed a strong enhancement, up to 2.5 times the steady flow value, for both amplitude and frequency.

Goldman and Xieu (1981) and Libbis and Goldman (1992) investigated a new method for achieving pulsating combustion based on a double chamber system using symmetrical injectors for the self-feeding of air. They found that entrance conditions, distance between nozzles, the conical angle of convergent-divergent nozzles, and the temperature of the jet strongly influence the injector efficiency. Location of a peak of entrainment ratio for a particular feeding device was defined. An investigation of a pulsed combustor with two opposed chambers, lined by refractory material, confirmed that the system can be adjusted to produce oscillating combustion also without acoustical tailpipes and that the nature of the pulsations and their frequency are controlled by flame propagation within each chamber. An optimization process was exploited to find the tailpipe geometry, giving the highest thrust. It was found that a straight tailpipe gives the best results.

## 4.3 The Pulsation Driving Mechanism

Goldman and Timnat (1983) and Libis and Goldman (1992) present an up-to-date discussion of the mechanism driving the pulsations. Figure 4.2 describes the flame development in the combustion chamber, showing the gas composition, as well as the temperature and pressure variations, as obtained experimentally, using fast photography, a suction pyrometer, pressure gages, and gas analysis. Air appeared as a fully transparent medium, while the rich mixture ($N_2$, $CO_2$, $CO$, $H_2O$, and $H_2$) was blue and the high temperature combustion products were yellow to transparent in color.

At time $t = 0$ the combustion cycle begins an air intake stroke. The temperature drops as the burned gases are exhausted and fresh air enters the combustor. The resonating effect from the pulsating combustor and this drop in temperature cause the pressure to fall until the combustion chamber is mostly filled with fresh air and incompletely

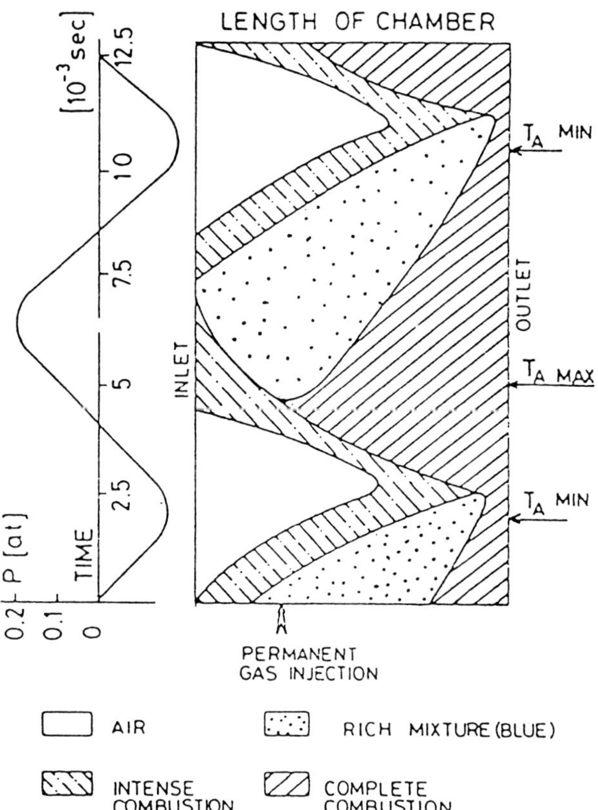


**Figure 4.2** The flame development in a pulse-jet (Goldman and Timnat 1983). *Copyright © 14th Int. Symp. on Shock Tubes and Waves. Used with permission.*

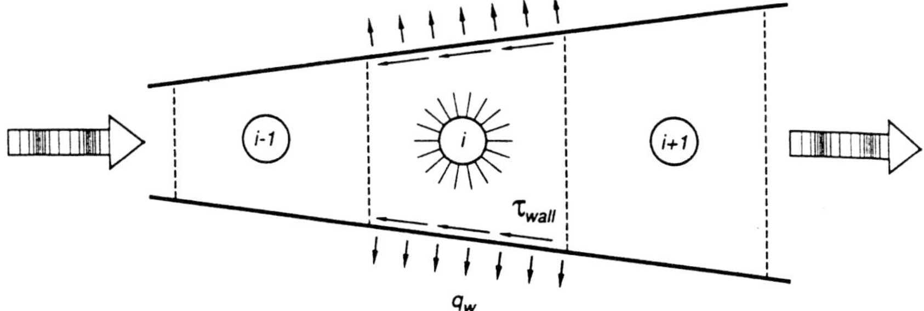

**Figure 4.3**  Typical control volume for pulse combustor (Barr and Dwyer 1991).
*Reprinted by permission of Kluwer Academic Publishers.*

burned gas at a minimum temperature. Then remnants of hot combustion products reignite the fuel-air mixture and cause the flame to propagate, thus increasing the temperature and driving the pressure above atmospheric.

The pressure cuts off the fresh air supply and the cycle is in the exhaust stroke. Maximum temperature occurs as all the air within the chamber is used in the combustion process when most of the chamber is filled with completely combusted products. Fuel continues to fill the combustion chamber behind the escaping gases. The temperature and the resonating effect within the combustor cause the pressure to increase until a maximum pressure is reached in the over-expanding chamber. The pressure then tends to decrease, and as it falls below atmospheric, fresh air rushes in to mix with the fuel and hot gases and the cycle begins again.

The average equivalence ratio measured over a cycle is 1.2 (gas rich). Supplementary computations show that it goes to 1.11 at the end of the exhaust stroke and reaches 1.27 at the end of the suction stroke. The corresponding gas temperature variations should be about 200°C and this explains the process of driving oscillations. The thermodynamic efficiency of the cycle and the mechanical energy gain are associated with the chemical energy release during the compression stroke and with the cooling of the gas volume by supplying fresh air during the suction stroke.

Libis and Goldman (1992) also developed a theoretical treatment of the pulsejet. It consists of two stages, one dealing with the operation of the combustion chamber, the other combining this model with another, adequate for the description of the pulse ejector operation. In an actual calculation one starts with the model for the nonsteady flow in the inlet and exhaust pipes, using a numerical method based on the hybrid approach, first advocated by Hartree (1958). This combines the best features of the characteristics analysis and the fixed grid methods. Then one models the combustion chamber, using energy and mass conservation and making simplifying assumptions, such as no heat transfer across the boundaries, fluid properties depending only on time, negligible kinetic and potential energy changes and ideal gases. Appropriate boundary conditions between the combustion chamber and the pipes must be prescribed, as

well as an ignition delay, which influences both the amplitude and the frequency of the pulsations.

## 4.4 Theoretical Study of a Pulsating Combustor

Numerical models for simulating the dynamics of a Helmholtz type pulse combustor have been described by various investigators. The first was proposed by Craigen (1976), who tracked the wave dynamics in the tailpipe by the method of characteristics. Ponizy and Wojcichi (1984) combined the method of characteristics with a well-stirred reactor to simulate the combustion process. Tsujimoto and Machii (1986) combined lumped parameters models with the method of characteristics to simulate the behaviour in more complex pulse combustor geometries. Barr, Dwyer, and Bramelette (1988) simulated pulse combustor dynamics with a one-dimensional finite difference model including losses due to heat transfer and wall friction. This model required experimental measurements of the cyclic energy release and injection processes as input data. We shall describe a recent model due to Barr and Dwyer (1991) which uses the finite difference model and includes the prediction of the flow through flapper valves.

The authors start by numerically simulating the one-dimensional, unsteady equations of gas dynamics. A typical control volume is shown in Figure 4.3 and the basic equations are:

(1) Continuity:

$$\frac{\partial}{\partial t} \iiint_V \rho \, dV + \iint_S \rho \mathbf{U} \cdot \mathbf{n} \, dA = 0$$

(2) X-momentum:

$$\frac{\partial}{\partial t} \iiint_V \rho \mathbf{U} \, dV + \iint_S \rho \mathbf{U} \mathbf{U} \cdot \mathbf{n} \, dA = -\mathbf{n} \iint_S p \, dA - \iint_W \tau_{\mathbf{w}} \, dA_W$$

(3) Energy:

$$\frac{\partial}{\partial t} \iiint_V \rho e \, dV + \iint_S [\rho e + p] \mathbf{U} \cdot \mathbf{n} \, dA$$
$$= -\iiint_V \dot{Q} \, dV + \iint_W q_W \, dA_W$$

where $e = c_V T + \mathbf{U}^2/2$.

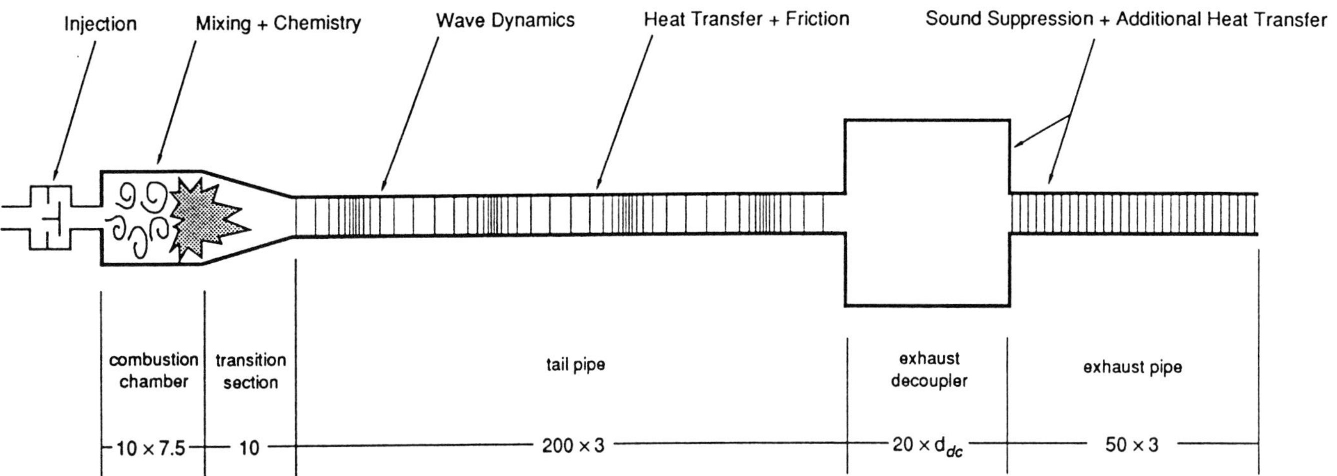


**Figure 4.4**   Exhaust decoupler system (Barr and Dwyer 1991). *Reprinted by permission of Kluwer Academic Publishers.*

(4) Equation of state:

$$p = \rho RT.$$

The local duct cross section is $A$ and the unit vector normal to it is $\mathbf{n}$. $A_w$ is the local duct wall surface area, $\tau_w$ the wall shear stress, and $q_w$ the local wall heat transfer. $\dot{Q}$ is the chemical energy release per unit volume. Subscripts $V$, $S$, and $w$ indicate integration over the cell volume, the cell face surface area, and the duct wall area, respectively.

This is a system of hyperbolic partial difference equations and the best method to solve it is due to McCormack (1969). He developed an explicit technique with the time step chosen to integrate the equations closely related to the local characteristic wave speed. With the addition of artificial viscosity it can give good resolution of shocks and contact surfaces. Incorporation of complex physics, chemistry, and multiphase models and phenomenologies are straightforward. Complex geometries and realistic boundary conditions can also be implemented. The basic method is to use a predictor step followed by corrector formulas.

The major additional numerical modelling issues introduced into the gas-dynamic equations by Helmholtz-type pulse combustor are injection of fuel and air into the combustor, wall friction due to viscous stresses, heat losses caused by the cooling of the duct walls, and energy release from the chemical reaction in the combustor. The injection of fuel and air into the combustor is treated as a boundary condition, so it is not necessary to include it explicitly in the flow equations. The other three terms are evaluated at cell centers, since they do not involve directly wave dynamics. The influence of wall friction, wall heat transfer, and chemical energy release involve multidimensional viscous and heat transfer effects as well as chemical kinetics. They can be introduced only in an approximate way in a one-dimensional analysis, either by quasi-steady approximations or by experimental measurements from similar pulse combustion systems.

The authors then describe the expressions used for the friction factor, the film coefficient and the energy release. Finally the system of equations is closed by imposing appropriate boundary conditions at the end of the cells in the combustor and in the tailpipe.

The paper is concluded by the simulation of two Helmholtz-type pulse combustors—one consisting only of a combustor and a tailpipe, which the authors call a "bare bones" pulse combustor, which is used primarily as a research tool to study the fundamental operation of such a device, containing the minimum features to operate. The other includes an exhaust decoupler system shown in Figure 4.4 which terminates the acoustic wave. The reader is referred to the original paper for the numerical results which include calculations of pressure, velocity and temperature profiles, pulse frequency, and efficiency.

# Chapter 5

# Test Facilities

Test facilities suitable for ramjet and scramjet research and development include many types that are common to other flight vehicles. Among them the most important are shock tubes, shock tunnels, and wind tunnels. The first two are particularly well suited for experiments on small models and testing of materials that can withstand high temperatures. Wind tunnels can simulate conditions encountered in flight at subsonic, transonic, and supersonic velocities and different pressure levels. As these facilities are well covered in the literature (for example, Bueteflisch 1989), we shall only describe some new hypervelocity wind tunnels; apart from this we will mainly treat devices, which were developed especially for ramjets and scramjets.

There are two unique types of facilities associated with ramjet research—the air-connected pipe and the freejet; for scramjets the range of existing installations has been extended and new ones have been erected reaching Mach 12.

**Table 5.1**  Ramjet development testing (Dunsworth and Reed 1979). *Copyright © AIAA 1979. Used with permission.*

*Components*
    Inlet
    Fuel system
    Combustor
    Nozzle
    Rocket booster
    Complete engine

*Objectives*
    Component performance
    System performance
    Operating characteristics
    Structural integrity
    Heat transfer
    Functional compatibility

*Facilities*
    Wind tunnel
    Rocket test cell
    Fuel system
    Test bench
    Direct-connect test cell*
    Freejet test cell*

*Unique ramjet simulation requirements.

Above this speed there are two specialized installations, the free-piston shock tunnel and the expansion tube, which can provide information on relatively small models at very high speeds and enthalpies, corresponding to flight Mach numbers above 20. Although these facilities operate for short times (on the order of milliseconds) they can validate results obtained by CFD calculations.

## 5.1 Development Testing Facilities

Dunsworth and Reed (1979) discuss in detail ramjet engine testing and simulation techniques used in the United States, with particular attention to integral rocket ramjet systems. The first step is component development, followed by further testing of complete systems, culminating in flight tests. Table 5.1 indicates the principal components, the major objectives of development testing, and the type of test facilities employed. As stated before, unique simulation requirements are necessary for the development of the combustion chamber and the complete engine and are performed in direct-connect and freejet facilities. The parameters simulated are given in Table 5.2.

Connected pipe tests are conducted primarily to demonstrate combustor thermal-structural design integrity and combustion performance (see Anderson 1975 and Beach 1979) using direct thrust stand measurements or total pres-

**Table 5.2**  Parameter simulation (Dunsworth and Reed 1979). *Copyright © AIAA 1979. Used with permission.*

    Stagnation pressure
    Stagnation temperature
    Mach number
    Pressure ratio
    Air mass flow
    Trajectory conditions
    External heating
    Maneuver loads
    Angle of attack
    Forebody flowfield
    Transition sequence

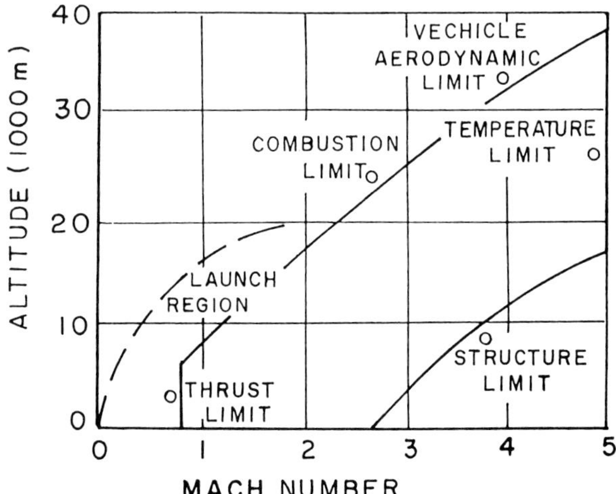



**Figure 5.1** Ramjet operating regime (Dunsworth and Reed 1979). *Copyright © AIAA 1979. Used with permission.*

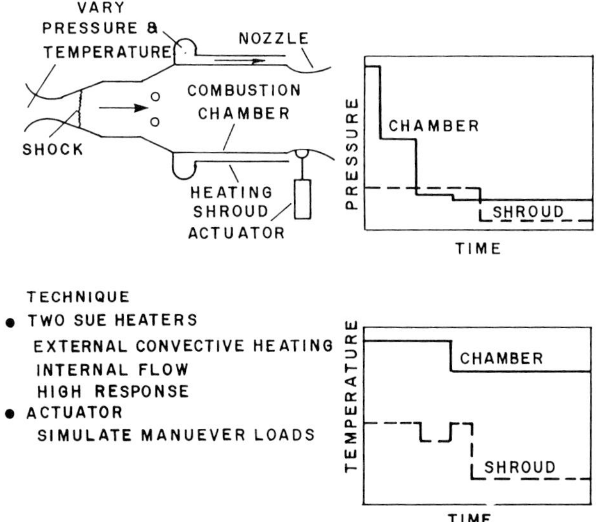



**Figure 5.2** Trajectory simulation in air-connected test facility (Dunsworth and Reed 1979). *Copyright © AIAA 1979. Used with permission.*

sure and temperature nozzle values. Precise input conditions of airflow, inlet temperature and fuel flow can supply accurate combustor performance data; in advanced facilities it is possible to simulate an entire trajectory time profile.

Freejet tests are conducted primarily to verify operation and performance of the integrated propulsion system in a supersonic or hypersonic field. It is desirable that the model should incorporate as many flight design systems as possible (i.e., vehicle forebody equipment and telemetry systems, pyrotechnical and ignition devices, inlet fairings, fuel management, and auxiliary power systems, booster/combustor units, etc).

Power requirements for continuous flow are extremely high, therefore such facilities are of the blowdown type. The Arnold facility can store 200 tons of air at a maximum pressure of 25 MPa, while the recently enlarged Marquardt complex, (Birch 1989) could handle already in 1980 about 80 tons of air at 19 MPa (Marquardt 1968). Table 5.2 lists the important parameters simulated in a test facility. Figure 5.1 indicates the operating regime for a subsonic combustion ramjet pointing out the various limits.

One of the major simulation problems is presented by the low-altitude boundary of the flight envelope, since here acceleration is high and the structural environment particularly severe. An important technique developed is the sudden expansion vitiated air burner with oxygen replenishment, called by Marquardt the SUE heater (Dunsworth and Reed 1979).

## 5.2 Air Connected Test Arrangements

A substantial portion of the engine development can be performed using a direct-connect test arrangement. Here the inlet is eliminated and the combustor is connected to the facility air supply (see Figure 5.2a). The air flow requirement is much lower than the freejet testing so the run time is relatively long and allows to simulate time-dependent flight trajectories, as shown schematically in Figure 5.2b. The inlet is replaced by a duct with a choked throat, followed by a supersonic expansion and a diffuser normal shock train similar to the internal structure of the inlet. This allows one to vary the fuel flow at constant air flow, simulating supercritical operation of the ramjet air inlet. A heavy-duty combustor can be used to investigate the fuel injector, flame holder, and exit nozzle. Adding an external heating shroud allows one to extend the technique to flight-weight combustors and combustion chamber insulators. One can match the calculated external heat for representative flight trajectories by flowing outside heated air through the shroud under controlled temperature and pressure conditions. Simulated maneuver loads can be applied to the hot structure during a test by hydraulic actuators. A photograph of a rocket ramjet with ducted intakes taken in the Modane S4 wind tunnel is shown in Figure 5.3 (Marguet and Cazin 1985).

One of the important parameters in solid fuel ramjets is the regression rate of the fuel, and knowledge of its behaviour is mandatory for a good understanding of these motors and for performance predictions.

In general, the regression rate may depend upon the location in the fuel grain while usually the regression rate will also vary in time. Therefore, it is customary to apply average values for the regression rates with respect to location and time. Most performance data that are related to the regression rate are given for these averaged regression rates. This

**Figure 5.3** Photograph of rocket-ramjet with ducted intakes (Marguet and Cazin 1985). *Courtesy of O.N.E.R.A. Copyright © AIAA 1985. Used with permission.*

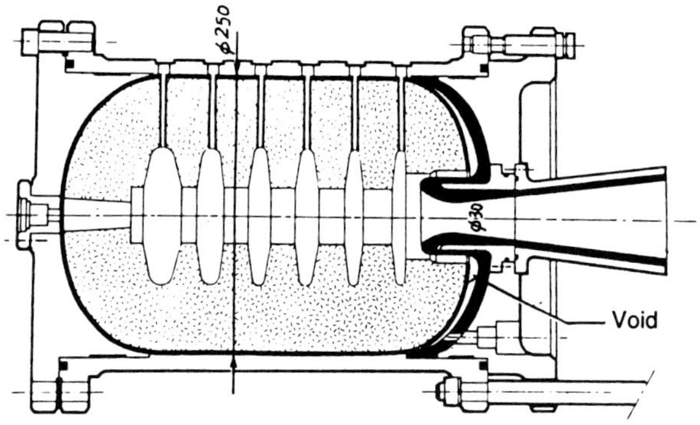


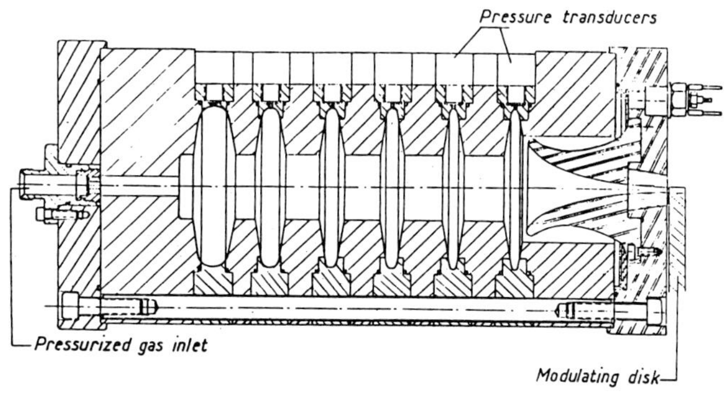


**Figure 5.4** Reference engine for ultrasonic technique (Kuentzmann and Lengellé 1977). Figure 1 used in AGARD CP 259, Solid Rocket Motor Technology, paper 22, fig. 1.

is understandable because it is these data that are most easily obtained from experiments. However, to obtain more insight in the process taking place in solid fuel ramjets (or hybrid rocket motors), a precise knowledge of local and instantaneous regression rates is of a great help.

An accurate and simple method to obtain local and instantaneous regression rates is by high speed video or motion pictures. However, these methods can only be applied to two dimensional fuel grains between two quartz windows. A grid system painted on one of the windows helps to obtain accurate regression rate data. The method has been applied successfully by the NLR in the Netherlands, but has a distinct disadvantage of analyzing the pictures frame by frame.

The most common techniques, however, are to determine an averaged regression rate from the weight loss of the grain or from the change of its inner dimensions. Although these latter techniques are the simplest ones in practice, much information may be lost because of the effect of averaging. If one wants to correlate the regression rate with parameters such as mass flux, combustion pressure, geometry, etc., then

more involved and more expensive techniques may be required.

Some of the drawbacks associated with the above mentioned techniques can be avoided by using an ultrasonic regression rate analysis technique.

### 5.2.1 Determination of the Regression Rate in Solid Fuel Ramjets by Means of the Ultrasonic Pulse Echo Method

This technique, which was developed by Korting and Schöyer (1985), is based on work performed by Kuentzmann and Lengellé (1977) for solid propellants. They developed it in a reference engine represented in Figure 5.4. The ultrasonic sensor excited by an electric pulse emits a mechanical wave that propagates towards the combustion surface, is reflected at the solid-gas interface because of the acoustic impedance discontinuity, and comes back towards the detector that then delivers an electric signal. They measured the mean travel time $\tau$ of the wave within the material, and deduced, assuming as a first approximation that the sound velocity is constant, the instantaneous thickness of

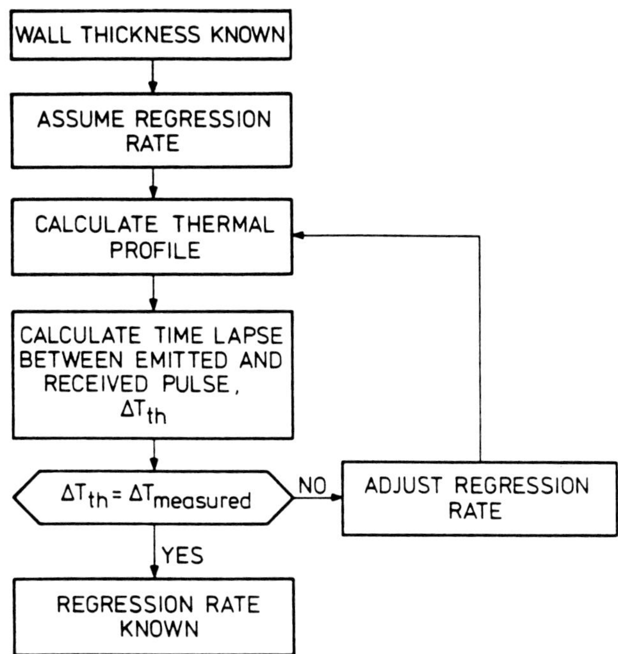

Figure 5.5 Iterative scheme for regression rate measurements (Schöyer and Korting 1984). *Used with permission.*

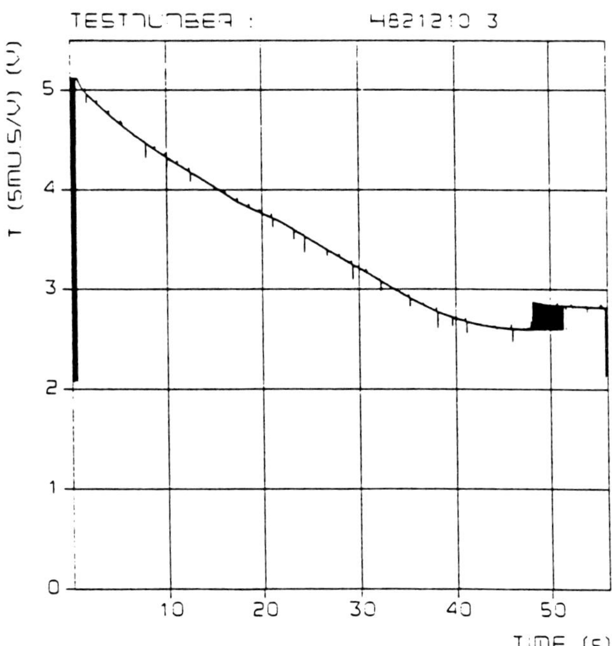


Figure 5.7 Time lapse history of emitted and received ultrasonic signal (Schöyer and Korting 1984).

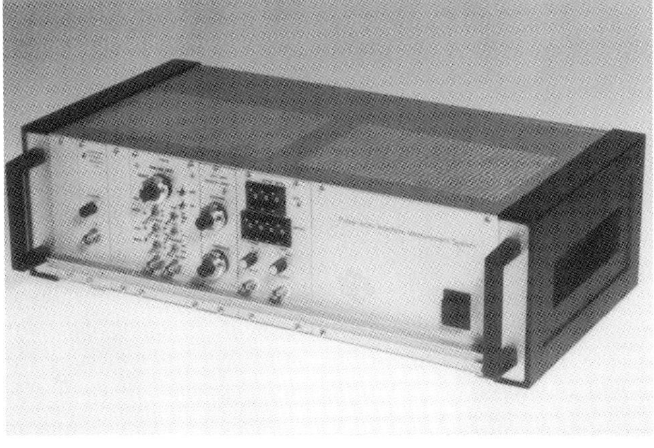

Figure 5.6 Ultrasonic regression rate analyzer (Schöyer and Korting 1984).

the propellant; since time is of the order of 50 μsec, this operation may be repeated up to 10,000 times per second. After filtering, the thickness signal is derived with a view to obtain the combustion rate. A coupling element is installed between the ultrasonic detector and the propellant grain; it makes it possible to ensure the measurement up to the complete combustion of the propellant which makes easier the calibration of the device. The maximum thickness of the propellant grain is limited to 40 mm for composite propellants. The operating pressure of the engine may exceed 10 MPa. The results obtained, when pressure is almost constant in time, are of a precision comparable to that of classical measurements. Korting and Schöyer developed an it-

erative procedure that takes into account the fact that there is a temperature profile in the fuel, which depends on the burning rate and the thermal diffusivity. This is shown in Figure 5.5. One of the probes they used for the emission and reception of sound also receives the reflected pulse. The probe and the ultrasonoscope are commercially available equipment. Simultaneously the emitted and reflected pulse are also sent to an ultrasonic regression rate analyzer (see Figure 5.6), they developed. It determines the time lapse between the emitted and received pulse and transforms this lapse into a suitable output signal, which is registered. After a test run the registered data are computer processed and the instantaneous and local burning rates are obtained.

A typical example of the time lapse as registered during an experiment is shown in Figure 5.7. Initially the signal is strongly disturbed, which was attributed to the ignition by a spark plug and the subsequent build-up of the thermal layer. After a short time a rather smooth continuous signal develops. They showed that there is a very good agreement between the mean regression rate as obtained by the ultrasonic technique and the mean regression rate based upon the increase in interim diameter. Figure 5.8 shows a pressure history and the corresponding regression rate history for an experiment with plexiglas as a fuel and a mixture of oxygen and nitrogen.

The very high regression rate observed during the first second is believed to be due to two causes: they ignited with a mixture of hydrogen and oxidizer which causes very high temperature initially. This also explains the high pressure peak during the first two seconds of the test run. The second cause is associated with the data reduction method.

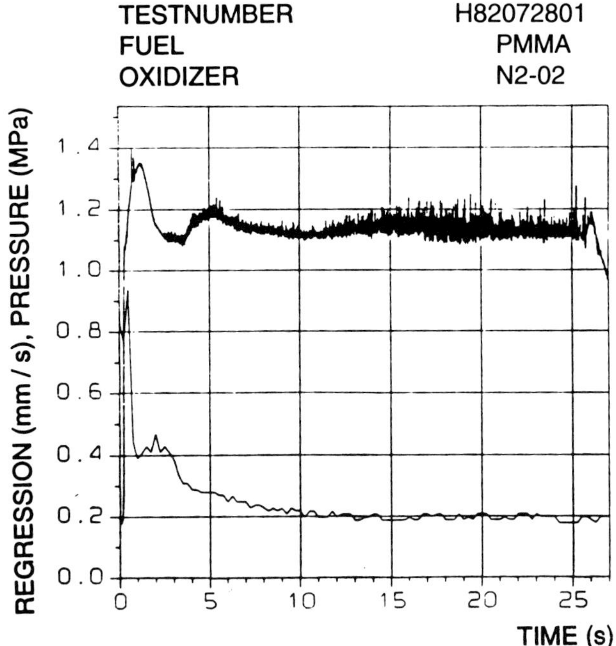



**Figure 5.8** Pressure and local regression rate history (Schöyer and Korting 1984).

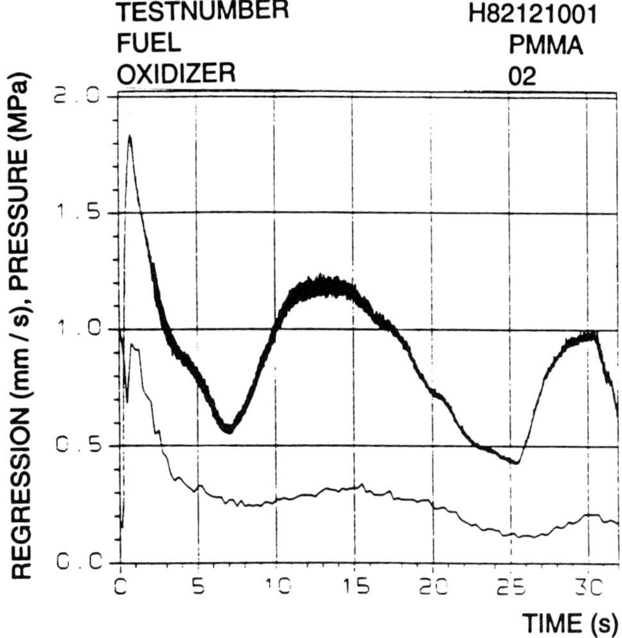



**Figure 5.9** Pressure and local regression rate history in SFCC Test with strongly varying pressure and mass fluxes (Schöyer and Korting 1984).

Although the program allows for nonstationary temperature profiles in the solid fuel it is assumed that immediately at ignition the surface temperature has been achieved. This will cause initially some discrepancies between computed results and reality.

It is estimated that the characteristic time associated with a change in thermal layer is of the order of 1 or 2 seconds. After about 3 seconds a roughly exponentially decreasing regression rate develops. Because of consumption of the fuel, the mass flux decreases, so the decrease in regression rate can be understood. It is possible, for a test run, where the oxidizer mass flux varies between 95 kg/m²s and 33 kg/m²s to correlate the regression rate with the oxidizer mass flux using the following relationship:

$$r = a\,G_{ox}^{m} \cdot p^{n}, \text{ where in this case}$$
$$m = 0.7607 \text{ and } a.p^{n} = 1.21 \times 10^{-2}.$$

One should bear in mind that this is a relation between the local and instantaneous regression rate and the local and instantaneous oxidizer mass flux and therefore the high value of the exponent is perhaps not applicable for the overall regression rate.

It is known that the regression rate of PMMA is pressure dependent for pressures roughly below 1.5 MPa. So one would expect that if the combustion pressure varies during a run, the regression rate varies correspondingly.

Figure 5.9 shows a run where the combustion pressure varied during a test. This was obtained by varying the mass flow rate. So in fact, two effects are combined. One sees that indeed the regression rate follows the trend of the combustion pressure. In addition one notices that the regression rate curve has shifted about 2 seconds to the right in comparison to the combustion pressure. This agrees roughly with the characteristic time associated with a change in the thermal layer.

## 5.3 Freejet Test Facilities

A series of axisymmetric convergent-divergent nozzles is employed to simulate representative flight Mach numbers within the engines operating envelope. One defines the nozzle contour required to obtain a given Mach number by the method of characteristics with suitable corrections for the boundary layer thickness. Additional techniques extend the test capability of such fixed nozzles (see Figure 5.10).

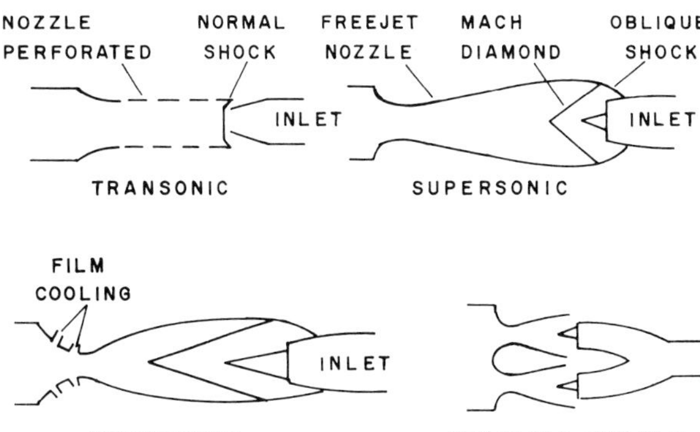



**Figure 5.10** Freejet nozzles (Dunsworth and Reed 1979). *Copyright © AIAA 1979. Used with permission.*

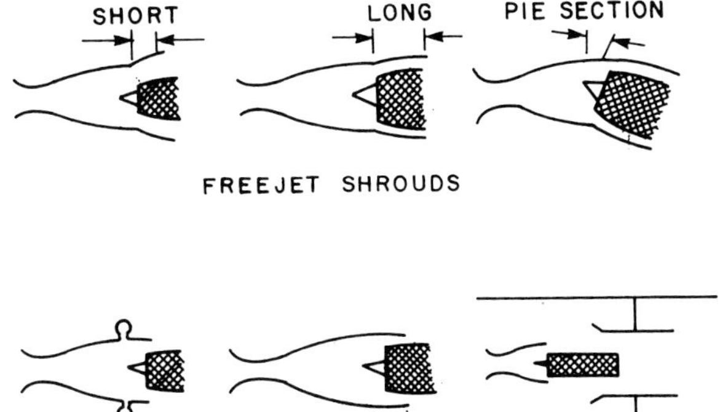



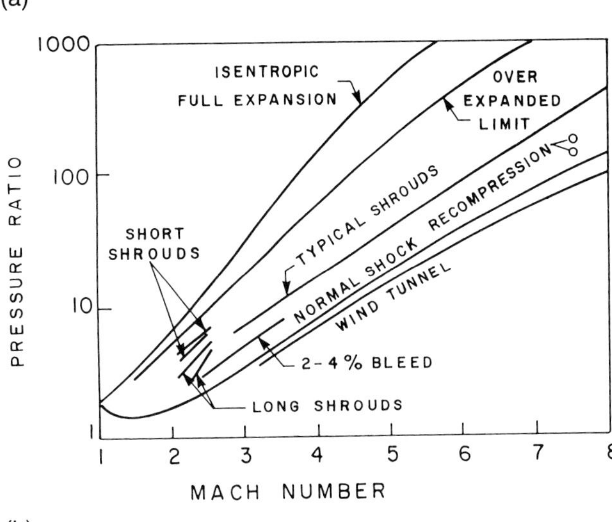



(a)



(b)

**Figure 5.11**   (a) Pressure ratio technique; (b) Pressure ratio performance (Dunsworth and Reed 1979). *Copyright © AIAA 1979. Used with permission.*

One must also protect the nozzle from high temperatures, say, by film cooling.

A simple but effective axisymmetric variable Mach nozzle for transonic free jet testing is obtained by welding a rolled up cylinder of perforated sheet metal to a sonic nozzle section. The desired test Mach number is obtained by operating at an appropriate pressure ratio, as is done in transonic slotted wind tunnels. The ratio of bleed air to test air increases from 0.18 at $M = 1.5$ to 0.69 at $M = 2$ and to 3.61 at $M = 3$. At higher Mach numbers this technique is not practical.

The ratio of total pressure to static pressure increases rapidly with Mach number. To simulate the static pressure at altitude one must reduce the cell pressure by connecting it to a vacuum tank or using an ejector or a powered exhaust system. A further reduction can be obtained by operating the freejet nozzle filled but over expanded. Since the limiting pressure ratio for a filled nozzle is a function of the

exit Mach number, it is possible to reduce the pressure ratio even more, if required, by bleeding the boundary layer near the nozzle exit or by designing the nozzle to recompress a portion of the flow, as shown in Figure 5.11. Other useful techniques illustrated are the use of a freejet shroud or an exhaust diffuser. The freejet shroud diffuses the flow from supersonic to subsonic speed through a shock train in the annular passage between the shroud and the test article. Exhaust diffusers use the freejet and engine test gases as the primary flow in an ejector-diffuser combination to reduce the cell pressure at the freejet nozzle exit. The ejector or exhausters draw off air downstream of a sealed bulkhead, which splits the cell. The pressure in the diffusers flow goes up through a series of shock waves and leaves the diffuser at subsonic velocity.

Figure 5.11 illustrates the effectiveness of the techniques described in reducing the operating pressure ratio below the isentropic ratio for a fully expanded nozzle. The curves are based on measurements; the pressure ratios associated with diffusing the air through a normal shock and typical wind tunnel operation are shown for comparison. Bleed flow at lower Mach numbers and recompression at hypersonic Mach numbers produce a substantial reduction in the operating pressure ratio.

Freejet testing at an angle of attack is facilitated by the use of articulated ducting, as shown in Figure 5.12. Rotating the air flow instead of the ramjet engine is beneficial for minimizing the problem of cooling the diffuser. Maintaining the alignment between the ramjet exit nozzle and the facility exhaust diffuser also deflects much of the external flow, as well as the internal engine flow, into the diffuser axially thereby maintaining high ejector pumping effectiveness. In a typical test setup instrumentation lines, fuel lines, and water-cooling lines are connected to the engine; if it is not necessary to rotate the engine for testing at various angles of attack, then one does not have to change the test installation between runs. This has the benefit of sim-

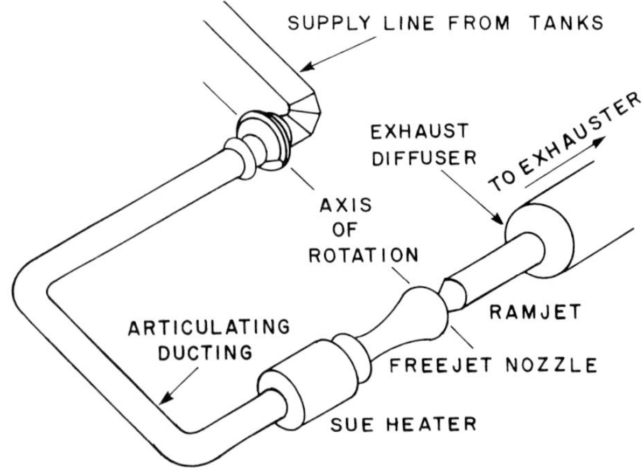



**Figure 5.12**   Angle of attack (Dunsworth and Reed 1979). *Copyright © AIAA 1979. Used with permission.*

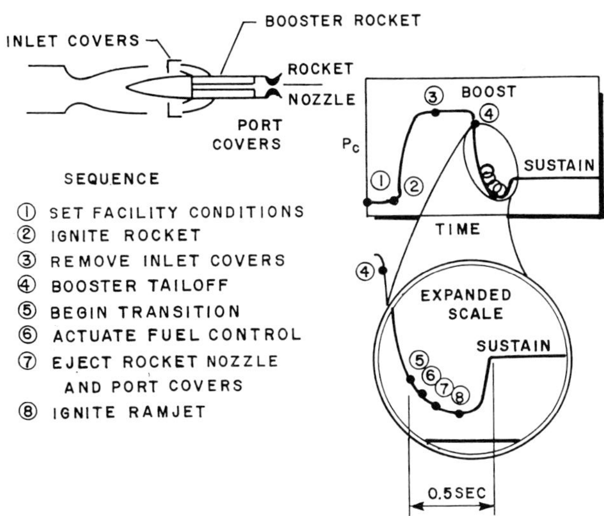


**Figure 5.13** Transition testing (Dunsworth and Reed 1979). *Copyright © AIAA 1979. Used with permission.*

plifying the procedure, the design of the thrust stand, and the method of mounting the engine in the test cell.

For ramrockets one must also demonstrate the capability of transition from the rocket to the ram-mode. Figure 5.13 illustrates schematically such transition testing. A freejet nozzle is selected to match the end-of-boost Mach number. The complete IRR system is installed in the cell and tested sequentially as indicated in the figure. Inlet covers isolate the rocket motors from the freejet air blast until the rocket ignites and the chamber pressure is established. Apart from possible secondary effects of axial acceleration the transition process is duplicated closely; it occurs within a few hundred milliseconds in order to minimize the deceleration process in flight between the two modes.

## 5.4 Hypersonic Test Facilities

Hypersonic test facilities have been greatly expanded as a result of the NASP program and of ESA activities (Muylaert et al. 1991). At the lower Mach number end, the following facilities should be mentioned: the engine test facility (ETF) at Aerojet, a Mach 8 vitiated propulsion facility, based on a storable propellant (nitrogen tetroxide—monomethylhydrazine) rocket gas generator (Hooper 1989); the 8 foot high temperature tunnel at Langley, a Mach 7 air-methane-oxygen direct combustion heated combustion facility (Puster et al. 1987); the high Re Mach 8 wind tunnel at the Naval Surface Warfare Center (Hedland et al. 1990), in which Reynolds numbers above 150 million/m and a 0.6 m diameter test core were demonstrated; the major European facility is the S4MA wind tunnel of ONERA at Modane, where the original Mach 6.4 capability has been extended to Mach 12 with a temperature limit of 1500 K. The NASA Lewis "Hot Gas Facility," which provides a high-enthalpy-high-heat flux environment, belongs also to this

category; it uses a hydrogen-oxygen rocket engine for this purpose.

At higher velocities we shall discuss the arc-heated hypersonic wind tunnels at the University of Texas (Wilson 1990) and ONERA Le Fauga (Muylaert et al. 1991), the low disturbance wind tunnel at Langley (Beckwith et al. 1990), the free piston shock tunnels (Stalker 1972 and 1989; Hornung and Belanger 1990; Muylaert et al. 1991), and the expansion tube (Tamagno et al. 1990).

### 5.4.1 The Hot Gas Facility

This is based on a hydrogen-oxygen rocket engine with a square combustion chamber, which provides hot combustion gases up to 3050 K and 4 MPa to the test specimen (Melis and Gladden 1990). By attaching a convergent-divergent nozzle to the chamber one can achieve supersonic flow up to Mach 2.5. The parts to be tested can be mounted in the combustion chamber and downstream of the nozzle, giving subsonic or supersonic flow.

The facility, shown during an experiment in Figure 5.14, generally operates in short 3 seconds bursts, but has been operated successfully up to 30 sec. Minor modifications allow the pressure to be increased to 6 MPa. Figure 5.15 shows the major components: the injector has 69 small holes in the center of the face plate for injecting the oxygen, while the hydrogen is forced through the remainder of the plate. The combustion chamber has a 34 cm$^2$ cross-section and can provide cooling by water or gaseous hydrogen. To the right one sees a test specimen spool, mounted on the rear of the combustor, forming a throat at the trailing edge of the specimen; this provides subsonic flow on the leading edge of the specimen. A sketch of the water-cooled nozzle and nozzle extension is shown in Figure 5.16. The main measurement system is a high-speed, analog-to-digital recorder with a capability of 100 data channels at 100 samples/sec/channel for obtaining time dependent data, such as pressures, temperatures, flow speeds, and strain. A 32 channel recording oscillograph provides immediate feedback of critical data. Photographic documentation is supplied by high-speed motion picture cameras, by video, and by 35 mm remotely operated cameras. Combustion experiments have shown that heat flux levels up to $9 \times 10^5$ kw/m$^2$/sec can be provided at a leading edge stagnation point, while side wall heat flux levels of the order of 2.5 to $5 \times 10^4$ kw/m$^2$/sec can be achieved in a supersonic flowfield. Thus the facility has been useful to the NASP program in providing Reynolds and Prandtl numbers, enthalpies, and heat flux similar to those expected in hypersonic flight.

### 5.4.2 Hypersonic Wind Tunnels

The two arc heated wind tunnels, although working on the same principles, are different in character. The smaller facility (Wilson, 1990) can operate continuously, while the

**Figure 5.14**   The hot gas facility (Melis and Gladden 1990). *Copyright © AIAA 1990. Used with permission.*

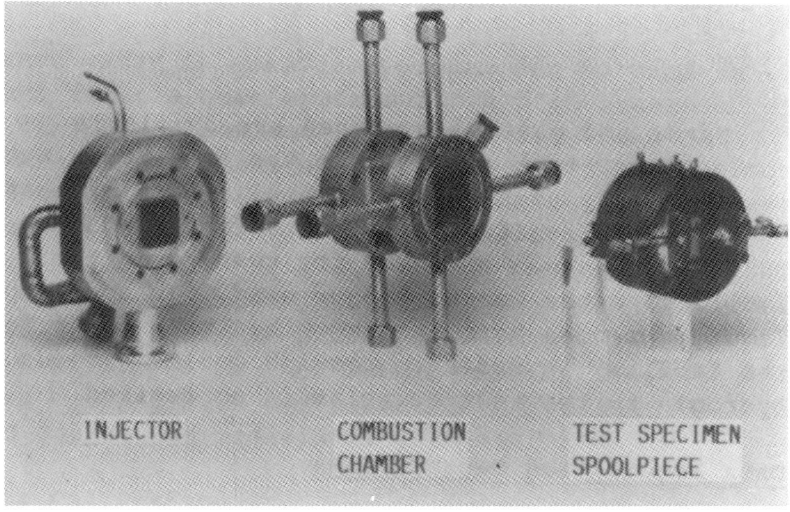


**Figure 5.15**   Major components of hot gas facility (Melis and Gladden 1990). *Copyright © AIAA 1990. Used with permission.*

larger F4 of ONERA, which was built primarily for Hermes, has a 20 to 100 msec test time. The University of Texas tunnel, shown schematically in Figure 5.17, employs a 2 MW DC electric arc heater, with a maximum gas flow rate of 0.2 Kg/sec. The nozzle throat region is water cooled for continuous operation; the test section is a standard freejet design, with a 203 mm exit diameter nozzle and a freejet section length of 610 mm, from the nozzle exit to the diffuser. The diameter of the test cabin is 762 mm. Diagnostic probes or models can be inserted through a 250 mm diameter porthole on top of the test cabin or a 406 by 610 mm support plate on its bottom. Optical ports are provided on both sides of the test cabin for holographic interferometry or schlieren photography. A closed jet section, 914 mm long, is also available. The diffuser has a 254 mm capture diameter; it reduces to a 229 mm constant diameter diffuser pipe,

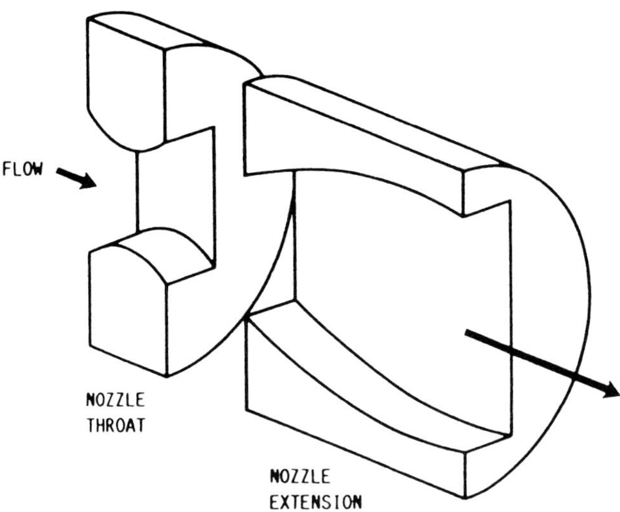



**Figure 5.16**  Water cooled nozzle extension (Melis and Gladden 1990). *Copyright © AIAA 1990. Used with permission.*

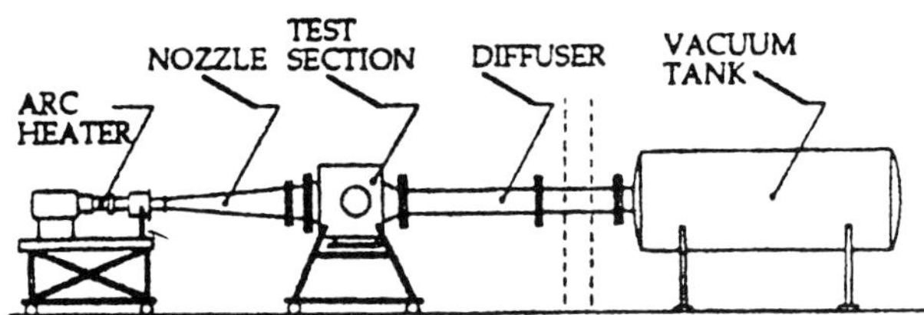



**Figure 5.17**  Hypersonic continuous flow arc heated wind tunnel (Wilson 1990). *Copyright © AIAA 1990. Used with permission.*

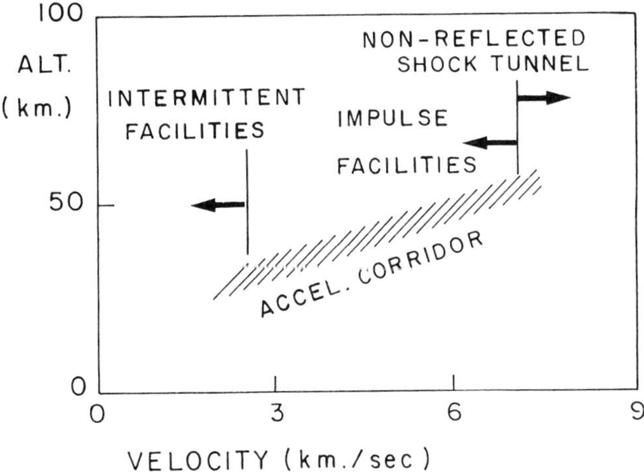



**Figure 5.18**  Binary scaling parameter vs. speed (Hornung and Belanger 1990). *Copyright © AIAA 1990. Used with permission.*

2.44 m long, which ducts the flow into a large vacuum tank outside the building. Apart from aerodynamic and aerodynamic heating studies, it is possible to investigate flow phenomena associated with hypersonic propulsion, in particular nozzle-airframe flow field interactions. By injecting an appropriate mixture of secondary gases into the arc

heater plenum chamber, a realistic simulation of the exhaust products is possible.

The F4 La Fauga installation is a 160 MW impulse generated facility; its total enthalpy goes from 40 to 160 MW/kg; the reservoir pressure is between 50 and 200 MPa, with a nozzle Mach number of 7 to 20. There are four different nozzles with exit diameters from 0.3 to 1 m. The test gas is nitrogen or air and the arc rotates at high speed to enhance homogeneous heating and reduce pollution. Here too it is possible to simulate chemical kinetics. Instrumentation includes schlieren spectroscopy, infrared thermography, and emission spectroscopy (to detect the presence of metals such as tungsten or copper in the throat or shock region). Holographic interferometry, CARS, and LIF are also planned to be added.

The NASA Langley Mach 18 quiet helium tunnel (Beckwith et al. 1990) is a helium blowdown tunnel fitted with a conical nozzle in an open jet configuration. Its main purpose is to study the effect of pressure gradients on the stability of laminar boundary layers in hypersonic flows, in order to validate existing prediction codes. It uses a conical slotted nozzle, 2.025 m long from the 17.2 mm diameter throat to the 356 mm diameter exit, with a 5° half-angle. The Mach number design range is from 12 to 18.

### 5.4.3 The Free-Piston Shock Tunnel

In order to obtain a good simulation of the combustion process taking place in a scramjet, the strong dependence on temperature of both equilibrium and nonequilibrium phenomena requires that combustion zone temperatures be reproduced relatively closely. In addition, either Mach number or velocity must be reproducible in order to properly represent wave interaction phenomena and mixing processes in the engine.

Only impulse facilities can fill this need at high flight speeds (see Figure 5.18), but their test time is too short to allow the establishment of steady flow temperatures at the surfaces and within the structures of an engine. For very high flight velocities impulse facilities can simulate gas dynamic pressure and heat loads, while for material limits, response to flight loads or cooling system effectiveness one must use other simulation methods, including numerical techniques.

To simulate exactly thermochemical equilibrium, both pressure and temperature should be matched. This requirement can, however, be relaxed in the combustion zone without introducing large errors. For example, raising the pressure of a stoichiometric hydrogen-oxygen mixture initially at 2400 K from 0.01 MPa to 1 MPa maintaining the enthalpy per unit mass constant, will raise the temperature by about 7%. This shows that the relation between temperature and specific enthalpy is not very sensitive to pressure; therefore, for most scramjet experiments involving vigorous

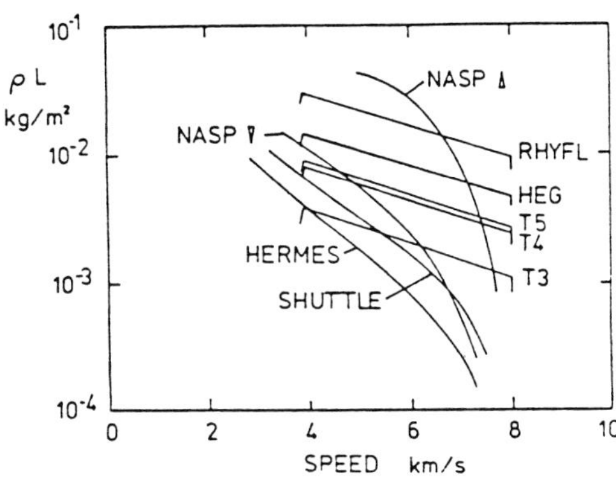

**Figure 5.19**  Facility stagnation enthalpy boundaries (Stalker and Morgan 1987). *Copyright © AIAA 1987. Used with permission.*

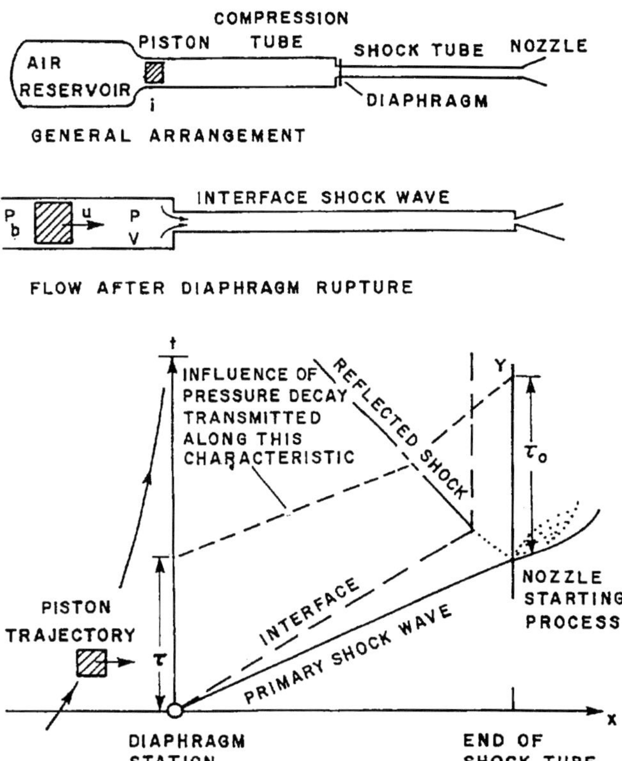


**Figure 5.20**  Operation of free-piston shock tunnels (Stalker 1972). *Courtesy of the Aeronautical Journal of the Royal Aeronautical Society.*

combustion, it can be expected that the relative magnitude of pressure loads will not be an important combustion simulation condition. The product of density and a typical linear size, $\rho L$, is an important scaling parameter. One speaks of binary scaling because it simulates correctly those chemical reactions that include two collision features. Figure 5.19 presents a plot of this parameter as a function of flight speed, showing the relevant values for some vehicles and impulse test facilities.

On the other hand, pressure levels may be important for thermochemical simulation of the expanding thrust nozzle flow. For hydrogen fuel the recombination processes involve three body collisions and therefore the associated chemical rate dependent phenomena may be expected to scale with the square of density. Thus, for models twenty times smaller than the flight article, pressures up to 4.5 times full scale values are required for simulation of recombination thermochemical phenomena. For a point in the acceleration corridor corresponding to 6 km/s at 45 km altitude (see Figure 5.18) this would involve nozzle reservoir pressures on the order of 450 MPa. This is a high value, but could be reached in a reflected shock tunnel.

In attempting to increase test section stagnation enthalpies, the reflected shock tunnel has been operated with both helium and hydrogen driver gas, heated in a number of ways. The most successful has been free-piston compression heating (Stalker 1972). Its operation is shown diagrammatically in Figure 5.20. The driven gas is contained initially in a large compression tube and a free piston is projected along it to initiate a test. The inertia of the piston causes an overshoot in its motion, producing a high pressure and temperature transient in the driver gas. By choosing the strength of the diaphragm at the entrance of the shock tube so that it ruptures close to the peak pressure, high shock speeds and associated high stagnation enthalpies can be produced in the shock tube.

A photograph of the free-piston shock tunnel T-3 at the Australian National University is shown in Figure 5.21. The 3 m reservoir and the 6 m compression tube are forged as a simple stainless steel tube, the first has an i.d. of 0.34 m and the second of 0.3 m. The high pressure end of the compression tube is reinforced by a steel sleeve of approximately 2 m with a 0.76 m overall diameter, which is a shrink fit onto the compression tube, so that the internal pressure can reach 270 MPa. The compression tube and the shock tube are free to move on rollers, thus isolating the system dynamically from its surroundings. Since the center of gravity must remain fixed, the large axial movement of the piston as it traverses the compression tube is balanced by a relatively small recoil movement (about 35 mm) of the compression-tube-shock-tube assembly. The compression tube, shock tube, and hypersonic nozzle are coupled rigidly together, and the movement is taken up at the joint between the nozzle and the dump tank. After a test pneumatic actuators return the assembly to its original position. The shock tube is 6 m long with a 76 mm i.d., while the dump tank is 3 m long and has a 1.2 m diameter with a volume of about 5 m³. The hypersonic nozzle is conical with an included angle of 15°. Throat size and test section diameter can be varied giving Mach numbers between 4.7 and 20 as shown in Table 5.3. The test section stagnation enthalpy corresponds approximately to 9 km/s.

More recently, a number of larger facilities of this type has been built; for example, the T-4 at the University of Queensland, the T-5 at CalTech, the Rocketdyne RFHYL in Santa Susana, California (Anderson et al. 1990), the TCM-2 at the University of Marseille, and the HEG at

**Figure 5.21**   Free-piston shock tunnel T-3 (Stalker 1972). *Courtesy of the Aeronautical Journal of the Royal Aeronautical Society.*

**Table 5.3**   Available test section Mach numbers (Stalker 1972). *Courtesy of the Aeronautical Journal of the Royal Aeronautical Society.*

| Throat size (cm) | Test section diameter (cm) | Test Mach number |
|---|---|---|
| 1.2 | 15 | 4.7 |
| 1.2 | 30 | 7.6 |
| 0.63 | 30 | 12.3 |
| 0.32 | 30 | 20.0 |

DLR, Goettingen (Muylaert et al. 1991). The two biggest facilities are the 122 m long RFHYL and the 60 m long HEG; the other dimensions are comparable (compressor tube diameter 61 vs. 55 cm, shock tube diameter 20.3 vs 15 cm), as are the maximum pressure (200 MPa), stagnation enthalpy (40 vs. 45 MJ/kg), and maximum attainable speed (8 km/sec). Test times are short (1–6 msec). Advanced measurement techniques, similar to those mentioned for F4 and the ones described in Section 13.6, are employed.

### 5.4.4 The Expansion Tube

The expansion tube was developed at NASA Langley during the 1960s and 1970s in order to generate hypervelocity flows with negligible levels of dissociation of the test gas and was used for hypersonic real gas studies of short duration (300–400 msec). The last operating tube, with a 152.4 mm i.d. and 36.6 m length was decomissioned in 1983. With the renewed interest in the field created by the NASP project the unit was refurbished and reactivated by GASL (Tamagno et al. 1990) and renamed Hypulse.

The unique feature of an expansion tube, aside from its high velocity capability, is that the test gas is delivered in a chemically "clean" state. That is, it is not contaminated by vitiation products or a large percentage of dissociated species. Therefore, it becomes an ideal facility with which to study strongly chemically-dependent hypervelocity flows such as scramjet combustion processes or external aeroheating. The available test time is brief, on the order of 350 $\mu$sec (in a reflected shock tunnel of comparable dimensions test time can be on the order of 2–3 msec, but the dissociation is very high (e.g., in the nozzle reservoir of the HEG all the oxygen is dissociated and 73% of the nitrogen). However, one can obtain useful data through the use of state-of-the-art fast-response instrumentation and by choosing models which are sized to permit the establishment of steady flow in the available test time.

The operation of an ideal expansion tube is described with reference to the distance-time $(x - t)$ diagram shown in Figure 5.22. An expansion tube is physically similar to a shock tube with the addition of a secondary diaphragm which separates the driven tube into two section—an inter-

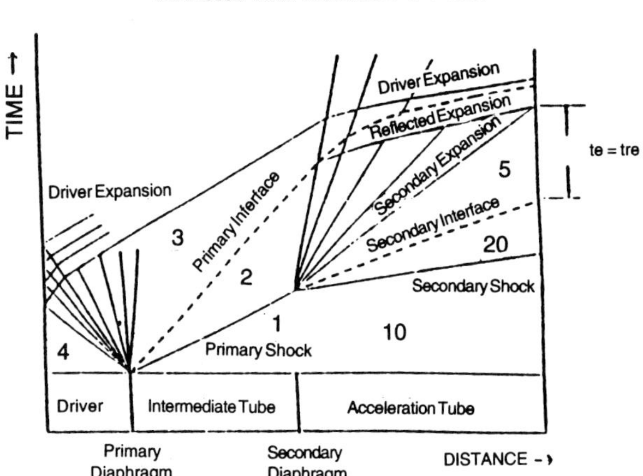



**Figure 5.22** Expansion tube operation (Tamagno et al. 1990). *Copyright © AIAA 1990. Used with permission.*

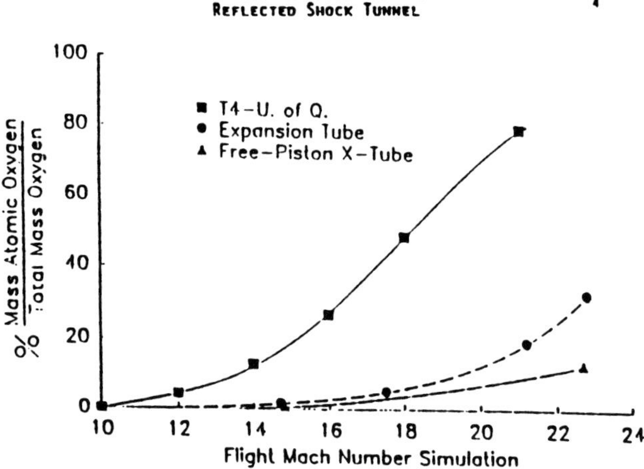



**Figure 5.23** Fraction of atomic hydrogen in HYPULSE (Tamagno et al. 1990). *Copyright © AIAA 1990. Used with permission.*

mediate tube, closest to the driver, and an acceleration tube further downstream. The secondary diaphragm is typically very thin mylar or cellophane having a low burst pressure. Initially, the driver tube (region 4 in the $x - t$ diagram) is filled to high pressure, preferably with a low molecular weight gas such as helium (the facility performance, as measured by the attainable total pressure and total enthalpy level, improves with increasing driver-to-intermediate tube sound speed ratio). The intermediate tube (region 1) is filled with the desired test gas to a pressure which is generally subatmospheric. The acceleration tube, denoted as region 10, is filled with a third (acceleration) gas at a still lower pressure. The same gas typically is used in both the intermediate and acceleration tubes.

The flow is initiated by bursting the primary diaphragm, causing the primary shock wave to propagate down the intermediate tube, producing flow in region 2. Upon striking and rupturing the secondary diaphragm, the primary shock acquires a higher Mach number as it enters the acceleration tube, leading to the flow in region 20. To equilibrate the pressure and velocity from region 2 to 20, a system of unsteady expansion waves propagates into region 2 accelerating the test flow to the high velocity region 5. The test section, which receives the test gas, is located at the exit of the acceleration tube.

The available test time is the period between arrival of the acceleration gas/test gas interface and the first wave which disrupts the uniform test flow. The three most significant waves which limit the test time are also shown in Figure 5.22. First, the arrival of the tail of the unsteady system of expansion waves, denoted as the secondary expansion wave, truncates the test time at $t_e$. Second, when the leading wave of the unsteady expansion wave system intersects the driver gas/test gas interface, a reflected expansion

is generated which arrives at the test section at time $t_{re}$. An optimum configuration of the $x - t$ diagram would have the secondary and reflected expansions arriving at the test section simultaneously (see Figure 5.22). The test time is also potentially limited by the arrival of the first reflection of the primary expansion system off the driver end; however, in practice this occurs later than either $t_e$ or $t_{re}$. Adjustments to the actual $x - t$ diagram to achieve a particular operating condition for the expansion tube are made by varying driver gas sound speed and pressure, as well as intermediate and acceleration gas pressures, and by changing the relative lengths of the intermediate and acceleration sections by moving the secondary diaphragm.

The Hypulse facility can be operated in the expansion-tube mode or in the expansion-tunnel mode by adding a nozzle downstream of the acceleration chamber. In the first mode one can achieve real gas hypervelocity flow conditions with arbitrary test gases. The plug of uniform test gas produced at the exit of the expansion tube is about 2 m long and 10 cm in diameter; the length of models that can be tested is between 0.7 and 1 m. Adding a free-piston driver and providing a large test core will allow one to duplicate flight conditions at speeds above 6 km/sec and altitudes as low as 40 km, thus allowing the use of larger model frontal area and reducing post-test loads. An important feature of Hypulse is that it generates high enthalpy conditions without stagnating the flow, thus minimizing dissociation. Figure 5.23 shows that at Mach 17 the fraction of atomic oxygen is less than 6%, while the T4 facility would have 40%. Adding the free-piston drive would reduce the dissociation level even more, because of the higher pressure. The test flows obtainable with the free-piston driver will reach speeds close to 9 km/sec, total temperatures of 14,000 K and total pressures of $10^4$ MPa.

# Chapter 6

# *Status of Hypersonic Combustion Technology*

Both experimental and computational simulations are very important for achieving progress in hypersonic combustion technology (Anderson et al. 1990), with flight tests giving the final confirmation. It is of interest to represent, in an alternative way, the enthalpy requirements, which are shown in Figure 6.1 for combustor entrance conditions corresponding to Mach 10, 15, 20, and 25. The static pressure is in the range of 0.05 to 0.1 MPa, while the stagnation pressure rises exponentially from 10 MPa at Mach 10 to 10,000 at Mach 25. Figure 6.2 compares the simulation capability of selected pulse facilities, including two reflected shock tunnels, in addition to those mentioned in section 5.4 Figure 6.3 shows the binary scaling parameter for such installations, together with the values for NASP ascent and descent and shuttle descent. The advantage of free piston driven shock tunnels is evident from this figure.

CFD predictions of complex three-dimensional hypersonic combustion flow fields are invaluable, since they allow one to perform parametric studies and to simulate two and three-body reactions simultaneously, but they must be calibrated with experimental data obtained at the appropriate conditions.

## *6.1 Critical Issues for Combustor Design*

The important factors involved are the thrust production potential of the flow leaving the combustor, the peak local and overall cooling requirements as well as the integration of the combustor with the inlet compression and the exhaust expansion processes. Losses due to shock waves, wall friction, and shock-boundary layer interactions are very important. The design is, therefore, dominated by the trade-off between accomplished mixing and combustion and the losses necessary to achieve mixing together with other pressure losses (see Swithenbank et al. 1989).

One way to deal with the importance of thrust production is to use combustor effectiveness, $\eta_{ce}$ (McClinton 1990) instead of combustion efficiency as a figure of merit. It is defined as

$$\eta_{ce} = \frac{(1 + f/a)I_{er} - I_i}{(1 + f/a)I'_{er} - I_i}$$

where

$I_i$ = combustor entrance vacuum specific impulse
   ($V/g + PA/w$), sec

$I_{er}$ = actual combustor exit vacuum specific impulse
   expanded isentropically to ambient pressure

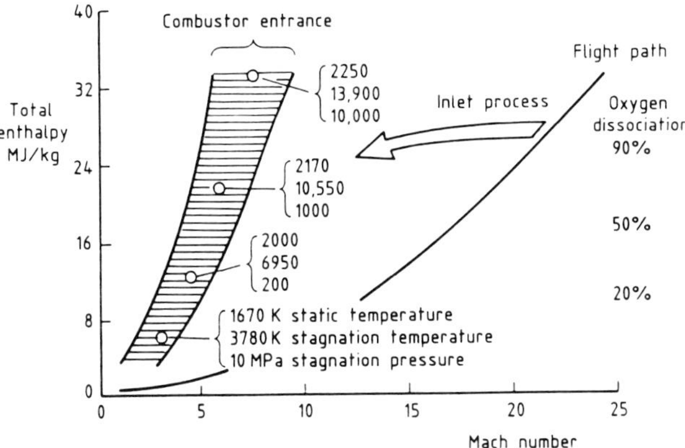



**Figure 6.1** Enthalpy requirements for hypersonic combustion facilities (Anderson et al. 1990). *Courtesy of NASA. Copyright © AIAA 1990.*

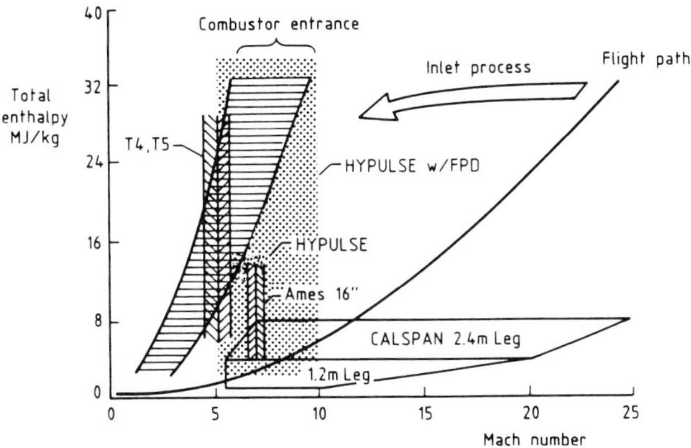



**Figure 6.2** Total enthalpy simulation capabilities of selected pulse facilities (Anderson et al. 1990). *Courtesy of NASA. Copyright © AIAA 1990.*

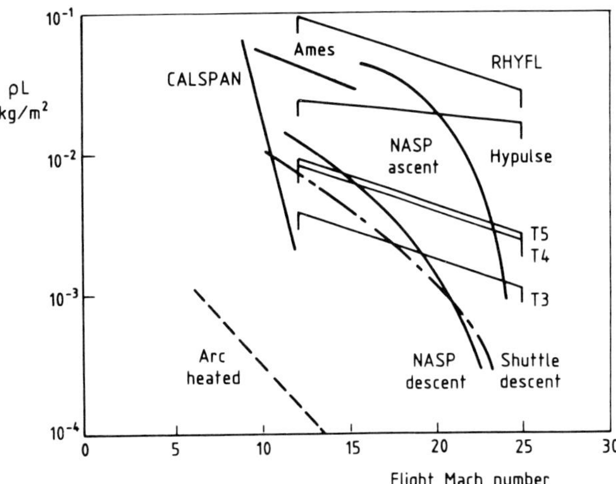

Figure 6.3   The binary scaling parameter (Anderson et al. 1990). *Courtesy of NASA. Copyright © AIAA 1990.*

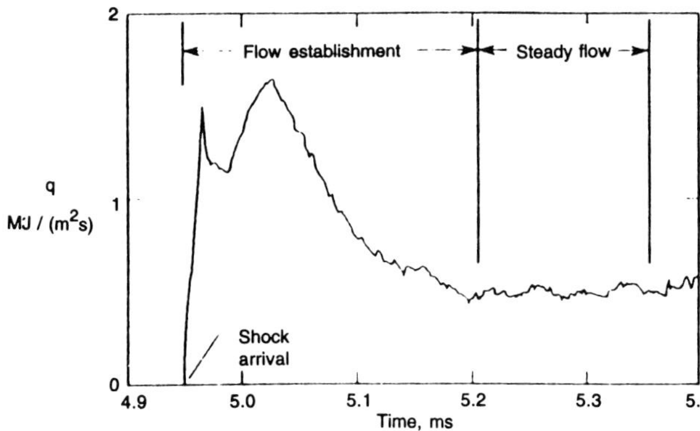

Figure 6.4   The flow establishment process (Anderson et al. 1990). *Courtesy of NASA. Copyright © AIAA 1990.*

$I'_{er}$ = ideal combustor exit vacuum specific impulse for ideal heat release at constant pressure with zero velocity combustion expanded isentropically to ambient pressure.

Combustor effectiveness thus represents the thrust production potential of the combustor exit flow.

Wall cooling considerations may dominate the design process, because the peak local heat transfer level must be compatible with the thermal-structural capability of the regeneratively cooled combustor wall. Flow distortion at the combustor inlet can be minimized by matching the fuel distribution created by the injector system with the airflow distribution.

Finally one should remember that the high local Mach number at combustor exit (5 or more) will cause pressure changes to propagate at relatively shallow angles to the flow direction. A profile in flow properties may exist there and could possibly be utilized for thrust control.

## 6.2 Experimental Considerations

Two important techniques used in the development of combustors are flow visualization and heat transfer measurements. Wall heat transfer rate is considered the best diagnostic tool to track the time required to reach steady state. In pulse facilities it can be measured accurately thanks to the combination of high stream stagnation enthalpy, cold model walls, and abrupt flow starting processes. Figure 6.4 is an example of the flow establishment process as shown by a laminar heat transfer trace obtained in the GASL Hypulse facility. The model, an open cylinder 89 cm long and 3.8 cm inside diameter with a sharp leading edge and smooth wall, was tested in air at 4.9 km/sec. The local internal Mach number was 5 and the pressure 0.17 bar. Shock arrival and the initial relaxation process were followed by arrival of the test gas and the subsequent relaxation of the heat transfer rate (as the boundary layer grows) to a sensibly steady state during the last 150 μsec of the test.

In many experiments contaminants, such as diaphragm material, driver gas, trace hydrocarbons from vaporized mylar may be present; the high temperature may cause the formation of atomic oxygen and nitric oxide.

Flow diagnostics can be performed by introducing uncooled probes for measuring wall pressure, heat flux distribution, and instream pitot pressure (no cooling is required because of the very short test time) and by a variety of optical methods. The preferred ones are interferometry, in particular laser holographic interferometry and planar laser induced fluorescence, while filtered Doppler-shifted Rayleigh scattering appears promising.

## 6.3 Computational Tools

Computational tools available for the analysis of the combustor flow field range from simple quasi-one-dimensional cycle codes to complex three-dimensional Navier-Stokes codes with detailed chemical kinetics and diffusion modeling. The relatively simple tools gain their simplicity and speed by sacrificing representation of the details of the flow to an integral or global approach. However, they can be very useful engineering tools with some of the major characteristics of the combustion process retained. These codes can provide the proper trends of flow behavior and set an upper bound on the combustor performance.

Computational techniques were first applied to scramjet combustor flow fields during the 1970s. A considerable amount of pioneering work on the solution of the combustor flow fields of hydrogen-fueled scramjet engines utilizing multistep/multicomponent chemical kinetics models was done under the direction of Ferri (1973). A viscous characteristics approach was used to split the analysis into strongly coupled parabolic mixing and hyperbolic wave solutions. Upgrades to the solution technique were made by Spalding and his colleagues (1974) which resulted in a two-dimensional/axisymmetric program CHARNAL, and a three-dimensional program, SHIP.

Drummond (1979) developed a two-dimensional Navier-Stokes program, TWODLE, based on MacCormack's explicit algorithm that included an equilibrium hydrogen-air chemistry model. This code was extended to include detailed finite-rate chemistry and transport models based on kinetic theory. This version of the program was named SPARK (Drummond et al. 1986). As scramjet technology evolved, a critical need developed for a three-dimensional combustor code. Venishi and Rogers (1986) extended the three-dimensional inlet code of Kumar (1986) to include a two-step hydrogen-air reaction model. The numerical algorithm was modified to include implicit evaluation of the chemical source term.

Because of the need for a more basic modeling capability in three-dimensions, Carpenter (1988) extended the two-dimensional SPARK code to three dimensions. He improved the algorithm to fourth-order compact based on the work of Abarbanel and Kumar (1988) and added a generalized chemistry package (i.e., equilibrium or finite-rate chemistry) for hydrogen-air. In addition, Kamath (1989) developed a parabolized version of the three-dimensional SPARK code using Gielda's explicit, space-marching algorithm (Gielda and McRae 1986) which is based on the MacCormack explicit algorithm. The SPARK series of codes represent the state-of-the-art in combustor analysis for the scramjet.

## 6.4 Examples of Computation and Experiment Applied to Hypersonic Combustion

Here, two examples of the combined application of computational methods and hypervelocity test facilities are presented. The first deals with the injection and combustion of hydrogen at Mach 10 flight conditions in a conventional reflected shock tunnel and compares the wall data with computations. The second discusses a Mach 14 shock induced combustion study carried out in a free-piston shock tunnel and compares it with calculations.

An example of the application of a conventional shock tunnel to the investigation of hydrogen injection, mixing, and combustion is provided by a cooperative experiment conducted by GASL and Calspan in the Calspan 96-inch reflected shock tunnel. Figure 6.5 shows an example of an LHI visualization of the density in the injection region of this experiment. The geometry is plane symmetric with slot injection along both walls at the conditions indicated in Figure 6.6. Figure 6.7 shows the measured wall static pressure data compared with pressure computed by the SHIP code. Since this code uses a "partial equilibrium" reaction scheme, mixing is allowed to occur without reaction in the computation until 60 cm where pressure rise begins. The

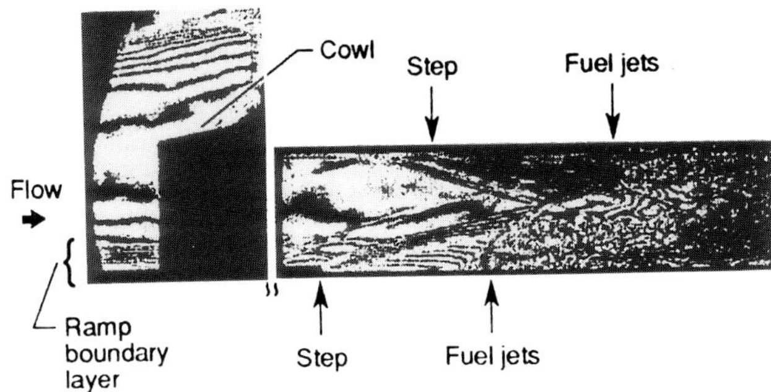


**Figure 6.5**  Density visualization in injection region (Anderson et al. 1990). *Courtesy of NASA. Copyright © AIAA 1990.*

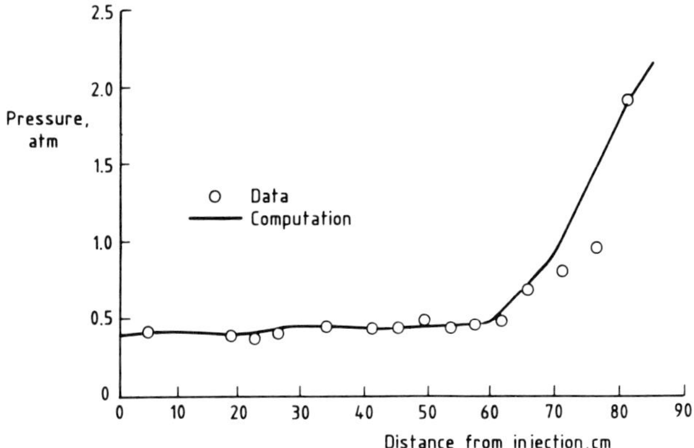


**Figure 6.6**  Measured and computed wall static pressure (Anderson et al. 1990). *Courtesy of NASA. Copyright © AIAA 1990.*

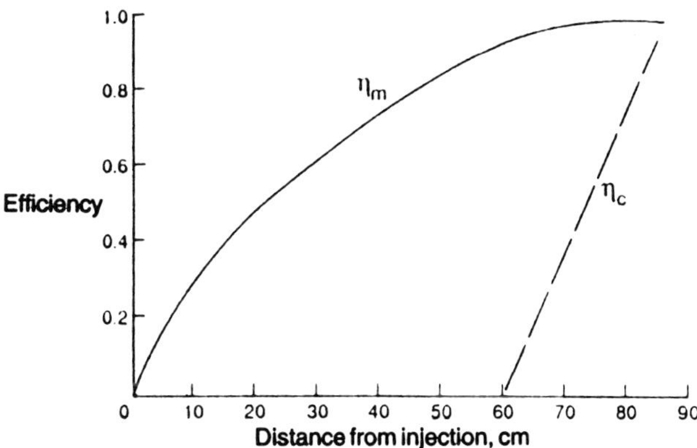


**Figure 6.7**  Mixing efficiency (Anderson et al. 1990). *Courtesy of NASA. Copyright © AIAA 1990.*

computed mixing is shown in Figure 6.8 in terms of an integral "mixing efficiency," which is defined as that fraction of the total fuel (both reacted and unreacted) that would be consumed in a given distribution if the reaction were locally complete without further mixing. As shown in Figure 6.8, the reaction is assumed to proceed linearly with distance from 60 to 84 cm and is nearly in local equilibrium (i.e., essentially complete) from 84 cm downstream. With the reaction keyed to the observed pressure distribution in this way,

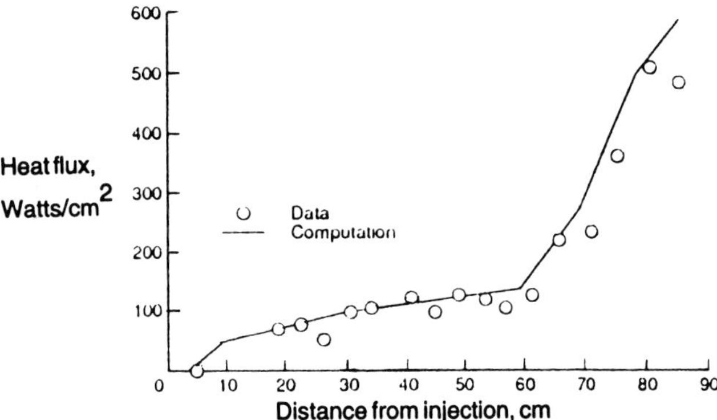

**Figure 6.8**  Measured and predicted wall fed flux (Anderson et al. 1990). *Courtesy of NASA. Copyright © AIAA 1990.*

it is interesting to compare the measured and predicted wall heat flux in Figure 6.8. The trend of the heat flux data is well represented by the computation.

The other example treats a two-dimensional flow field with strut injection of hydrogen into air at Mach 14 flight conditions investigated by Stalker. The local Mach number is 5 and the static pressure is sufficiently low that combustion does not occur until initiated by a shock which is generated by the inclined wall of the duct (see Figure 6.9). Analysis of the data was conducted by Rogers et al. (1987) with a two-dimensional version of the SPARK code which includes a 9-species, 18-reaction finite rate kinetics scheme for hydrogen air. Figure 6.9 shows the measured and computed pressure along the lower wall of the duct for cases with air-only (circular symbols, solid curve) and with hydrogen injected into the air (triangle symbols, broken curve). The section at the top of the figures shows the geometry of the duct and isobars in the computed flow with hydrogen injected into the air. Generally speaking, the data follow the computed pressure trends reasonably well. Injection and combustion (evident from production of water shown by other plots of computed results included in Rogers et al. 1987) tend to raise the level of pressure in the flow as expected.

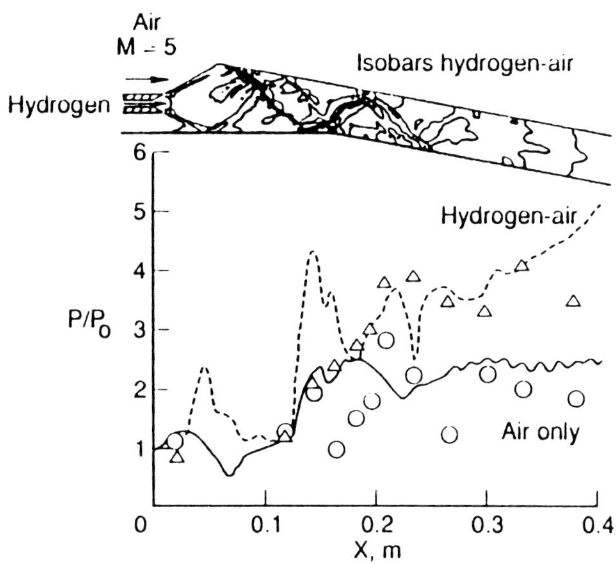



**Figure 6.9**  Pressure distribution for shock induced combustion experiment (Anderson et al. 1990). *Courtesy of NASA. Copyright © AIAA 1990.*

However, closer inspection shows that the details of the wave structure in the computed and real flow are somewhat different. The computation for hydrogen into air shows a definite pressure peak at about 15 cm, where the incident shock reflects from the lower wall before the expansion corner. In the experiment, two pressure transducers immediately adjacent to the corner do not show this peak. In fact, examination of the measured pressure with air only suggests that the incident wave reflects from the lower wall slightly downstream of the corner in the experiment rather than slightly upstream of the corner as in the calculation.

Many possible explanations come to mind, such as the difference in geometry between experiment and calculation, the effect of oxygen dissociation (near 20% according to Figure 5.23), the nonuniformity of the tunnel flow and others, but only additional measurements, better defining the flow field, will provide the answer.

# Part II

# *Modern Turbojets*

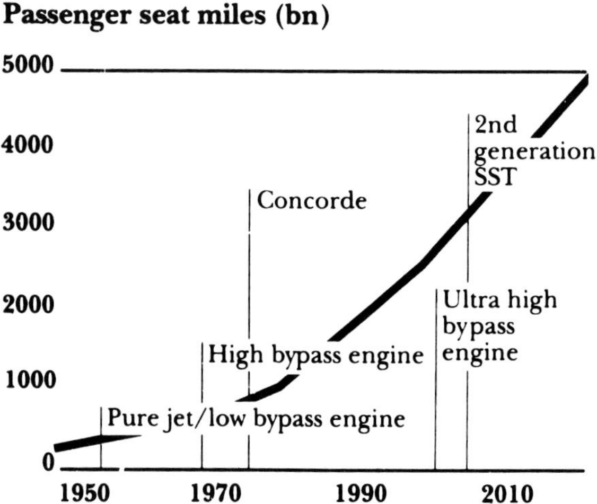


**Figure 7.1** Evolution of civil transport engines (Rolls-Royce 1989). *Used with permission.*

# Chapter 7

# Modern Turbine Engine Development

## 7.1. Introduction

The second section of this book deals with modern turbojets.

This chapter starts with an evaluation of development trends in engines which are intimately connected with the evolution of aircraft. Chapter 8 deals with energy development, a program which was started in the 1970s continued in the 1980s, and is now being applied to large modern engines such as the GE90 and the 4000 series of P&W, as well as to smaller engines. The following chapter treats civil engines for subsonic aircraft, starting with large engines with two and three shafts, going over to medium size powerplants, differentiating them according to thrust magnitude, with particular attention to the propfan. It is concluded with a discussion of the technology of small engines.

Chapter 10 is concerned with civil engines for supersonic aircraft and emphasizes innovative concepts, such as double bypass, variable stream control, and turbine bypass. We then treat military engines, stressing their requirements and giving as examples the Rafale and the EJ200. Particular attention is given to stealth requirements and thrust vector control. Chapter 12 treats V/STOL technology, comparing it to conventional aircraft, discusses the different options, and then describes existing powerplants, such as Pegasus for the Harrier and the combination of cruise and lift engines in the Forger. The chapter concludes with the prospects for supersonic operation.

An extensive discussion of combustion chambers includes the definition of their requirements, a treatment of pollutant formation and the methods adopted to reduce it, and measurements of performance and durability. Special types of combustion, such as fuel prevaporization and premixing, hot wall, recuperative cooling, and catalytic converter combustors are the next subject.

The chapter is completed by a discussion of chamber cooling and a short review of combustion diagnostics. The last three chapters deal with components and materials (this includes also those used in hypersonic powerplants),

noise problems, and the importance of maintenance and reliability.

The emphasis is on modern developments, showing how they came about and what the contemporary status and the foreseable trends are, stressing the growing role of international cooperation and providing up-to-date examples.

## 7.2. Trends in the Development of Airplanes and Engines

The general trend in the 1990s is to improve performance by employing a higher pressure ratio and lowering fuel consumption, to facilitate maintenance by dividing the airframe and the engine into a small number of replaceable subsystems and to minimize pollution and noise. There are considerable differences between civil and military applications. In the first, economy is the overriding consideration, and as a consequence, high bypass ratio turbofans and propfans will dominate the scene (see Figure 7.1). In the second, fast response and low vulnerability are the controlling factors, with stealth properties very important. It is, therefore, natural to treat the two categories separately.

### 7.2.1. Civil Aircraft and Engines

Sir John Chainley, writing in 1982, predicted the success of the Boeing 757 and 767 and the Airbus 310 as medium sized airplanes, with continued improvement being made on the much larger Boeing 747. Indeed, while these aircraft have all been very successful, new planes are also being developed; for example, the A320, A321, A330, and A340 by Airbus, covering the short, medium and long haul portions of the passenger market, the Boeing 777, a two-engine widebody with a capacity of 300–400 passengers for medium and long distances, as well as the MD11 and MD12 of McDonnell Douglas, aimed at a similar market.

The Russians are developing two large civil airplanes: the Tu-204 a medium range airliner designed for 190–214 passengers, with the basic Tu-204-100 version having a range of 5300 km and the larger Tu-204-200 reaching up to 7180 km. Their second aircraft, the Il 96-300, a long-range aircraft, using advanced structural materials, can carry 300 passengers 7500 km; with a 30 ton payload it can reach 9000 km and with a 15 ton payload, 11,000 km. Originally it was to enter service in 1991; an enlarged version, seating 350 passengers and having a range of 7000 km (high capacity medium haul) should enter service in 1993. Both planes use the PS-90A turbofan, the Tu-204 has two engines, the Il 96-300 has four.

According to Pierson (1987) of Airbus Industries, technology is the moving force of the new projects. When conceiving a new aeronautical program he points out three main objectives that must be taken into consideration when evaluating a new technology: (1) Safety and reliability which are a "sine-qua-non"; (2) Profitability, so the user can reduce cost and maximize revenue and (3) Compatibility with the environment.

He distinguishes between two types of innovation; on one hand there are completely new technologies, such as electric-electronic flight commands (fly-by-wire) and the propfan; on the other hand (and this is more common) there is an evolution rather than a revolution. This is exemplified by the gradual introduction of composite components, which started from secondary flight commands, has arrived to whole wing structures, and may in the future include the fuselage. The decisive factor is economic: the new technique must result in lower initial cost of the aircraft, cheaper maintenance and lower fuel consumption. He cites as an example the A320, a narrow body aircraft for 150 passengers using fly-by-wire commands, utilizing composite materials for the wings and incorporating second generation computers.

As regards the engines, Airbus is participating in the development of various types of propfans, while, at the same time, equipping the newer aircraft with high bypass, high pressure engines, such as the CFM56, the IAE V2500, the CF6-80, the PW4000 series and the Rolls-Royce Trent engines.

In the 777, Boeing will use either the Trent or the new GE90 engine which will have a bypass ratio of 10 and will exhibit technologies developed in the $E^3$ program, to achieve low pollution and a 10% improvement in fuel consumption. It is interesting to note that the first models of the A330 will fly with CF6-80E1 engines (303kN), going over later to the GE90 or Trent powerplants. The pressure ratio of the CF6-80E1 is already above 30, while the new engines will have an overall pressure ratio exceeding 40.

The PS-90A engine is named after Pavel A. Soloviev, who designed it. After his retirement in 1989, the design bureau was renamed the Perm Scientific Manufacturing Association. This engine has a takeoff thrust of 156.9 kN and an overall pressure ratio of 35 with a bypass ration of 4.4 and

a fuel consumption of 0.595 kg/hr/kg. The PS-90A is in the same class as the Rolls-Royce RB535E4 and the Pratt & Whitney PW 2000.

Future supersonic commerical transports continue to generate great interest and a group comprising the major manufacturers (Aerospatiale, Boeing, BAC, Deutsche Airbus, MDD, Alenia and Japanese Aircraft Industries) is studying technical, environmental, and business issues to determine if it is possible or desirable to initiate a formal international program to design and build a second generation Super Sonic Transport (SST).

## 7.2.2. Military Aircraft and Engines

The main U.S. fighter planes, which will continue their service during the 1990s, are the F14, F15, F16, and F/A18, together with the F117A stealth fighter, which operated very successfully during the Gulf War. This plane will continue service after the year 2000 together with the Advanced Tactical Fighter (ATF), for which the Lockheed F22A was chosen and is expected to enter service in the mid 1990s.

The Russians will employ the Mig 25, 27, 29, and 31 and the Sukhoy 24, 25, 26, and 27. The Europeans will continue to use the Tornado, while introducing the EFA (European Fighter Aircraft). The French, who prefer to go it alone, will fly the Mirage 2000 and the Rafale, while Sweden has decided to use the Gripen.

With regard to bombers, the United States will supplement the B-IB with the B-2, while the Russians will employ the Tu-26 Backfire and the Tu-160 Blackjack. In the V/STOL field the Harrier II or AV-8B will be further developed in the West, with an eye on supersonic performance, while the Russians will continue to use various Yakovlev models (Yak 28, 38, and 41).

As regards engines, modern versions of the GE F404 power the initial Rafale prototype (thrust 71.2 kN), the F/A 18 (79 kN), and the Gripen (F404/RM12, 80 kN). The GE F110 provides power for the F14D (120.2 kN) and the F16 (129 kN), the GE F118 is the engine of the B-2 bomber (84.5 kN). P&W offers the F-100 for the F15 and F16 (106 kN thrust), while an improved design engine, the F-100 PW-229, can supply 129 kN. For the ATF GE is developing the variable cycle F 120 and P&W the F119, which will provide supersonic performance without afterburner.

In the area of V/STOL aircraft, Rolls-Royce is developing the Pegasus 11-61 for the Harrier II, which will provide 106 kN thrust with improved reliability and maintainability and low life-cycle cost. They are also studying vectored thrust engines, either with plenum burning combustion or with other types. The Russians use a turbojet and two lift engines for their Yak 38 aircraft.

# Chapter 8
# *Energy Development Programs*

As was mentioned in the previous chapter, an important goal of the engine designer is to optimize the utilization of the available energy, which will, in general, correlate with minimum fuel consumption while obeying environmental constraints. In the United States, a number of programs toward this goal have been sponsored by NASA. One of the first was the Experimental Clean Combustion Program (ECCP), conducted in parallel by General Electric (Gleason and Bahr 1979) and by Pratt and Whitney (Roberts et al. 1977). This was succeeded by the Energy Efficient Engine or $E^3$ program, which is well described by Sokolowski and Rhode (1981) and summed up by Ciepluch et al. (1987). After describing the goals of the $E^3$ program, we shall see how GE and P&W proposed to achieve them and will discuss the results obtained by the two companies. The success of the program can be judged from the fact that the new GE90 engine incorporates many of the techniques developed in the course of the $E^3$ program and implements the recommendations presented there.

## 8.1 The Energy Efficient Engine

The $E^3$ program was initiated in the mid 1970s by NASA as one of the major activities under the NASA Aircraft Energy Efficiency Program, whose aim was to improve fuel efficiency (Nored et al. 1979). The $E^3$ project goals, which took into account fuel savings, and economic and environmental improvements, were to reduce the SFC by 12%, to reduce SFC performance deterioration by 50%, to meet FAA noise regulations and the (then) proposed EPA emission standards.

Before starting the project P&W and GE were contracted to study advanced propulsion systems for aircraft. Boeing, Lockheed, and McDonnel Douglas also provided guidance for these studies, which examined the size and the cycle of the turbofan engines, as well as advanced component technologies. Two then existing airlines, Eastern and Pan Am, were contracted to review the overall benefits of the engine concepts and the proposed technologies.

### 8.1.1. The GE Engine

The GE configuration and the major components are illustrated in Figure 8.1. The engine is sized for 160 kN takeoff thrust; it has two counter-rotating spools supported by five bearings in two main frames. The cycle and performance characteristics are compared to those of the GE CF6-50C which was then the state of the art large engine, for maxi-

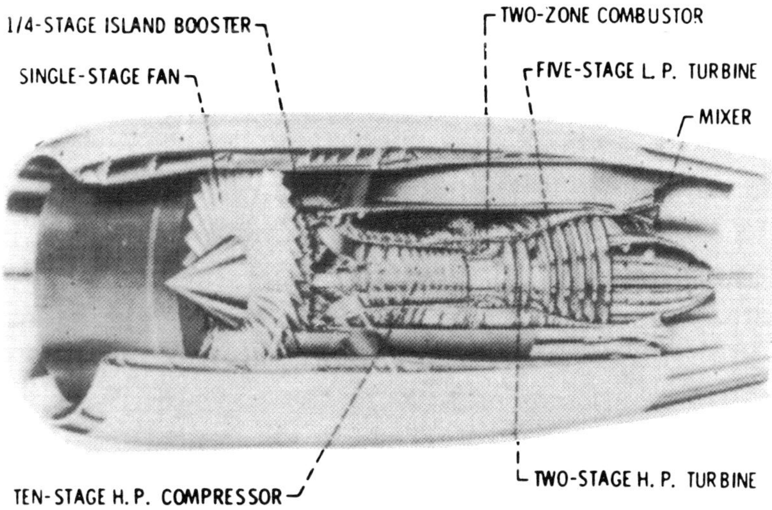


**Figure 8.1**  G.E. energy efficient engine (Ciepluch et al. 1987). *Copyright © AIAA 1987. Used with permission.*

**Table 8.1a** Comparison of cycle and performance characteristics for GE CF6-50C and GE E$^3$ EPS (Ciepluch et al. 1987). *Copyright © AIAA 1987. Used with permission.*

|  | GE CF6-50C | GE E$^3$ FPS |
|---|---|---|
| Bypass ratio | 4.3 | 6.9 |
| Fan pressure ratio | 1.73 | 1.61 |
| Overall pressure ratio | 30.1 | 36.1 |
| Compressor pressure ratio | 12.5 | 22.6 |
| Turbine rotor inlet temperature |  |  |
| For hot day takeoff | 2160 K | 2200 K |
| For maximum cruise | 1905 K | 1975 K |
| SFC | Reference | −14.2% |

**Table 8.1b** Emission levels of CF6-50E2 engine with production and low emission combustor (Bahr 1992). *Reprinted by permission of Kluwer Academic Publishers.*

| Category | Applicable ICAO regulatory limit | Status-original production engine | Status-low emission production engine |
|---|---|---|---|
| Smoke (SN) | 18.8 | 6.5 | 12.5 |
| HC | 19.6 | 57.8 | 3.4 |
| CO (g/kN) | 118.0 | 97.3 | 29.8 |
| NO$_x$ | 100.0 | 58.2 | 51.6 |

mum cruise conditions at 0.8 Mach and 10,670 m altitude in Table 8.1a, while the emission levels of the more modern CF6-50E2 engine are shown in Table 8.1b; here the ICAO regulatory limits are compared with the original production engine and a low emission production model.

The nacelle is slender and has low drag characteristics. Bulk-absorber type acoustic treatment is used in the inlet, the inner and outer walls of the fan duct, and after the low pressure turbine at the end of the core flow passage. There is a single stage fan consisting of solid titanium blades, while the vanes are integrated with the support struts, minimizing the number of airfoils and reducing the fan-frame weight and cost. To provide adequate support while avoiding excessive blockage, an equal number of vanes and blades is recommended. This is achieved by increasing the axial spacing between vanes and blades to about two blade chord widths. A quarter-stage island booster is placed behind the fan for automatic core flow matching; it also acts as a foreign object separator. The 10-stage compressor gives a pressure ratio of 23; it has four variable vane stages and a variable inlet guide vane. The last five stages have active clearance control.

The double annular combustor, designed for low emissions, is an outgrowth of the high-pressure combustor developed in the ECCP program shown in Figure 8.2. A segmented liner provides increased life and reduces main-

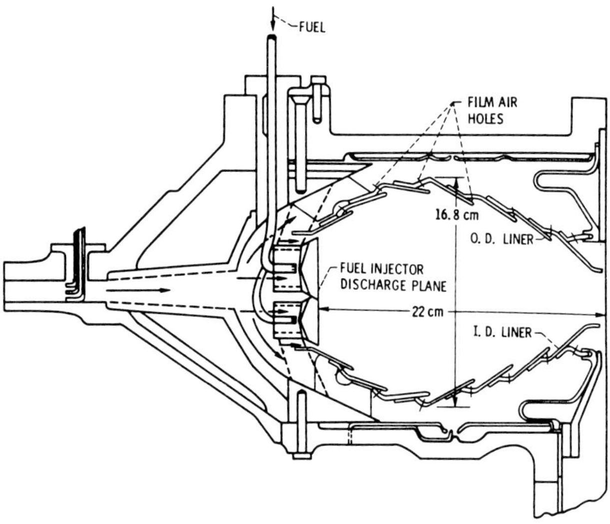


**Figure 8.2** G.E. double annular combustor (Sokolowski and Rhode 1981). *Copyright © AIAA 1981. Used with permission.*

tenance, while the split duct diffuser divides the flow for the two concentric burning zones thus shortening the combustor length. The segmented design should increase burner life by 30%, thanks to reduced thermal stresses during rapid heating and cooling cycles.

The two-stage high-pressure turbine is cooled by means of compressor discharge and interstage air for the vanes and blades respectively. Advanced directionally solidified material in the airfoil, a ceramic shroud in the first stage rotor, and advanced powder metallurgy discs make it possible to reach a turbine entry temperature (TET) of 2200 K. The five-stage low-pressure turbine is acoustically tuned to reduce noise. The hot core engine stream is mixed with the cooler fan bypass air before the engine exhaust by a 12 chute mixer. A full authority digital electronic control (FADEC), which incorporates failure indication and correction action (FICA), coordinates all engine variables. Therefore, normal engine operation can be sustained even with the loss of one or more input parameters.

### 8.1.2. The P&W Engine

The P&W engine, shown in Figure 8.3 is also sized for 160 kN takeoff thrust. A substantially higher pressure ratio and bypass ratio than for the reference JT 9D-7A engine was chosen, as is shown in Table 8.2. This counterrotating two-spool engine has a five-bearing design with two main support frames and two main bearing compartments. An advanced, low drag, single nozzle nacelle system accommodates the fan and core exhaust mixer. A low loss acoustic liner in the inlet and exhaust ducts attenuates noise coming from the fan and the core.

Two approaches were taken for the fan design; the more advanced one was a single-stage, shroudless, 24 hollow blade design, which promises higher performance. Since it was not

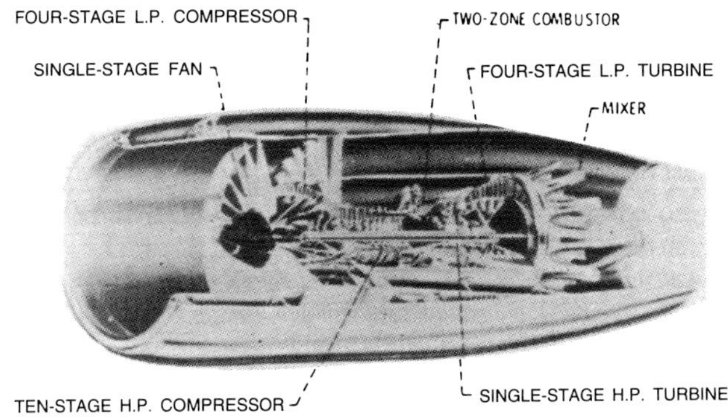



**Figure 8.3**  P&W energy efficient engine (Ciepluch et al. 1987). *Copyright © AIAA 1987. Used with permission.*

**Table 8.2**  Comparison of cycle and performance characteristics for P&W JT9D-7A and P&W $E^3$ FPS (Ciepluch et al. 1987). *Copyright © AIAA 1987. Used with permission.*

|  | P&W JT9D-7A[a] | P&W $E^3$ FPS[a] |
|---|---|---|
| Bypass ratio | 5.1 | 6.6 |
| Fan pressure ratio | 1.58 | 1.71 |
| Overall pressure ratio | 25.4 | 37.3 |
| Compressor pressure ratio | 10 | 14 |
| Turbine rotor inlet temperatures |  |  |
|    For hot day takeoff | 2070 K | 2225 K |
|    For maximum cruise | 1840 K | 2000 K |
| SFC | Reference | −15.1% |

[a]Parameters at 35,000-ft altitude, Mach 0.8, maximum cruise power setting.

clear if the required technology would be completely developed during the course of the $E^3$ program, an alternative more conventional design had a shrouded fan rotor consisting of 36 solid blades. The fan rotors are acoustically matched to the duct exit guide vanes by spacing the blades and vanes apart by three blade chord lengths.

The compressor section is configured as two short, stiff, drum-rotor low and high pressure sections, the latter producing a 14:1 pressure ratio in ten stages. The four front stator vanes are variable, the six rear stages of the compressor have actively controlled clearance.

The two-zone combustor, shown in Figure 8.4 was designed to meet both low and high power EPA emission standards, as proposed at the time. The zones are axially aligned as primary and secondary combustion regions. The liner was designed as a segmented structure to allow for axial and thermal expansion. In the advanced cooling system the compressor discharge air first cools convectively the back fans of the liner segments, then reverses direction and film-cools the combustor side of the segments.

The single stage high-pressure turbine is a transonic design using a powder/metal/nickel alloy for the disk, while the airfoils consist of high-temperature single crystal nickel alloy; a ceramic outer air seal is employed over the rotor. The 24 vanes require only 6.4% of the core engine inlet flow for cooling, while the 54 blades utilize only 2.75%; both vanes and blades are designed for advanced cooling. The low-pressure four-stage turbine includes advanced aerodynamic blading concepts, high-temperature materials (to eliminate cooling), active clearance control, and counter-rotates with

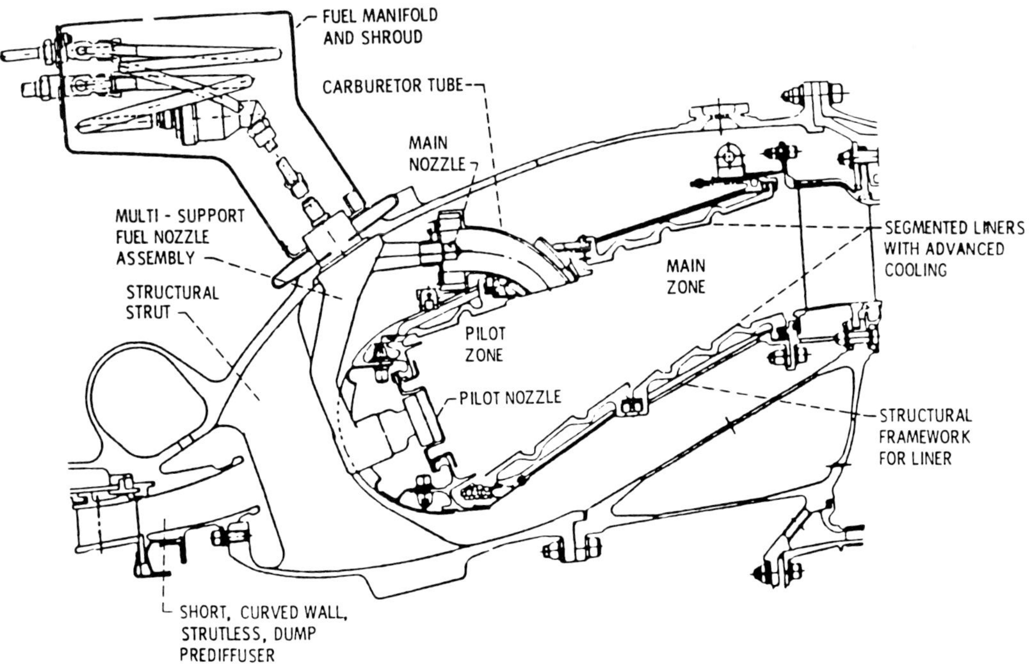



**Figure 8.4**  P&W $E^3$ combustor design and features (Sokolowski and Rhode 1981). *Copyright © AIAA 1981. Used with permission.*

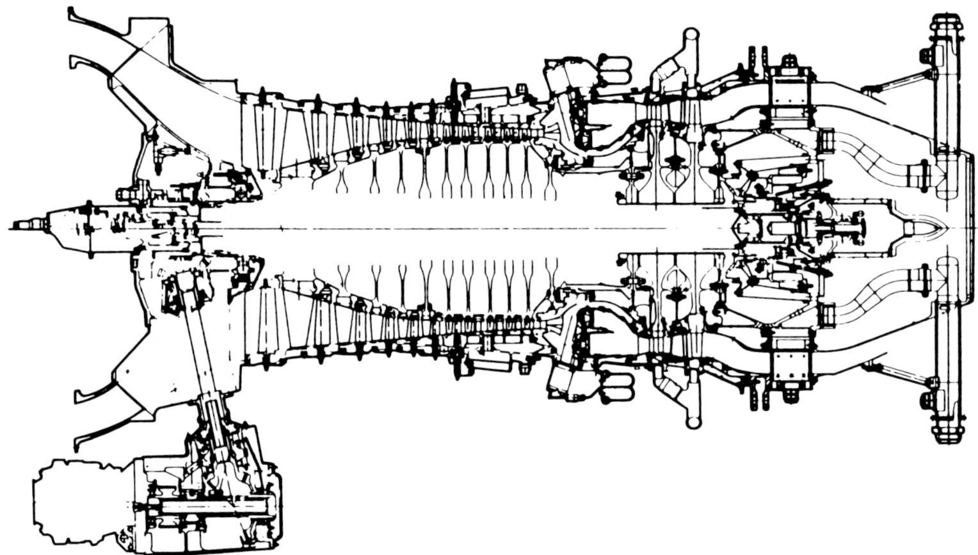

**Figure 8.5**   G.E. engine core cross section (Ciepluch et al. 1987). *Copyright © AIAA 1987. Used with permission.*

respect to the HP turbine. The hot LP exhaust gases are mixed with the fan discharge air in an 18 lobe mixer. Control is similar to that of the GE engine.

### 8.1.3 GE's Results

Technology development activities started in 1978, and their results were incorporated in bench, subcomponent, and full-scale component tests. This stage of the work was common to both programs; however, the final verification of the technology, obtained by evaluating in a real engine environment the advanced components through core (HP compressor, combustor and HP turbine) and integrated core-low pressure spool (ICLS) tests was limited to the GE engine, due to budget limitations.

#### 8.1.3.1 Component Tests

In the GE six year program over 2000 hours of component rig testing were performed. A full scale component test provided data (See Table 8.1a) on the efficiency, the stall margin, and blade vibratory response. For takeoff conditions the stall margin was 2–4% over the 16% goal; blade vibratory responses were low. The high-pressure compressor tests optimized aerodynamic design and variable geometry; they included 270 intentional stalls for performance and stall mapping. The takeoff stall margin was 21% with an efficiency of 0.849 at 10,125 m, Mach 0.8, maximum cruise.

Both sector and full-annular, full-scale combustor tests were performed, which demonstrated a circumferentially averaged temperature profile entering the turbine rotor for simulated high power operation at

$$0.125\,(T_{\text{circ. avg. at given radius}} - T_{\text{comp. inlet}})/$$
$$/T_{\text{comb. outlet}} - T_{\text{comb. inlet}})$$

The goals for smoke, carbon monoxide (CO), and unburned hydrocarbons ($C_xH_y$) emissions were surpassed, while the $NO_x$ goal was not achieved, although a considerable reduction with respect to the reference engine was obtained. High-pressure turbine rig tests demonstrated a 0.925 efficiency at standard day 10,625 m, Mach 0.8, maximum cruise, while the LP turbine achieved 0.915 efficiency.

#### 8.1.3.2 Core Tests

The $E^3$ core was assembled (see cross-section in Figure 8.5) and run in a test cell for 45 hours, using over 1400 sensors, which provided over a million parameter readings. The tests included sea level and ram inlet mechanical checkouts, HP compressor stator optimization, active clearance control evaluation, wind-milling characteristics, starting optimization, operating line migrations, and transient evaluations.

Overall performance was better than predicted. High-pressure compressor efficiency was up 0.8% relative to rig test results because of additional aerodynamic adjustments incorporated after rig test completion. High-pressure test efficiency was very close to that predicted based on rig tests. Transitioning the combustor from single- to dual-annular burning and back was accomplished without problems. Combustor emissions and smoke results were low, as shown in Tables 8.3 and 8.4.

#### 8.1.3.3 ICLS Tests

The core engine, fan, low pressure turbine, mixer, and nacelle were combined into the ICLS turbofan test vehicle and installed in an outdoor test facility. All aspects of the $E^3$ propulsion package were evaluated: mechanical and aeromechanical characteristics and integrity, performance, acoustics, engine control, rapid accelerations and decelera-

**Table 8.3**  Combustor emission results vs.goals (Ciepluch et al. 1987). *Copyright © AIAA 1987. Used with permission.*

|  | Emission index in g/kg of fuel | | |
|---|---|---|---|
|  | CO | $H_xC_y$ | $NO_x$ |
| Required to meet goal | 20.75 at idle | 2.75 at idle | 17.50 at T/O |
| Measured (corrected to turbofan cycle) | 19.13 | 1.45 | 20.1 |

**Table 8.4**  Smoke results vs. goals (Ciepluch et al. 1987). *Copyright © AIAA 1987. Used with permission.*

| Combustor inlet temp., °K | SAE smoke number | |
|---|---|---|
|  | Goal | Measured |
| 544 | 20 | 2.2 |
| 588 | 20 | 2.1 |
| 808 | 20 | 0.6 |

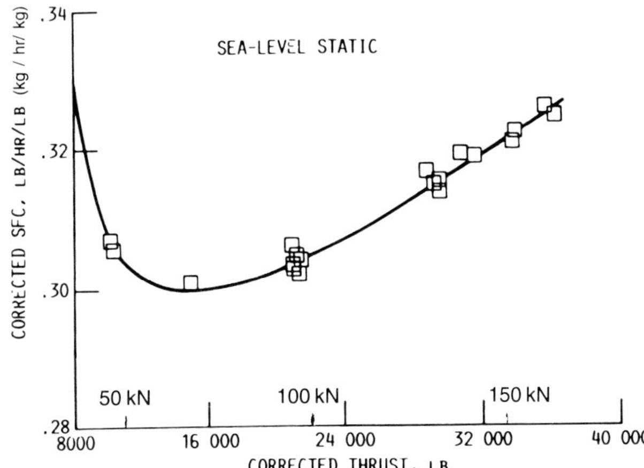


**Figure 8.6**  Measured G.E. $E^3$ SFC as function of thrust (Ciepluch et al. 1987). *Copyright © AIAA 1987. Used with permission.*

tions, and starting. No aeromechanical or mechanical problems were found over the full engine operating range, up to the maximum thrust achieved: 166.4 kN.

The measured sea level static takeoff SFC is shown in Figure 8.6. An SFC of 0.326 at 162 kN thrust was achieved, representing the lowest SFC ever demonstrated for a high-bypass-ratio turbofan engine.

Starting was accomplished in 45 s. There were no stalls and bleed was not used. The control system operated flawlessly. Bursts from 10–90% thrust were made in 5 s with no turbine temperature overshoot. All overrides held, FICA control strategy was demonstrated, and accurate variable stator and speed control were demonstrated, as was FADEC control of clearances.

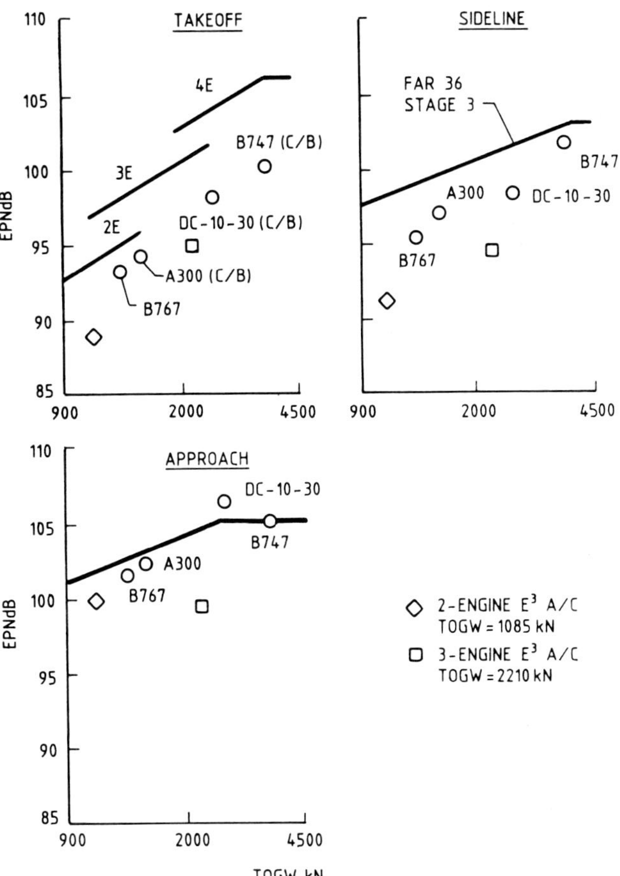


**Figure 8.7**  Project noise level of $E^3$ powered aircraft (Ciepluch et al. 1987). *Copyright © AIAA 1987. Used with permission.*

Fan performance paralleled that of the fan rig test with stresses and efficiencies being very close to those of the rig test. Core performance was also very close to that demonstrated during the core vehicle test.

The low-pressure turbine, in a full-scale test for the first time, performed very close to prediction. Mechanical integrity was excellent over the full range of operation, with blade vibratory stresses never exceeding 25% of limits. All temperatures were within limits and close to predictions. Active clearance control was effective.

Mixer performance was better than predicted, achieving a mixing effectiveness of 78% at 0.64% pressure loss for a 2.4% SFC improvement at the standard day, 10,675 m, Mach 0.8 maximum cruise condition. Figure 8.7 shows the analytically projected aircraft noise levels of $E^3$-powered aircraft, based on engine acoustic measurements, compared to measurements made on existing aircraft. As can be seen, two- and three-engine $E^3$-powered aircraft meet the FAA requirements with at least a 2-dB margin.

Altitude performance projections of the $E^3$ ICLS sea level static data were made using data from component tests covering a range of altitude, corrected speeds, Reynolds

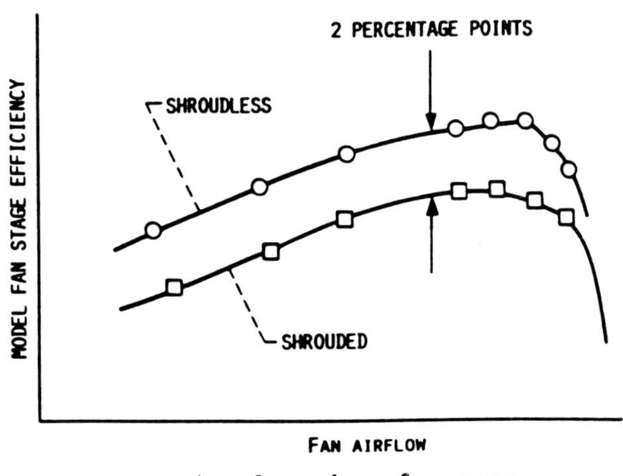


**Aerodynamic performance.**

**Figure 8.8** Results of shroudless hollow fan technology (Ciepluch et al. 1987). *Copyright © AIAA 1987. Used with permission.*

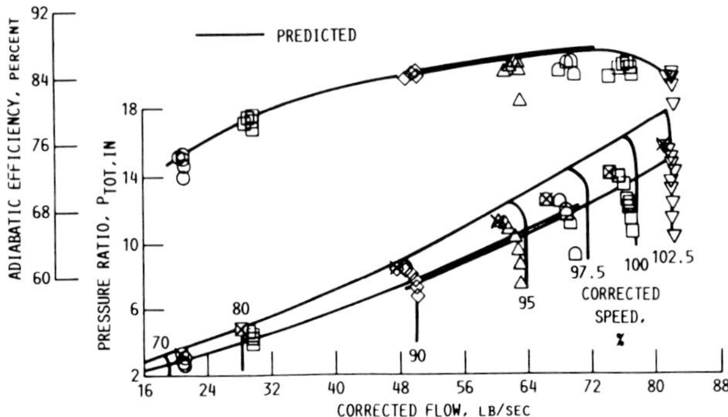


**Figure 8.9** P&W HP compressor performance (Ciepluch et al. 1987). *Copyright © AIAA 1987. Used with permission.*

numbers, etc. This projection indicated that the 10,625 m, Mach 0.8, standard day, uninstalled maximum cruise SFC would be 0.55. This is a SFC reduction of 1.5 percentage points more than the program goal of a 12% decrease in SFC.

In terms of fuel savings, results from component and ICLS (full engine) tests indicate that the fuel savings goal of 12% was exceeded with experimental hardware. Fully developed engines based on $E^3$ technology would show even greater fuel savings. Attainment of the fuel savings goal along with the flight propulsion system weight estimates also indicate that the goal of a 5% reduction in direct operating costs should be met when applied to various transport aircraft. In terms of environmental program goals, projection of ICLS ground test noise measurements to flight conditions shows that all FAA noise requirements could be attained with at least several dB margin for both two- and three-engine aircraft. Also, emissions measurements indicate that all the original EPA goals were met except for $NO_x$ as shown in Tables 8.3 and 8.4.

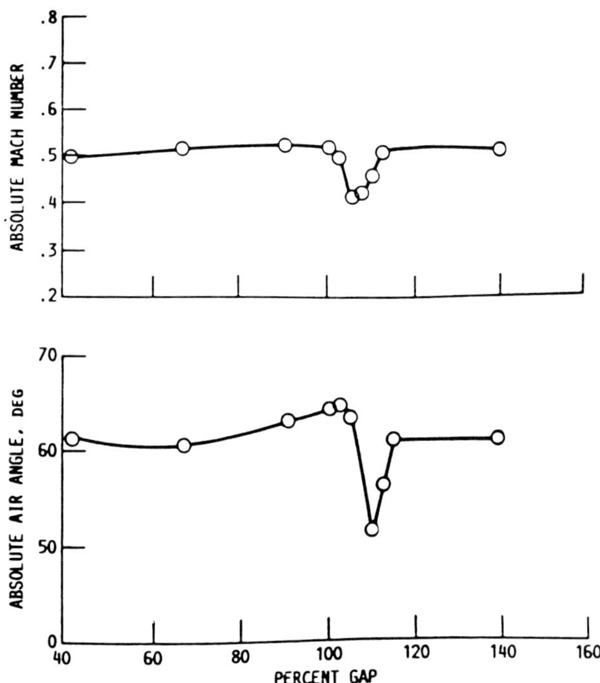


**Figure 8.10** LDV measurements behind ninth stage stator (Ciepluch et al. 1987). *Copyright © AIAA 1987. Used with permission.*

### 8.1.4. P&W Results

Pratt and Whitney concentrated their efforts on the main sections of the engine, with particular emphasis on the fan and core components. For the fan a rotor bladed with solid titanium shroudless airfoils was tested; dynamic tests gave satisfactory results: the shroudless configuration was two percentage points more efficient than a shrouded configuration (Figure 8.8), of which 1.5% was attributed to shroud removal. Different approaches for fabricating hollow fan blades were investigated, such as superplastic forming/diffusion bonding techniques.

Efficiency and performance potential of the HP compressor were demonstrated, as shown in Figure 8.9. A highlight of this program was the use of LDV techniques to measure flow angles and velocities (see Figure 8.10). Combustor tests confirmed that significant improvements in combustor life can be achieved with cast metallic segments and improved cooling methods. Goals for emissions of smoke, CO, and UHC were comfortably exceeded, while $NO_x$ levels were slightly in excess of the original goals.

The HP turbine underwent experiments in an uncooled and a cooled component rig. Stage reaction levels of 35% and 43% were investigated and showed that the higher reaction level improves stage efficiency. A three-phase mixer model test program, using one-tenth scale mixer/tailpipe models was conducted. The following major conclusions reached were: 1) an 18-lobe configuration provides best overall per-

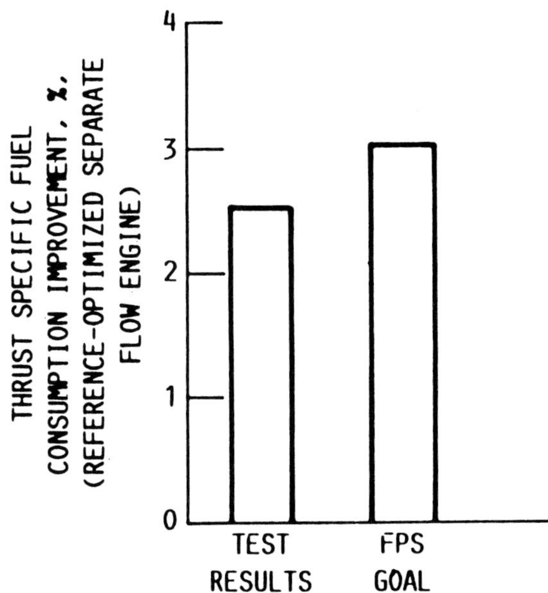



**Figure 8.11** Results of experimental mixer model research program (Ciepluch et al. 1987). *Copyright © AIAA 1987. Used with permission.*

formance; 2) inlet swirl produces large mixer losses; 3) scalloping of lobes provides significant mixing improvements; 4) pylon fairings can be integrated with the mixer with little loss; 5) longer mixers have lower internal loss; and 6) duct stream upstream lobe fairings (or hoods) improve performance. The mixing goal was nearly met and substantial progress was made toward achieving the specific fuel consumption goal, as indicated in Figure 8.11.

# Chapter 9

# *Civil Engines for Subsonic Aircraft*

The engines for civil subsonic aircraft constitute today the largest jet engine market. They are classified here according to their size (large, medium and small), with further subdivisions reflecting their architecture and specific tasks. In the West there are three manufacturers, which cover the whole spectrum of engine sizes: General Electric, and Pratt & Whitney in the United States, and Rolls-Royce in Great Britain. The Russians have introduced recently the three-shaft Lotarev D-18T engine, a second generation turbofan developing a thrust of 230 kN. Their most advanced engine is the two-shaft PS-90A with a thrust of 160 kN designed by Soloviev.

In the medium and small engine market there are also other manufacturers. In the United States the most important are Allison, Garrett, Teledyne, and Textron-Lycoming. In Western Europe there is a tendency toward collaboration, represented by CFM International (SNECMA-GE), Turbomeca—RR, MTU taking part in International Aero Engines (IAE), which manufactures the V-2500, FIAT-AVIO cooperating with both GE and P&W. Volvo has agreements with all the "Big Three," and this is also the case for Japan. Finally, one should mention the Chinese manufacturers, now working mainly under licence from Russia and from the West, who are expected to enter the aeroengine market sometime in the 1990s.

## *9.1 Large-size Engines*

The first large-size engine to enter service was the P&W JT9D-3A, at the beginning of 1970, on the Boeing 747-100. This two-spool unit was the first high-bypass ratio engine, with a value of 5.17 as compared to about 1 for the JT8D first generation turbofans; it could develop a maximum thrust of 200 kN at takeoff, with an SFC of 0.346 kg/hr/kg (to be compared with around 0.6 for the JT8D models). It was followed in 1971 by GE's CF6-6D, the two-shaft engine of the DC-10-10, and by RR's three-spool RB 211-22B, the powerplant of the Lockheed 1011 Tristar, which started airline service in 1973.

The advantages of the high-bypass, high-pressure ratio

(over 20) engines with respect to the first generation turbofans was a 25% reduction in cruise SFC, and their modular construction. This feature permits rapid change of engine parts and enabled service life to be set up for each module, rather than for the complete engine, as was done before. We shall describe in some detail two families of large powerplants: the CF-6 family, which is followed by the third generation GE90, as an example of two-shaft engines; and the RB211 family, which is going over to the Trent, as representative of three-spool engines.

### 9.1.1. Two-shaft Engines

The CF-6 family of engines was first announced by GE in 1967 for the Lockheed Tristar and the MDD DC-10 in a thrust range from 142 to 160 kN. It developed up to 233.5 kN in the various CF6-6 and the CF6-50C types, which power also the Airbus A300 and some versions of the Boeing 747. These are two-shaft engines having a so-called 1-1/4 stage fan, driven by a five-stage LP turbine, while the core initially had a sixteen-stage HP compressor, with a pressure ratio of 24, an annular combustor, and a two-stage HP turbine. The 1-1/4 stage fan consists of a single stage fan having a bypass ratio of 5.7 with an integrally mounted single stage LP compressor that acts as a booster for the airflow into the core engine. In the HP compressor the inlet guide vanes and the first six stator rows have variable incidence. The annular combustor has comprehensive film cooling. The two-stage aircooled HP turbine has a TET of 1600K with rotor blades film and convection cooled. These are cast from René 80 with the discs and the shafts of Inconel 718. The takeoff SFC of the engine was around 0.35 kg/hr/kg.

A major redesign in the 1980s resulted in the CF6-80C2 (Figure 9.1), which, at 5.05, has a slightly lower bypass ratio. Here, the fan is followed by a four-stage LP compressor; the HP compressor has fourteen stages and the overall pressure ratio has risen to 30.4. The blades of the first stage HP turbine are directionally solidified. The SFC has gone down to 0.33 kg/hr/kg. There are different models, covering a thrust range from 231.5 to 268 kN, for various aircraft (A300, A310, Boeing 767, 747, and MD-11). A further development

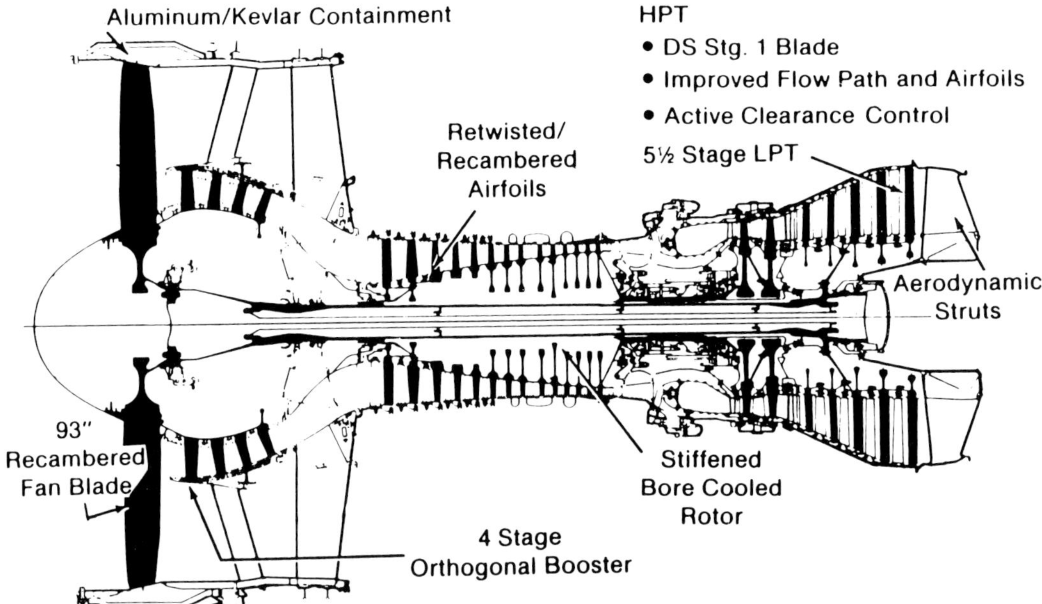



**Figure 9.1**   The CF6-80C2 engine (Jane's, 1990–91). *Reproduced with permission of Jane's Information Group.*

**Figure 9.2**   The GE90 engine (Jane's, 1991–92). *Reproduced with permission of Jane's Information Group.*

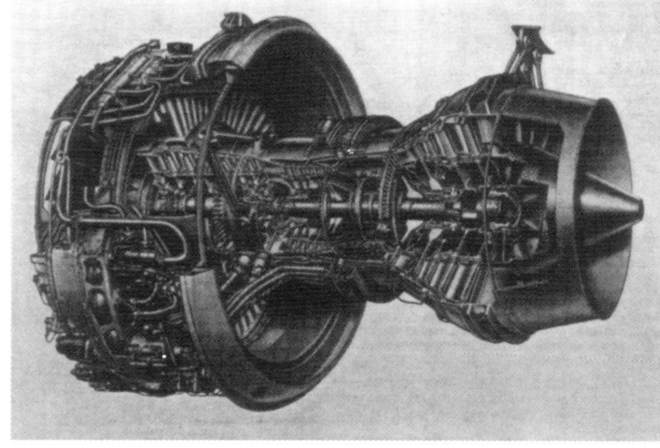

**Figure 9.3**   The RB 211-22B engine (Jane's, 1973–74). *Reproduced with permission of Jane's Information Group.*

is the CF6-80-E1, which has a larger diameter fan, a somewhat higher pressure ratio (up to 34.6), and improvements in materials and cooling techniques in the turbine. As a result of this, the thrust is up to 303.5 kN for the A330.

One should note that the development of the CF6-80 series is done together with SNECMA, MTU, Volvo Flygmotor, and Fiat Aviazione. As mentioned previously the next step is a completely new engine, the GE90 (Figure 9.2) with a bypass ratio of 10 and an overall pressure ratio of 40 and more. This engine is planned for a thrust range of 333 to 422 kN and will be capable of powering the new widebody aircraft entering the market in the 1990s. The engine is scheduled for certification in 1994 at 355.5 kN and is expected to enter service in 1995 on the twin engine Boeing 777. It is

based on a compressor scaled up from the Energy Efficient Engine program (see Section 8.1). The fan diameter will be 3.05 m with wide-chord blades. The higher bypass ratio should lower the SFC by about 10% with respect to the engines it will replace, as well as reducing emission levels by one-third, and lowering the noise level.

The propulsion system will integrate the engine with a composite, load sharing nacelle, resulting in weight savings, improved performance, and reduced maintenance. The engine will have a three-stage LP low speed compressor, a ten-stage HP compressor, a double annular combustor (see Section 8.1.1), a two-stage HP turbine with monocrystalline blades and powdered metal discs, and a seven-stage LP turbine with relatively low loading.

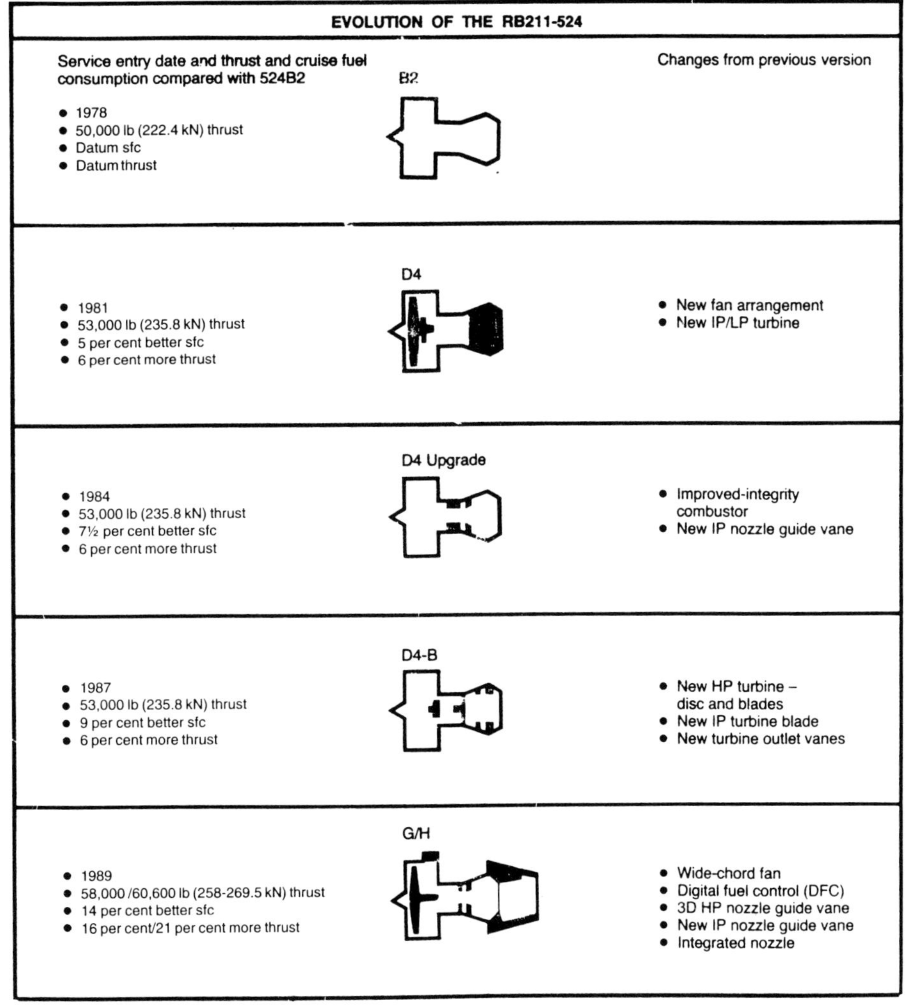


**Figure 9.4** Evolution of the RB 211-524 (Rolls-Royce Magazine No. 38, 1988). *Courtesy of Rolls-Royce plc.*

### 9.1.2 Three-Shaft Engines

Rolls-Royce decided in the late sixties on a modular three-shaft design for the Tristar engines, which, according to RR, gives greater structural strength, smaller size, and lower weight, making transport and maintenance easier. This was implemented in the RB-211.

The architecture of this engine, shown in Figure 9.3 consists of an LP single-stage fan, powered by a three-stage LP turbine, followed by the seven-stage IP compressor driven by the single stage IP turbine; then comes the six-stage HP compressor, also driven by a single-stage turbine and the annular combustion chamber which has 18 airspray burners with annular atomizers. The original RB 211-22B was rated at 187 kN thrust with a bypass ratio of 5 and a pressure ratio of 25. Over the years, the RB 211 has developed in two directions: the RB 211-524, which has grown to 270 kN, and the RB 211-535 for medium range aircraft at a thrust level of 166 to 192 kN.

The RB 211-524 started in 1977 with 222 kN (version B2) and has undergone successive improvements, shown in Figure 9.4, reaching 270 kN in version H. The main advances were in the fan, which in 1989 became a wide-chord fan with turbine discs and blades, in the compressor's guide vanes, and in the introduction of digital fuel control. Other changes made include a slight increase in length (3.175 vs. 3.033 m) and fan diameter (2.192 vs. 2.154 m), considerably higher pressure ratio (33 vs. 25), and, as a consequence, higher cruise thrust (52.1 vs. 42.2 kN, corresponding to a 23.5% increase).

The Trent engine (Figures 9.4 and 9.5), which was previously known as the RB 211-524L, has a similar bypass ratio (5.1), but is in many aspects a new engine; for example, the fan diameter has grown by 15% to 2.477 m diameter (these blades are manufactured by superplastic forming and diffusion bonding), the increased flow IP compressor has eight stages vs. seven, and the improved six-stage HP com-

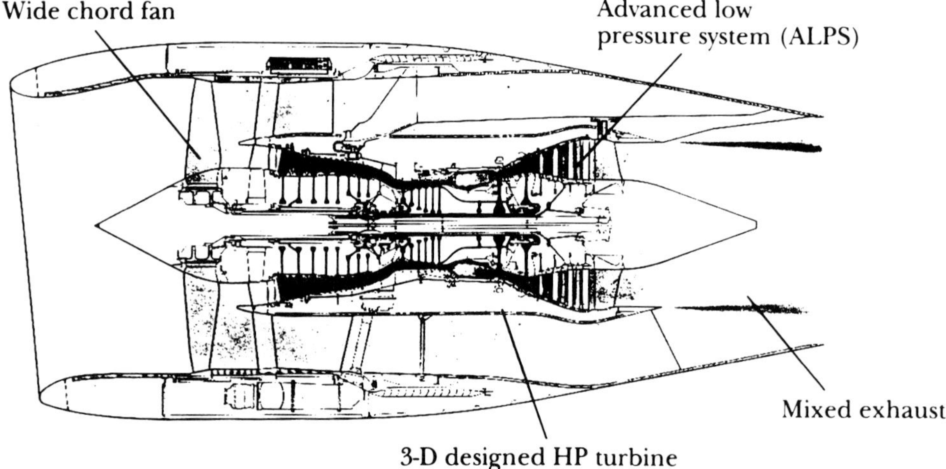


**Figure 9.5**    General features of the Trent engine (Rolls-Royce Magazine No. 42, 1989). *Courtesy of Rolls-Royce plc.*

pressor is based on V2500 technology. The overall pressure ratio has risen to 35 from 33, and the HP turbine is fitted with blading of the latest 3D design. The IP turbine has an increased annulus with single crystal uncooled blades (these required cooling in the 524), and the four-stage (vs. 3) LP turbine has 3D design blading. The SFC has been lowered by 4% relative to the RB 211-524G, to 0.543 kg/hr/kg at cruise. The certification of this engine is planned for 1993 at 308.7 kN thrust (for use in the MD-11), growing to 320.8 kN in 1994 (for the Airbus 330), then to 385 kN for the Boeing 777, superating 400 kN in the future, for the MD-12 and later versions of the B-747. The higher thrust engines will have a diameter of 2.79 meters.

The second branch of the RB 211 has two entries; namely, the RB 211-535C and the RB 211-535E4 (see Figure 9.7). The 166.4 kN thrust 535C was the launch engine of the Boeing 757; it has a HP module based on the RB 211-22B (i.e., six-stage HP compressor, annular combustion chamber, and single-stage cooled turbine), a six-stage (down from seven previous stages) IP compressor module, and a scaled down fan module. The 535E4 is an improved version of the powerplant, offering a reduction in SFC (from 0.646 to 0.607 kg/hr/kg at cruise, higher at climb).

## 9.2 Medium-size Engines

One can subdivide this class into two groups: an upper thrust category, for short and medium haul airliners, and a lower thrust one, which includes commuter aircraft carrying about 50 passengers and high class business jets. At the higher thrust levels the West prefers international cooperation with the French-American CFM 56 and the International Aero Engines V2500, while the Russians have the PS-90A. In the same thrust bracket is the recently developed propfan, on which P&W, GE, RR, and the Russians are working; its greatest promise is in the anticipation of a considerably lower fuel consumption.

### 9.2.1 Higher Thrust Powerplants

The CFM 56 (Figure 9.8) is a joint venture of GE and SNECMA covering the 88–145 kN thrust class. The CFM 56-2 was launched in 1979 to reengine the DC-8; later versions power the Boeing 737 and the Airbus A320, 321, and 340.

The initial bypass ratio was 6 increasing to 6.6 in the CFM 56-5C version. The architecture is similar in the different models. There is a robust single-stage fan with Ti discs and blades with a three-stage LP compressor (in later versions, four). This is preceded by a forward cone which reduced ice accumulation. All rotating parts are designed for a 36,000 hour life cycle duration. The HP compressor has nine stages with the first four stators variable; the overall pressure ratio is in the 25:1 class. The fully annular combustion chamber has advanced film cooling and the fuel-air mixers are machined. In the single-stage HP turbine both stator and rotor airfoils are aircooled, with directional solidification used in the latest 56-5 models; the TET is 1530 K. The core is kept short and rigid. The cruise SFC (given at 10.67 km, $M = 0.8$) diminishes from 0.67 kg/hr/kg in the 56-2 series to 0.567 kg/hr/kg in the 56-5C series for a reduction of 15%. The thrust is expected to increase to 151 kN by 1994 and eventually to 160 kN. CFM International is also investigating the possibility of increasing the bypass ratio to the range of 10–15 with further fuel consumption gains.

The V2500 engine is an example of international cooperation, with the design responsibility divided between RR (30%) and P&W (30%), which are the leaders, while the Japanese Aero Engine Corporation (the appropriate divisions of IHI, KHI, and MHI) is responsible for 23%, the German MTU for 11%, and FIAT Avio for 6%, as presented in Figure 9.9. This engine provides outstanding performance and advanced technology at low risk. It incorporates the modern technology developed for, and proven in other engines with additional refinements for the V2500.

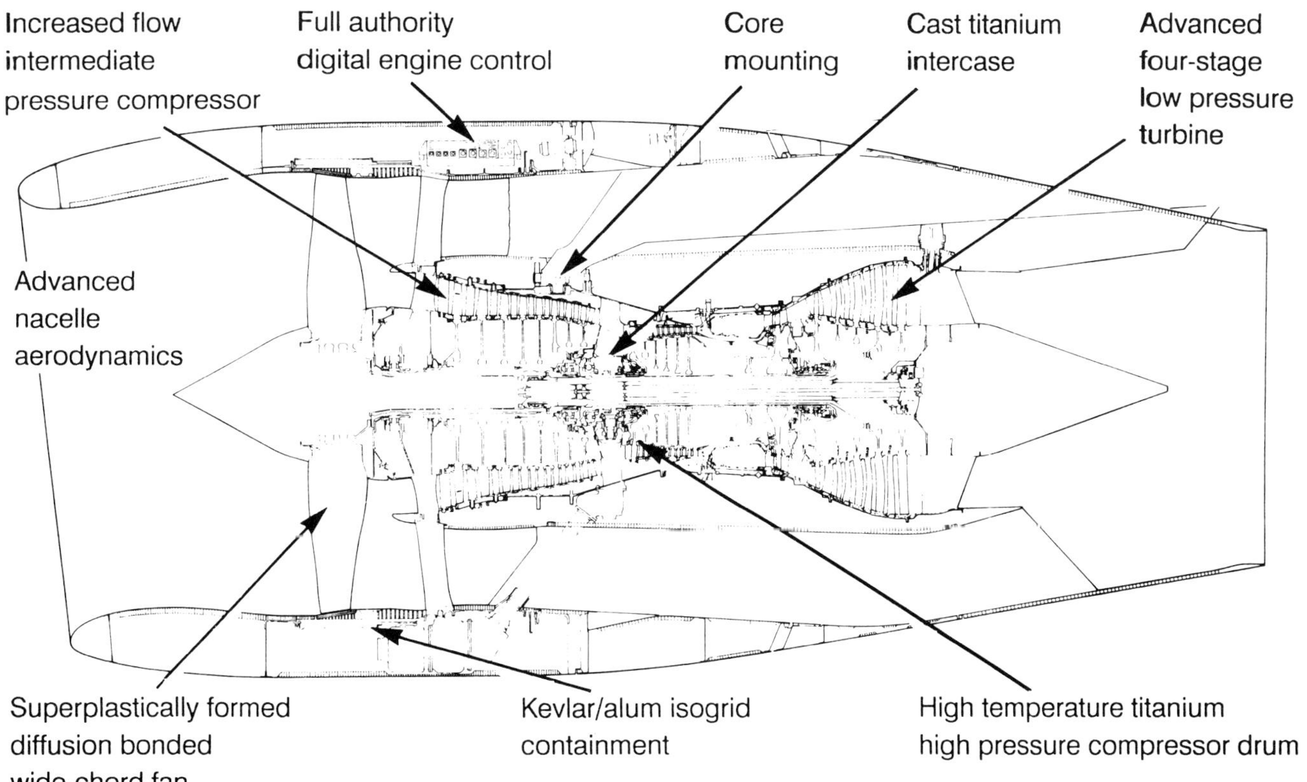


**Figure 9.6** The Trent engine. *Courtesy of Rolls-Royce plc.*

**Figure 9.7**  The RB 535-E-4 engine. *Courtesy of Rolls-Royce plc.*

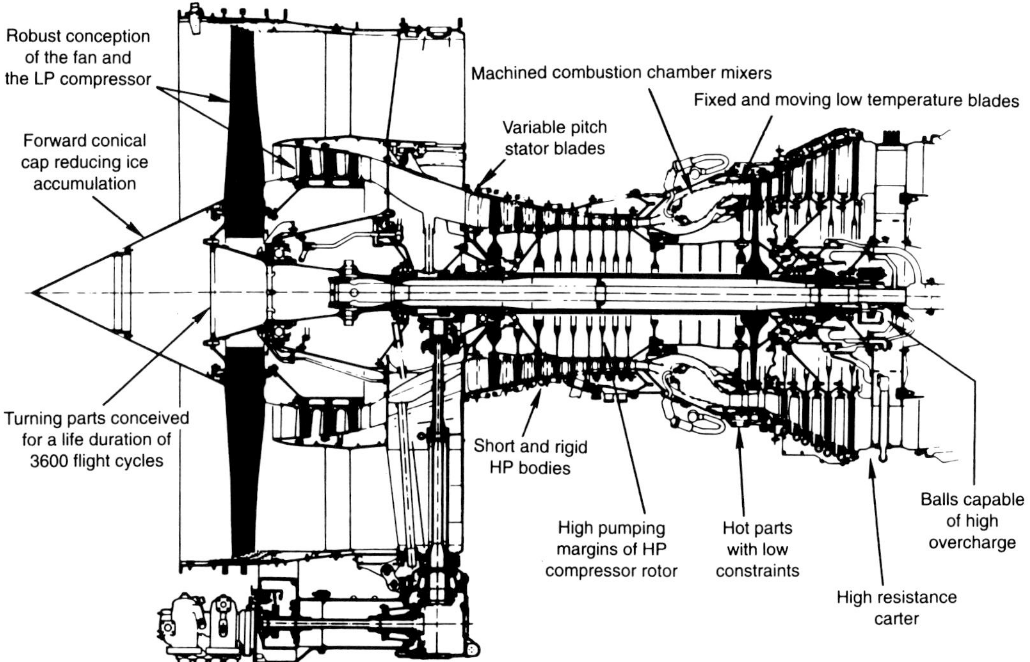


**Figure 9.8**  The CFM 56 engine (Jane's 1991–92). *Reproduced with permission of Jane's Information Group.*

The Japanese are responsible for the 3D wide-chord fan ($\beta$ = 5.42, PR = 1.7) designed with 3D aerodynamics, which provides maximum fuel efficiency and low noise; the wide-chord fan blade and the axial spacing from the core give ingestion protection to the HP compressor for excellent performance retention. The three-stage booster provides core supercharging and a uniform flow profile to the HP compressor, while the fan exit design meets structural stiffness requirements at low noise levels.

The 10-stage high pressure ratio HP compressor, manufactured by R R, gives high efficiency at low weight; the inlet guide vanes and the first three stators are variable. The overall pressure ratio of the engine is 29.4. The stiff casings,

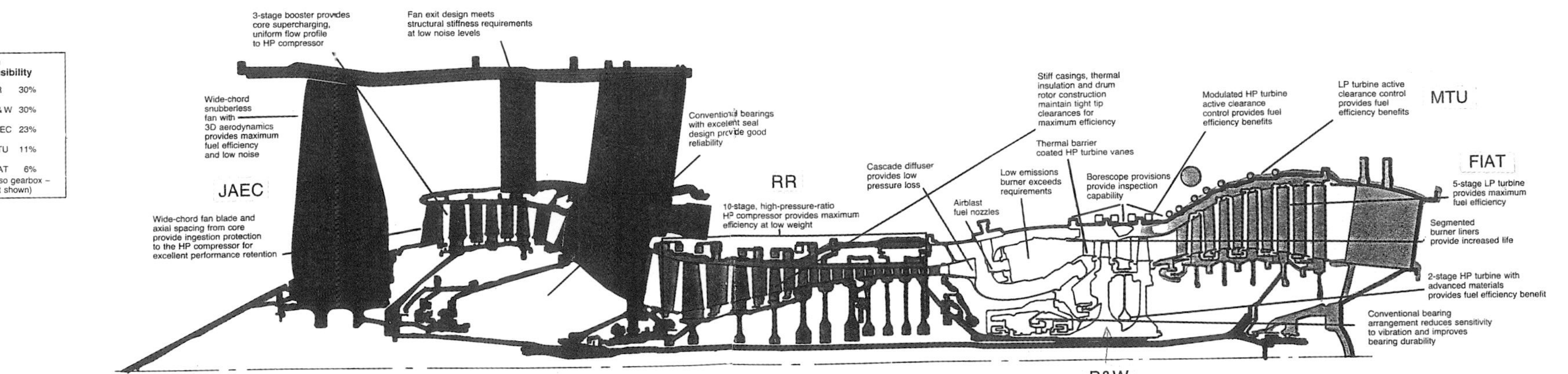


**Figure 9.9** The V-2500 engine (Rolls-Royce Magazine No. 41, 1989). *Courtesy of International Aero Engines A. G., Rolls-Royce plc.*

## How the PS-90A shapes up

The 16 tonne (35,275 lb) thrust of the Soviet Union's PS-90A engine puts it in the same class as Rolls-Royce's 535E4 and the Pratt & Whitney PW2000.

Its design weight of 2,800 kg is also highly competitive, and makes it a slightly smaller engine than the E4 and the 2000.

The general designer of the Perm Scientific Manufacturing Association (PNPO), Yuri Yevgenyevich Reshetnikov, says that his team has not yet brought the cruise specific fuel consumption (sfc) below 0.595. To supercharge the core further, a third stage was being added to the low pressure booster. This will be the definitive engine, to be certificated in 1993.

Such a change would be difficult to arrange as a retrofit, so it may be that as many as 50 Tupolev Tu-204s and a few Ilyushin Il-96-300s will be re-engined. This is not to suggest any failure in today's engine, merely that the PNPO is determined to reach its cruise sfc target of 0.58. Even today's figure of 0.595 is better than the E4, so the PS-90A is already world class.

How the Tu-204 performs against the Boeing 757 will be an important test. The aircraft are similar to each other except that the 757 carries more fuel and payload. This is largely because the 204 has to meet more severe field requirements. *Bill Gunston*

### TECHNICAL DETAILS

**Configuration:**
Two-shaft turbofan with full-length bypass duct and integral fan thrust reverser.

**LP system:**
Single-stage fan and two-stage booster compressor, four-stage turbine. HP system; 13-stage compressor, can-annular combustor with 12 front-section liners plus annular aft-section liner, two-stage turbine.

**Exhaust system:**
Multi-lobe exhaust mixer and single exit nozzle.

**Performance data:**
Take-off thrust 16,000 kg (35,275 lb) flat-rated to 30 degrees C at ISA sea level static. Representative cruise thrust 3,500 kg (7,716 lb) at Mach 0.8 and 11,000 m (36,000 ft) ISA; sfc 0.595 kg/hr/kg (0.595 lb/hr/lb). Air mass flow 470 kg/sec (1,036 lb/sec); bypass ratio 4.4:1; fan pressure ratio 1.6:1; overall pressure ratio 35:1; maximum turbine entry temperature 1,565 degrees K.

**Dimensions:**
Fan tip diameter 1,900 mm (74.8 in); engine maximum width 2,270 mm (89.4 in); maximum height 2,396 mm (94.3 in); overall length 4,964 mm (195.4 in).

**Weight:**
Basic dry engine 2,900 kg (6,390 lb).

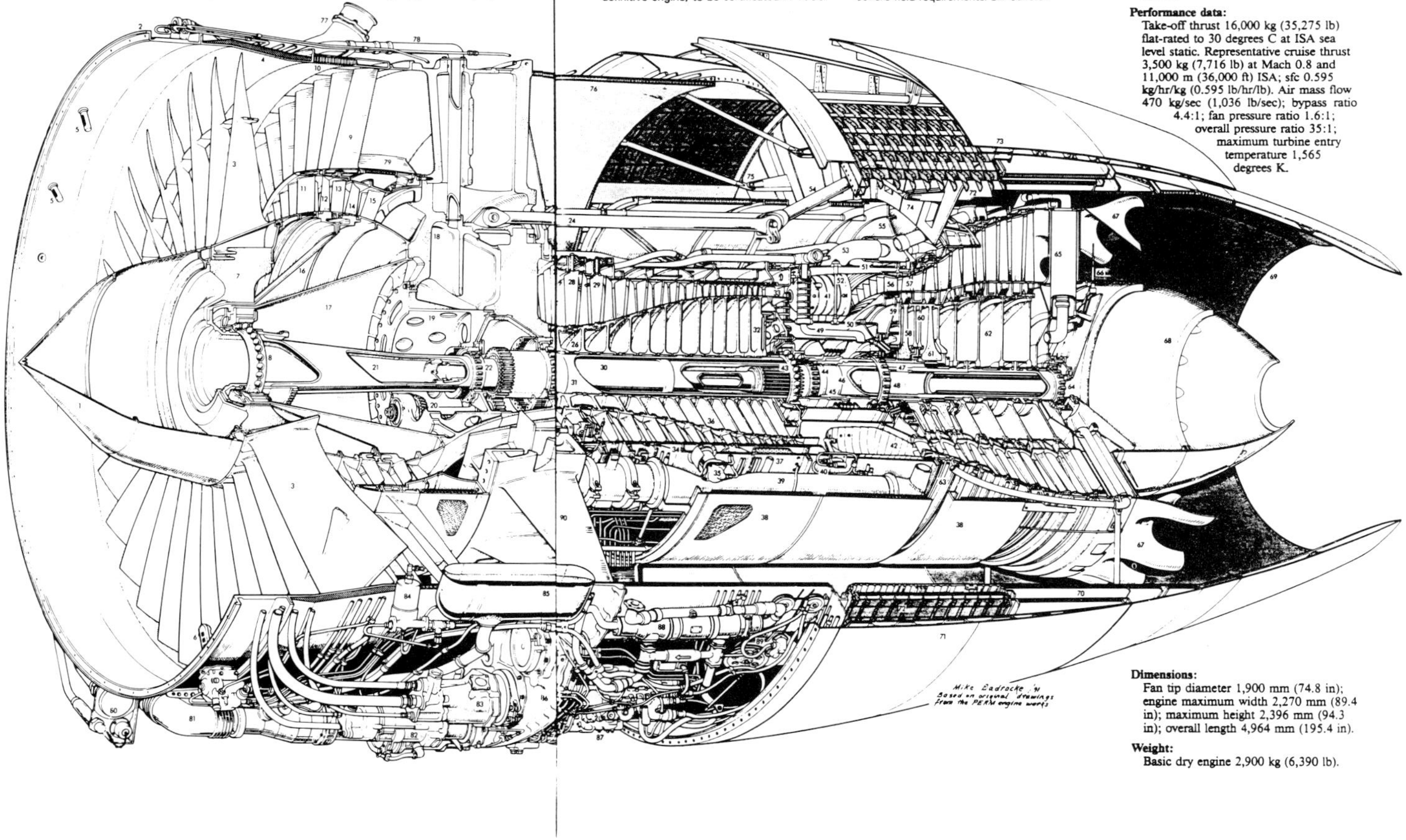

### CUTAWAY DRAWING KEY

1 Nose cone, ice shedding profile
2 engine/intake seal
3 Fan blades with snubbers [33 off]
4 Composite blade containment ring
5 Air intake temperature probe [electrical]
6 Air intake temperature probe [hydraulic]
7 Fan disc
8 Fan thrust bearing [ball]
9 Fan airflow straightening vanes
10 Fan casing honeycomb lining
11 Low pressure compressor inlet guide vanes
12 LP first stage rotor blades
13 LP first stage stator blades
14 LP second stage rotor blades
15 Compressor exit vanes
16 LP compressor inlet guide vanes
17 Conical support casing for overhung fan rotor
18 Intermediate casing
19 Bearing and gear drive support drum
20 Bevel gear drive to accessory gearbox
21 Fan front drive shaft
22 LP rotor system front roller bearing
23 HP rotor system front roller bearing
24 Engine support structure tie rod
25 HP compressor variable incidence inlet guide vanes
26 HP compressor 2nd stage rotor disc
27 HP compressor 1st stage rotor blade with snubber
28 HP compressor 1st stage variable stator blades
29 HP compressor 2nd stage variable stator blades
30 Fan drive shaft from LP turbine
31 HP compressor shaft
32 HP compressor 13th stage rotor disc
33 Compressor exit vanes
34 HP compressor three-segment titanium casing
35 HP compressor air bleed valve
36 HP compressor rotor assembly
37 HP compressor air supply for HP turbine tip clearance control system
38 Bypass duct honeycomb inner liner
39 Turbine tip clearance HP air duct
40 Combustor fuel injector
41 Can-annular combustor line [12 cans]
42 Combustor annular exit duct
43 HP rotor thrust bearing [ball]
44 HP/LP intershaft roller bearing
45 HP rotor shaft coupling
46 LP turbine shaft
47 HP turbine shaft
48 HP rotor system rear rotor bearing
49 Oil breather pipe
50 Combustor inner air casing
51 Combustor outer air casing
52 Inner air casing retaining bolts [12]
53 Turbine tip clearance air duct
54 Support link brace
55 Aft firewall
56 Turbine inlet guide vanes
57 HP turbine first stage rotor blades
58 HP turbine first stage rotor disc
59 Cooling air deflector plate
60 Air deflector and labyrinth seal support disc
61 HP turbine 2nd stage rotor disc
62 4-stage LP turbine rotor
63 Blade tip clearance air distribution pipes
64 LP rotor rear roller bearing
65 Rear bearing support structure
66 Turbine exhaust temperature thermocouple
67 By-pass air 18-lobe mixer duct
68 Turbine inner exhaust cone
69 Turbine exhaust nozzle
70 Honeycomb lined inner by-pass duct
71 Thrust reverser cowling, closed position
72 Multi-vane thrust reverser cascades
73 Thrust reverser cowling, open position
74 By-pass duct blocker doors
75 Actuator rods
76 Honeycomb lined outer by-pass duct
77 Bleed air duct to cabin conditioning
78 Oil breather pipe
79 Intermediate casing support strut with borescope port
80 Hydraulic system connectors
81 Starter air supply duct
82 Air starter motor
83 Hydraulic pump
84 Fuel shut-off cock
85 Lubricating oil tank
86 Accessory gearbox casing
87 Fuel pump regulator
88 Fuel cooled oil cooler
89 Engine control system cable pulleys
90 Intermediate casing support strut [12]



**Figure 9.10** The PS-90A engine (Jane's 1990–91). *Reproduced with permission of Jane's Information Group.*

the thermal insulation, and the drum rotor construction maintain tight tip clearances with consequent maximum efficiency.

The annular combustor by P&W has low pressure loss, thanks to the cascade diffuser; the segmented construction provides for low emissions and uniform exit temperature. The two-stage high-pressure turbine has single crystal air-cooled blades and powder metallurgy discs. These features allow high airflow through a low diameter turbine, improving cycle efficiency. Thermal barrier coatings of the vanes reduce the cooling air requirement. The turbine efficiency is improved by 3D aerofoils and by minimal tip leakage achieved by active clearance control.

The low pressure turbine, built by MTU, has five lightly loaded stages also with active clearance control, providing fuel efficiency benefits. Finally, FIAT Avio is in charge of the exhaust nozzle and the gearbox.

The engine and the nacelle have been designed integrally, drawing on the RB 211-535 engine experience; advanced composites are used in the nacelle to reduce weight. Noise from the fan and the LP booster system has been minimized by selecting the appropriate number of blades and vanes and optimizing their axial spacing, while acoustic lining is used in the ducts. Moreover, the bypass and turbine exhaust flows are mixed before leaving the nozzle.

The nominal takeoff thrust of the engine is 116 kN and the complete powerplant weighs 3400 kg; the cruise SFC is 0.575 kg/hr/kg. The launch aircraft is the Airbus A320 and it will be used, in slightly different configurations, in the A321 and the MD 90 series. Thrust will range from 100 to 133 kN, the by-pass ratio between 4.6 and 5.42, and the pressure ratio from 25 to 32.5. The engine is expected to remain in use well into the twenty-first century.

In conclusion, one can say that the main objectives of the engine, as stated below, have been achieved; namely, low fuel consumption throughout its life, low noise levels meeting airport rules, high reliability and integrity, good handling, modular design for easy low cost maintenance, good resistance to foreign object damage, growth potential, reliable, low cost operation over short stages, and low pollution emissions.

It is interesting to compare the Russian entry in this category, the PS-90A, developed by P. A. Soloviev (in his honour it is called PS) since 1982. We shall see a great similarity in the techniques employed. This engine powers two Aeroflot modern airliners, the medium range twin-engine Tu 204 and the long range four engine Il 96-300. This is a two-shaft turbofan with a full length bypass duct (bypass ratio 4.4) developing a takeoff thrust of 156.9 kN with a pressure ratio of 35:1, a cruise SFC of 0.58 kg/hr/kg, and a TET of 1565 K. The architecture, shown in Figure 9.10, comprises an LP system consisting of a single-stage fan and a two-stage booster compressor driven by a four-stage turbine. The HP

system has a 13 stage compressor, with the inlet guide vanes and the first two stators variable, a can-annular combustor with 12 front-section liners plus an annular aft-section liner (this is the main difference from the Western powerplants) and a two-stage turbine.

Interesting features of this modern engine include fan design for minimizing foreign object damage, and the fact that the HP turbine rotor blades are directionally solidified castings and the discs are produced by powder metallurgy. First and second stage blades are cooled with high and intermediate pressure compressor bleed air. The cooling of the first-stage blades is controlled to minimize the effect of air bleed on fuel efficiency at cruising altitudes between 7 and 12 km. Active tip clearance of LP turbine stages is provided by air bled from the forward end of the HP compressor. Noise attenuation has been achieved by optimizing the number of fan exit vanes and their spacing from the fan rotor blades, plus the use of perforated honeycomb material around the inner and outer wall of the bypass duct. Moreover, there is an 18 lobe exhaust mixer, which contributes to noise reduction and enhances propulsion efficiency. As a consequence, the PS-90A noise level conforms to international noise regulations (see ICAO Ch. 3 Annex 16 or FAA FAR Part 36 stage 3).

The engine has dual channel digital electronic control with hydromechanical backup. It is also equipped with a computer based monitoring system, providing continuous in-flight signals to a data recorder of the main operating parameters, flight cycles, and operating hours.

## 9.2.2 The Propfan

The energy crisis in the early 1970s rekindled an interest in the turboprop because of its fuel efficiency. Studies performed by NASA and the U.S. aerospace industry showed that the propeller offered larger gains than the turbofan, provided it could be operated more efficiently at Mach numbers above 0.6. In 1975 Hamilton Standard proposed an innovative propeller, the propfan, using a larger number of shorter, thin swept blades (8 to 12 instead of 4), which could provide high efficiency up to Mach 0.8. Since then, two types of propfans have been developed; one by P&W and Allison, based on the original Hamilton Standard proposal, the other by GE and SNECMA, known as the unducted fan or UDF. It is also known that Rolls-Royce and MKB in Russia are doing work in this direction.

A chart by Calmon (1988), shown in Figure 9.11, presents the evolution of the specific consumption of civil aircraft engines. One sees a decrease of 15% in SFC from the first turbojets to the first generation turbofans, then a further 18% improvement for the second generation (high bypass) turbofans; another 14% decrease occurred during the 1980s by improvements in component performance, better energy utilization, and more advanced controls. The propfan, whose feasibility was proven by 1989, could bring a further

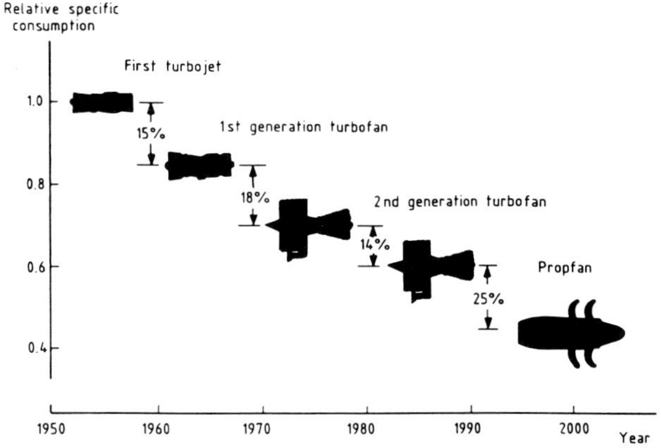



**Figure 9.11** Evolution of SFC in civil engines (Calmon 1988). *Courtesy L'Aéronautique et L'Astronautique.*

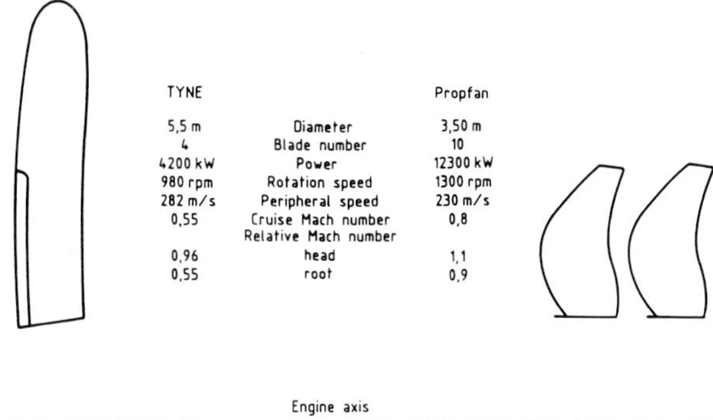



**Figure 9.12** Comparison of propeller of propfan with that of the Tyne engine (Calmon 1988). *Courtesy L'Aéronautique et L'Astronautique.*

**Figure 9.13** The GE36 unducted fan (Jane's 1990–91). *Reproduced with permission of Jane's Information Group.*

**Figure 9.14** The PW-Allison 578-DX engine (Jane's 1990–91). *Reproduced with permission of Jane's Information Group.*

reduction of 25%, thus improving (lowering) the SFC by over 60% since the beginning of the jet age.

Recent progress in CFD has demonstrated that a one row propfan could reach Mach 0.75 with acceptable efficiency, while adding a second row of blades, turning in opposite direction, increases the efficiency beyond 80%, as shown in Figure 9.13. A comparison of the propeller of a 10-blade propfan, which provides 100,000 N takeoff thrust, to that of the Tyne engine, shows that it supplies three times the power with a diameter smaller by 36%. The relative cruise Mach number varies between 0.9 to 1.1 according to the position (see Figure 9.12).

The GE-SNECMA unducted fan (UDF), known as GE 36, is shown in Figure 9.13. It has a bypass ratio of 36, uses an F404 core (110 kN thrust class) and has a four-stage LP compressor and a seven-stage HP compressor with single-stage turbines and a very compact FADEC regulated combustion chamber. This powerplant, which would be suitable for the MD-90 series of MDD, has eight scimitar-like blades in two rows. It was tested on a proof-of-concept demonstrator, accumulating 162 hours on a test bench, 97 hours on a B-727, and 366 hours on a MD-80 on the ground, with 25

and 93 flight hours respectively in the two planes. One of its features is the elimination of the drive gearbox, so that the hot gases pass through large multistage contrarotating turbines downstream. This requires that rotor blades be substituted for what normally would be the turbine stators, and the visible propeller blades are mounted on the rotors of the power turbine. The blade loading of the LP power turbine is light, and the speed is low with consequent low stress levels. The composite material blades change pitch from flight speed settings to feather, to reverse. The pitch is set to control speed and power output. Development was concluded in 1989 with completion to service entry within four years at an estimated cost of 1.3 billion dollars. GE and SNECMA are conducting studies on an UDF engine based on the core of the CFM 56-5 which could provide approximately 180 kN thrust with a gain of 6% to 10% in SFC.

The PW-Allison 578-DX, shown in Figure 9.14, was built as a flight test demonstrator using a geared counter-rotating propfan propulsion system. It is rated at a thrust of 89 kN, consisting of a three-stage LP compressor, a modified Allison 571 core engine with a 9689 kW (13,000 shp) differential planetary gear system. This drives two 3.53 m diameter six-blade Hamilton Standard propfans. A modified PW-Hamilton Standard FADEC regulates the fuel flow, the variable compressor stators, and the propfan blade angles. The first flight was on an MD-80 airliner in April 1989, and the program was completed during the same quarter.

The Russian propfan activity is conducted by two bureaus, Lotarev and MKB. Lotarev experimented with a contrarotating propfan, the D-236, based on their D-136 engines. It is interesting that they built an engine with eight blades in the front unit and six blades in the rear unit. The MKB bureau has studied advanced propfans derived from the PS-90A, paying particular attention to aft-fan derivatives with contrarotating shrouded fans having a bypass ratio of 30.

In the early 1990s the future of the propfan is unclear. On one hand, the world oil price is relatively low, while on the other, the high bypass fans, like the GE90 and similar units being developed in Russia, will decrease the SFC by about 10% below current values.

### 9.2.3 The Lower Thrust Engines

In this range we shall treat two interesting engines, the Garrett ATF3, which combines a three-spool design with a reverse flow combustion system and turbines, having a mixed flow exhaust, and the Allison T-406 with the related GMA 2100 and GMA 3000 engines.

The ATF3-6A, which powers the Falcon 200 business jet, has a takeoff thrust of 24.2 kN. Figure 9.15 shows the arrangement of the components, which makes it possible for the engineers to design the fan with little dependence on the gas generator, thus permitting it to operate at optimum fan speed. Omission of the inlet guide vanes, mixing the exhaust with the fan airflow, and the double reversal of the airflow considerably reduces the engine's length, as well as its infrared signature, which is important in military applications.

The architecture of the engine includes a single-stage titanium fan driven by a three-stage LP turbine with shrouded blades. The bypass ratio at takeoff is 2.8 with a mass flow rate of 73 kg/sec. The intermediate pressure system features a five-stage titanium compressor driven by a two-stage IP turbine, also with shrouded blades. The flow is fed to the rearward facing HP compressor via eight tubes feeding into an annular duct. The core airflow is 18.15 kg/sec. In the HP system a single stage centrifugal titanium compressor is driven by a single stage turbine with air-cooled rotor blades. The overall pressure ratio of the engine is 21 at takeoff and 25 at high altitude cruise. The combustor is a reverse flow annular type unit. The exhaust gases are turned by 180° through eight sets of cascades to mix with the fan bypass flow and are then discharged via an annular nozzle which surrounds the combustion section. The SFC is low (0.506 kg/hr/kg) at takeoff.

The other engines in this range which we will describe are based on the Allison T-406 and belong to the GMA 2100 (turboshaft and turboprop) and GMA 3000 (turbofan) series. This family may, in the future, include a propfan, if demand will justify its development. In this range the engines are common to military and civil powerplants. The first application was the T-406-AD-400 turboshaft, shown in Figure 9.16, developed for the V-22 Osprey, with a maximum power of 4586 kW. It has a 14-stage axial compressor in which the inlet guide and the first five stator vanes are variable. The pressure ratio is 14 and the air mass flow is 16.1 kg/sec. The annular combustor has film-cooling and is provided with sixteen airblast fuel nozzles. The gas genera-

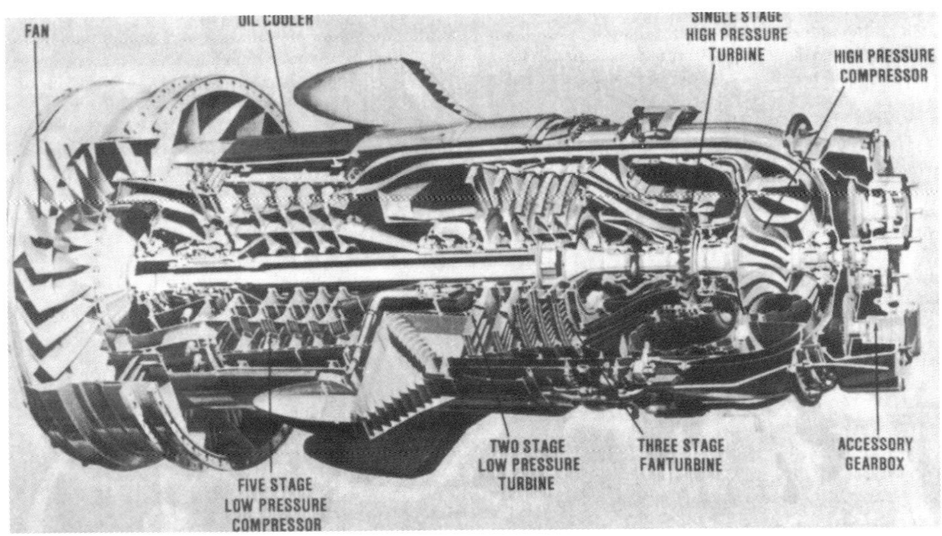


**Figure 9.15** The ATF3-6A engine (Jane's 1990–91). *Reproduced with permission of Jane's Information Group.*

**Figure 9.16**   The T406 engine (Jane's 1990–91). *Reproduced with permission of Jane's Information Group.*

tor turbine has two axial stages with aircooled single-crystal blades. The power turbine is also a two-stage device.

Turboprop applications are envisaged for the power range of 3000–5500 kW. The first such unit to be built is the GMA 2120, which, at 3394 kW, has been chosen for the Swedish fifty seat SAAB 2000 commuter plane. In this application, the engine is equipped with a six-blade Dowty propeller.

The turbofan derivatives, designated GMA 3000, could reach 55 kW thrust. The first application announced, the GMA 3007, has been selected for the Embraer EMB-145, a twin turbofan with 45–48 passengers. It will develop a thrust of 31.14 kN using a single-stage fan with wide chord blades, single crystal HP turbine blades, and FADEC control.

## 9.3 Technology of Small Engines

Engines in the power range up to 3000 kW for turboshafts and turboprops and 7500 N thrust for turbofans make up this category, which covers a wide spectrum of civil and military applications. These include, as shown in Table 9.1, helicopters, trainers, commuter aircraft, executive jets, light attack aircraft, maritime reconnaissance, and anti-submarine warfare. In this section we shall treat in some detail the Garrett F 109 turbofan engine (Krieger et al. 1988) and the RR-Turbomeca RTM 322 family, which is planned to include turboshaft, turboprop, and turbofan engines, as well as the recent MTR-390 a turboshaft "European" engine, developed jointly by MTU, Turbomeca, and RR.

Ruffles (1984) shows that, for a given standard of component efficiency and leakage, fuel consumption improves with overall pressure ratio and TET, as shown in Figure 9.17 for a 1500 kW class turboprop engine. If, however, one takes into account that scaling and cooling problems cause

greater losses with higher pressure and TET, the optimum cycle occurred at a lower pressure ratio (18:1 instead of 30:1) and lower TET (1550 K instead of 1650 K) in 1984. He foresaw future improvement, thanks largely to advances in technology resulting in a pressure ratio of 23:1 and a TET of 1600 K. The author states that similar results are obtained for helicopter engines, while in smaller engines (such as the 750 kW class) size effects are even more pronounced. In modern engines a small number of compressor stages is used, utilizing centrifugal stages, combined with a reverse flow combustor, a single- or two-stage aircooled turbine for the core and a two-stage LP or power turbine.

Mestre and Lagain (1984) in France performed experiments with a reverse flow combustor for a small gas turbine

**Table 9.1**   Small engine applications-aviation (Ruffles 1984).

|  | HELICOPTERS<br>Turboshaft | FIXED WING AIRCRAFT<br>Turboprop | Turbofan |
|---|---|---|---|
| Army/Air-force | • Light attack/ anti-tank<br>• Multi-role combat<br>• Tactical transport | • Primary trainer<br>• Transport<br>• Light attack | • Primary trainer<br>• Light attack |
| MILITARY<br>Navy | • Heavy lift<br>• Air search rescue<br>• Airborne early warning<br>• Anti-submarine warfare | • Maritime reconnaisance | • Anti- submarine warfare |
| CIVIL | • Off-shore oil exploration<br>• Executive<br>• Air taxi<br>• Medivac | • Commuter aviation<br>• Air taxi<br>• Executive<br>• Agriculture | • Executive<br>• Special cargo |

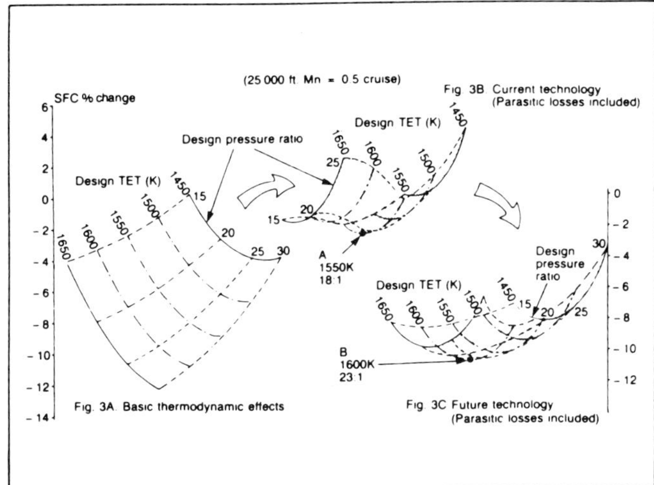



**Figure 9.17** SFC vs. $p_2/p_1$ and TET (Ruffles 1984). *Reproduced with permission of Aer. J.*

at atmospheric pressure and found that it is important to optimize the distribution of air among combustion, dilution, and cooling. They also conclude that prevaporization of the fuel allows one to obtain a satisfactory temperature distribution at the combustion exit, low emissions, and good efficiency for a large range of operating conditions.

The Garrett F109 turbofan engine was developed in response to the needs of a new primary trainer for the U.S. Air Force. The requirements were high reliability, fuel efficiency, and ease of control, to minimize the workload of the student pilot. The solution was a high bypass ratio turbofan employing a state of the art compression system driven by a turbine with relatively low TET and controlled by a FADEC unit. Two unique requirements from the manufacturer were that the engine should conform to the engine structural integrity program (ENSIP) and that it should satisfy a new four step qualification program, the accelerated mission test (Bauer 1982).

ENSIP is an organized and disciplined approach to the design, analysis, development testing, quality assurance, and life management of the engine. The benefits of ENSIP are many: rigorous design validation and verification, higher engine maturity at production, improved reliability/durability/availability, unparalleled safety, and lower operating cost. A key element of ENSIP is to establish damage tolerance design as a basic concept.

The design point flight condition selected was maximum continuum power at 7.7 km height, Mach 0.35, ISA. They also evaluated part power (for loitering) at 4.575 km, ISA; maximum power at sea level takeoff, ISA and at 1.525 km altitude, Mach 0.192 at 38°C.

In the process of selecting the core engine, SFC tradeoff studies for the core pressure ratio (CPR) and bypass ratio (BPR) were conducted for cruise and loiter flight conditions. Figure 9.18 shows the results for the cruise conditions; a CPR of 13 gives the best results, while a BPR of 5 was selected, which is best at loiter and only 0.5% off at cruise. After defining the core pressure ratio a fan optimization study was conducted, evaluating twelve design combinations of fan pressure ratio (FPR) between 1.5 and 1.7, and bypass ratio between 3.5 and 6. Although the optimum BPR is greater than 5 and the FPR slightly less than 1.6, these values were chosen, based on the past experience of Garrett with an FPR of 1.6 and small differences from optimum SFC.

The final engine, shown in Figure 9.19, is a two-spool, contra-rotating concentric shaft, fixed-geometry, turbofan engine. The low-pressure (LP) spool consists of a single-stage fan directly driven by a two-stage axial flow turbine. The high-pressure (HP) spool consists of a two-stage centrifugal compressor driven by a two-stage axial flow turbine having a cooled first stage. The combustion system incorporates a reverse-flow annular combustor and piloted airblast fuel nozzles. An accessory gearbox, driven by the HP spool, provides shaft horsepower for engine components and aircraft accessories. The engine exhaust geometry is arranged to operate with customer-furnished compound thrust nozzles for fan and core air flows.

In the development of the HP compressor, whose impellers utilize backward curved blading and integral machined titanium forgings, initial tests indicated that surge margin and performance did not meet design goals. As a result, an extensive development program was initiated to improve compressor performance. A single-stage program was conducted to establish an alternate second-stage design having higher efficiency potential and improved operating range in conjunction with the first-stage design. Additional tests were conducted to evaluate and develop diffuser performance, confirm design criteria, and provide proper matching with the alternate design impeller. Full-scale testing of the compressor established baseline performance levels and confirmed minimal performance sensitivity to classical inlet distortion patterns with distortion levels exceeding specification requirements. A total of 26 separate compressor rig tests were conducted. After introducing appropriate modifications compressor rig test results demonstrated that the initial production compressor configuration achieved its surge margin goals.

The moderately loaded, two-stage axial turbine, which must combine high performance with long life, also required modifications during the design and development phase. The overall performance and operability throughout the flight envelope were determined by tests conducted at the Arnold Engineering Development Center. These included altitude characteristics and component qualification tests, such as the fan foreign object damage test, which veri-

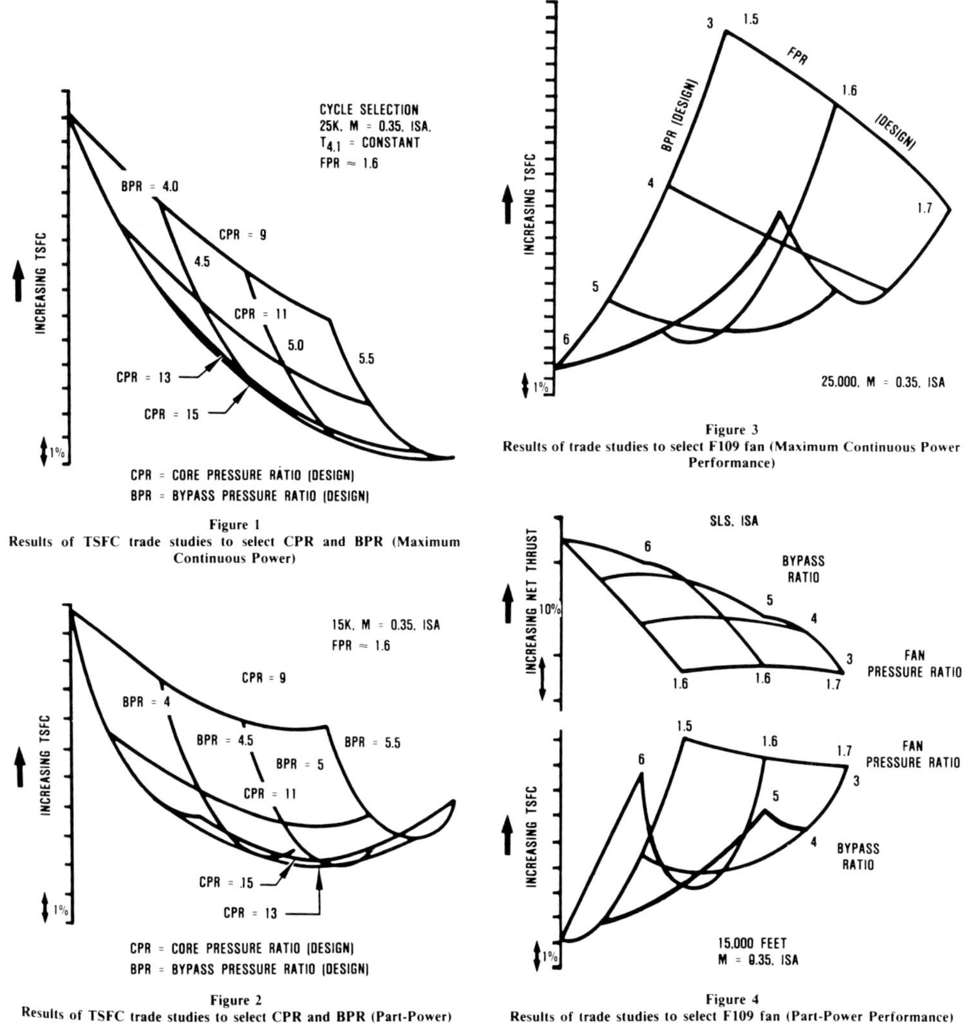



**Figure 1**
Results of TSFC trade studies to select CPR and BPR (Maximum Continuous Power)



**Figure 2**
Results of TSFC trade studies to select CPR and BPR (Part-Power)



**Figure 3**
Results of trade studies to select F109 fan (Maximum Continuous Power Performance)



**Figure 4**
Results of trade studies to select F109 fan (Part-Power Performance)

**Figure 9.18**   Selection of CPR and BPR for the Garrett F109 (Krieger et al. 1988) *Courtesy of Allied-Signal Aerospace Co.*

fied the ability of the fan rotor to continue to operate after this type of deterioration. Tests of tolerance to damage, conducted on rotating and static ENSIP components, verified the adequacy of the recommended inspection intervals.

The accelerated mission test, which is superior in many aspects to the previously used model qualification test, showed that the engine met all reliability, maintainability, safety and life cycle requirements. Finally one should mention, that by technology improvements, such as introduction of single crystal blades in the HP turbine and increasing the flow by addition of blade height in the HP compressor, the manufacturer expected to improve the maximum take-off thrust to 7.5 kN from the original 5.92 kN of the TFE109-1. As the U.S. Air Force cancelled its trainer, the original engine was used in the first prototype of the Pro-mavia Jet Squalus trainer. A more powerful TFE 109-2 (6.7 kN) powers the twin engine advanced trainer-tactical fighter (ATTA 3000) offered by the same Belgium-based company, and a 7.12 kN TFE 109-3 has been installed in the second Jet Squalus.

The launch engine of the RTM 322 family, built by Rolls-Royce and Turbomeca, with the participation of the Italian company Piaggio (10%) since 1986, is the RTM 322-01 turboshaft rated at 1560 kW with growth potential to 2250 kW. The family combines simple design, reliability, low fuel consumption, light weight, and low cost. Turboprops in the 1200–1500 kW range, with potential growth to 2100 kW, are under study for 35–70 passenger aircraft.

The RTM 322-01, shown in Figure 9.20, is equipped with FADEC and has options for an inlet particle separator and an infrared suppressor. It has potential civil and military applications in the 7 to 15 metric ton class for helicopters such as the EH101, the Sikorsky Black Hawk and Seahawk series, the Westland WS-70, and the AH-64 Apache. It has

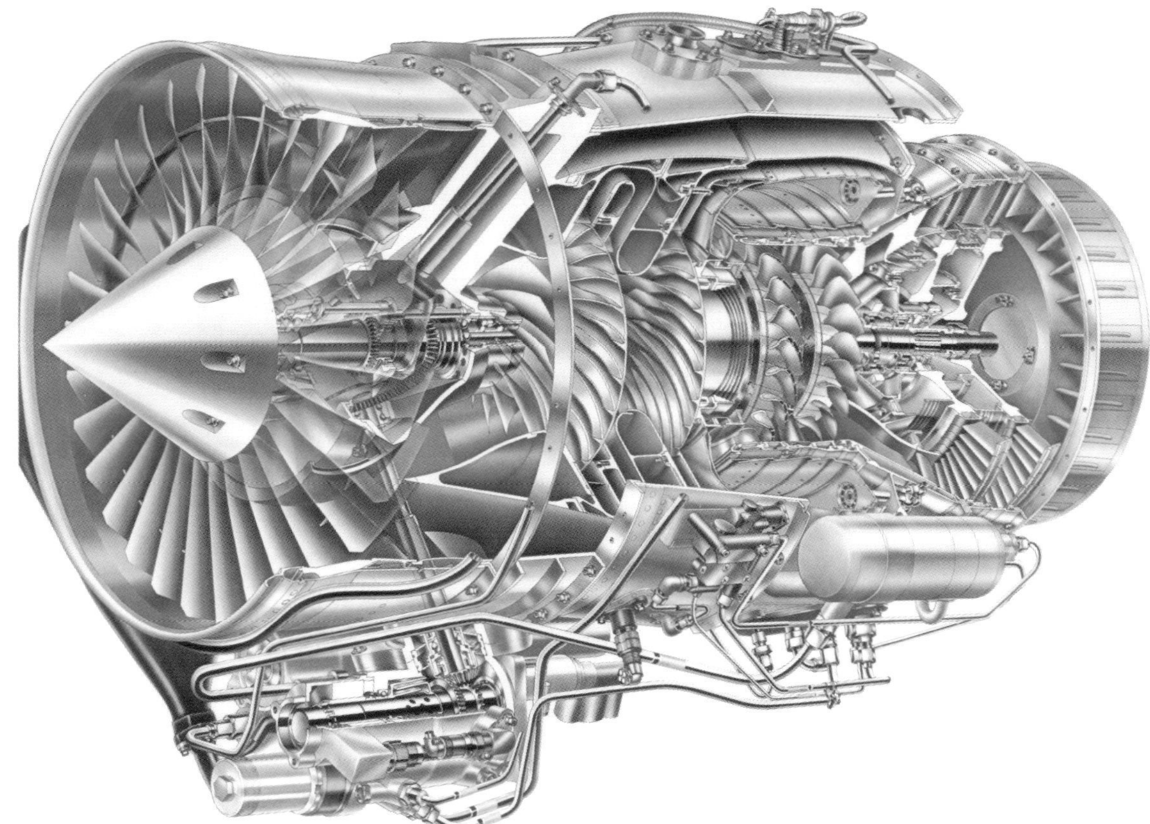

**Figure 9.19**   Final version of Garrett F109 engine (Krieger et al. 1988) *Courtesy of Allied-Signal Aerospace Co.*

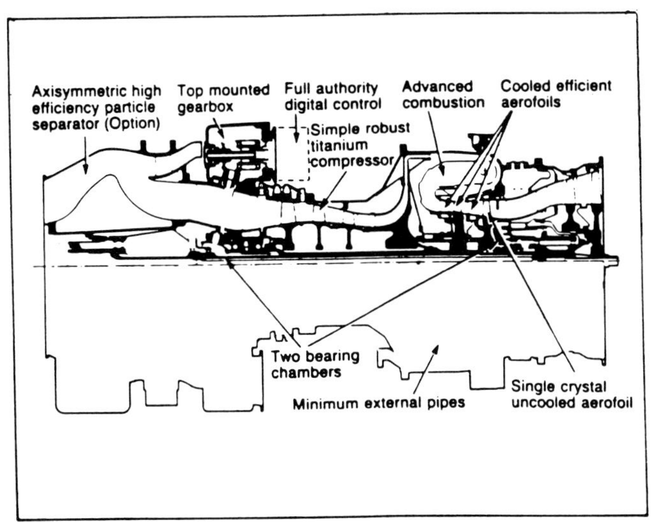


**Figure 9.20**   The RTM 322 engine (Jane's 1990–91). *Reproduced with permission of Jane's Information Group.*

demonstrated 1790 kW power. The compressor has three axial and one centrifugal stage, the combustor is annular with reverse flow, the two-stage gas generator turbine has an air-cooled first and second stage stator and first stage rotor, with an uncooled single crystal second stage rotor. The power turbine is also a two-stage unit.

The smaller MTR-390 is a turboshaft engine developed jointly by MTU, Turbomeca, and RR. Its first application is geared to an 8000 rpm output for the Eurocopter Tiger. The direct drive MTR 390T (27000 rpm output) has been proposed for the JEH Tonal, a derivative of the Agusta A 129. It has two centrifugal compressor stages, with an air flow of 3.2 kg/sec, a pressure ratio of 13 and a reverse flow annular combustor. The gas generator turbine is a single stage air-cooled design, while the free-power turbine has two stages. The maximum output power is 1160 kW.

# Chapter 10

# *Civil Engines for Supersonic Aircraft*

The number of engines built specifically for supersonic civil airliners is very limited, including the Anglo-French Olympus 593, which powers the Concorde, and the Nk 144, the engine of the Russian Tu-144. A third engine, which was tested on the ground, is the GE4, developed by General Electric for the American SST; it never flew because of program cancellation in 1971. On the other hand, a lot of research was carried out in the United States, first through the SCR Supersonic Cruise Research (SCR) propulsion program and, since 1985, under NASA's High Speed Research Program (HSPR) in which new engine concepts, such as the double bypass engine and the variable stream control engine (VSCE), were proposed and tested.

The major requirements for the viability of a supersonic aircraft are that it should meet acceptable standards of airport noise and sonic boom levels, that it does not have harmful effects on the atmosphere, and that it is economically competitive with new generation long-haul subsonic transports. So a high speed commercial transport (HSCT) should carry three times the Concorde passenger load over twice the distance at about 1/7 the cost per passenger-mile with a tenfold reduction in community noise (Wesoky et al. 1990). Rolls-Royce and SNECMA are studying variable cycle engines for SST applications (see Figure 10.1). RR also works with the Russian Lyulka design bureau on the definition of a powerplant for the proposed Gulfstream-Sukhoi supersonic business jet.[1]

The environmental problems will be discussed in Section 13.2 and Chapter 15. Here we shall describe briefly the three above mentioned engines, together with two advanced engine concepts (Morris et al. 1989) and then we shall treat some of the new concepts (Fishbach et al. 1982; Douglas Aircraft Co. 1989).

[1]It is interesting to note that the AL-31 may be used for prototype flight testing of the projected Gulfstream-Sukhoi Supersonic Business Jet; this would utilize two advanced technology RR/Lyulka non afterburning turbofans.

## *10.1 Existing, Experimental, and Conceptual Engines*

The Olympus 593 Mk 610 (Taylor 1978) is a two-spool turbojet with partial afterburning, which develops a maximum sea level thrust of 167.7 kN for a weight of 3380 kg (including the exhaust system). It has a mass flow of 180 kg/sec and a TET of 1355 K, with a pressure ratio of 15.5; the SFC at Mach 2 is 1.19 kg/hr/kg with an overall efficiency[2] of 41%.

The Kuznetsov Nk 144 is a three-spool design, with a two-stage titanium fan, a two-stage IP compressor, an 11-stage HP compressor, an annular combustor, a single-stage HP turbine with air-cooled blades, a one-stage IP turbine, and a two-stage LP turbine. This is a low bypass turbofan ($\beta = 1$); the pressure ratio is 15, the maximum air flow 250 kg/sec, the TET is 1350 K, and the weight (without jet pipe) is 2850 kg. Its thrust is rated at 196.1 kN.

The GE-4 is a single-spool turbojet with an afterburner, that produces 311 kN thrust and weighs 6012 kg. Its mass flow rate is 300 kg/sec, the pressure ratio 12.3, and the TET 1560 K. It was planned for cruise at Mach 2.7 with an overall efficiency of 42%.

The three first-generation engines show that different approaches are possible; all have, in common, fixed geometry and high cost. Presently, RR is conducting studies for an engine that could power a 250 passenger airliner at a speed of 2 to 2.5 Mach over a range of 9000 km. They favour a tandem fan engine, which would operate as a turbofan during takeoff and subsonic cruise, becoming a turbojet at supersonic speeds. This would be achieved by using an increased air flow, obtained by modifying air flow paths between the compressors (or downstream of them) with auxiliary fans providing the excess flow. At supersonic speeds the ducting and valve system can be stored within the intake area.

SNECMA is studying a three-spool mid-fan, the MCV

[2]This is the useful work provided by the propulsion system divided by the energy stored in the fuel consumed by it.

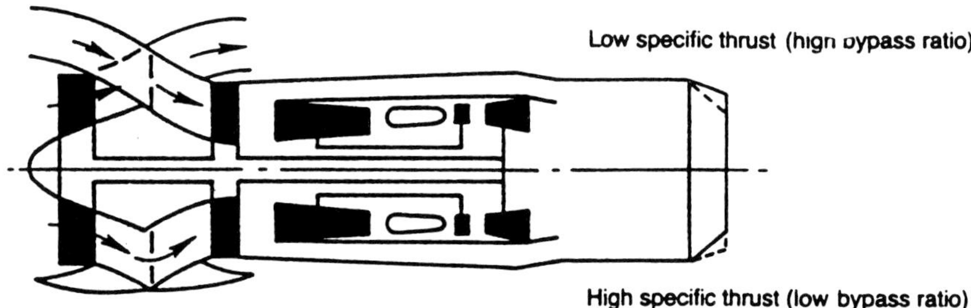



**Figure 10.1**  RR-SNECMA studies for SST (Air et Cosmos 1990:1). *Reproduced with permission.*

99 with inlets on each side of the nacelle feeding a large single stage fan driven by its own reverse flow turbine. This gives satisfactory performance at subsonic speeds. At supersonic cruise, the engine becomes a turbojet with no bypass; the fan is stopped and a secondary inlet, used at subsonic conditions, is closed. The engine features a clean combustion chamber with premixed, prevaporized combustion.

RR and SNECMA, which collaborated on the Olympus, have performed a joint market survey in order to identify possible engine concepts and determine the key technologies and the additional test facilities required.

In the United States, General Electric and P&W have investigated jointly the feasibility of developing a propulsion system for the Mach range between 1.5 and 3.5, although their technical approach is different. P&W proposes to use a variable geometry primary inlet as well as a variable geometry core inlet with an ejector assembly operating also as a thrust reverser. GE favors an augmented, double bypass, variable cycle powerplant, similar to the GE21, proposed in the framework of the SCR project for the AST-205. This planned 2.62 Mach airliner could carry 290 passengers over 8000 km (versus 100 passengers and 5400 km for Concorde or 140 passengers and 6500 km for the Tu-144). The GE 21 was planned as a two-spool turbofan with a compression ratio of 15 and a TET of 1800 K. Its performance is shown in Figure 10.2. The calculated overall efficiency of 47% is based on the use of more efficient components. Morris and coworkers at NASA Langley proposed in 1989 a more advanced conceptual engine, cruising at Mach 3, which could achieve an efficiency of over 54%. Such an engine, which would use technologies available in the period

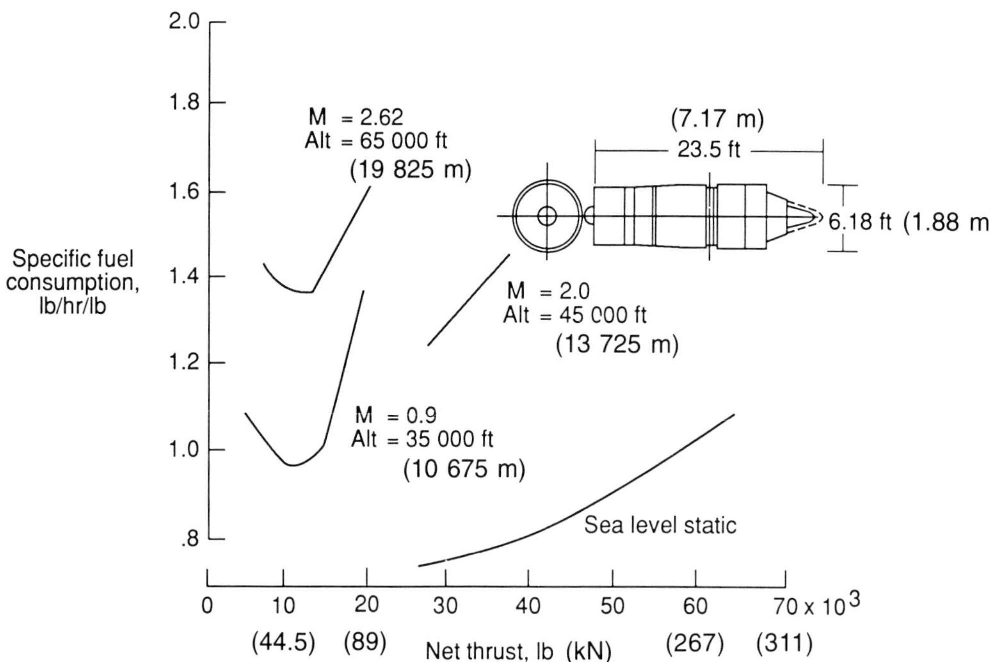



**Figure 10.2**  Proposed GE21 performance (Morris 1989). *Courtesy of NASA.*

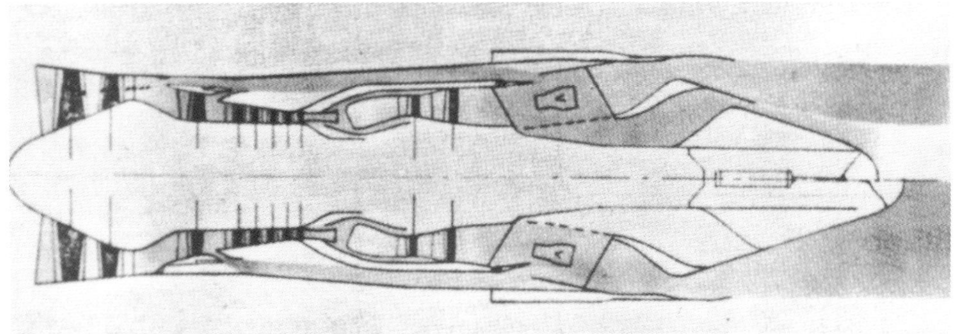

**Figure 10.3**   Double bypass engine (Fishbach et al. 1982). *Sponsored by NASA. Copyright © AIAA 1982. Used with permission.*

2005–2010, would have an overall pressure ratio of 38.6, a TET of 1945 K, an air mass flow of 184 kg/sec, and would operate at a height of 18.3 km. Two versions, shown schematically in Figure 10.4, are envisaged: a non-afterburning one supplying 159 kN thrust and a more powerful one, with reheat reaching a thrust of 272.5 kN.

## 10.2 Innovative Engines for Supersonic Propulsion

After the American SST program was canceled, it was recognized that significant technical progress was required to make a second generation of supersonic transport economically attractive. Consequently, in 1972 NASA started the supersonic research cruise (SCR) program, directed by NASA Langley, working closely with Boeing, MDD, and Lockheed on the airframe and with GE and P&W on the engine. The engine work was monitored by NASA Lewis and the results obtained, up to 1982, were reviewed by Fishbach et al (1982). Initially, there were a large number of engine concepts, but by 1976 two of them were selected: the GE double-bypass engine (Figure 10.3), and the P&W variable stream control engine (Figure 10.4). An additional candidate was proposed by Boeing; namely, the turbine bypass engine which was originally conceived as a two-spoof turbojet with an undersized HP turbine. It was later found, however, that a single-spool version could give the same performance in a simpler configuration and this was studied by P&W.

### 10.2.1. The Double Bypass Engine

This engine, shown schematically in Figure 10.3, is a variable geometry turbofan, in which the fan has been split into two blocks, each with its own bypass duct, allowing better control of the flow over a broad operational spectrum. The enlarged front block fan can accommodate all the airflow required for takeoff with reduced jet velocity, giving low jet noise in the double bypass operating mode. The lower capacity rear block fan is sized for the nominal single bypass, high specific thrust operating mode needed for transonic and supersonic acceleration and supersonic cruise.

**Figure 10.4**   The VSCE engine (Fishbach et al. 1982). *Sponsored by NASA. Copyright © AIAA 1982. Used with permission.*

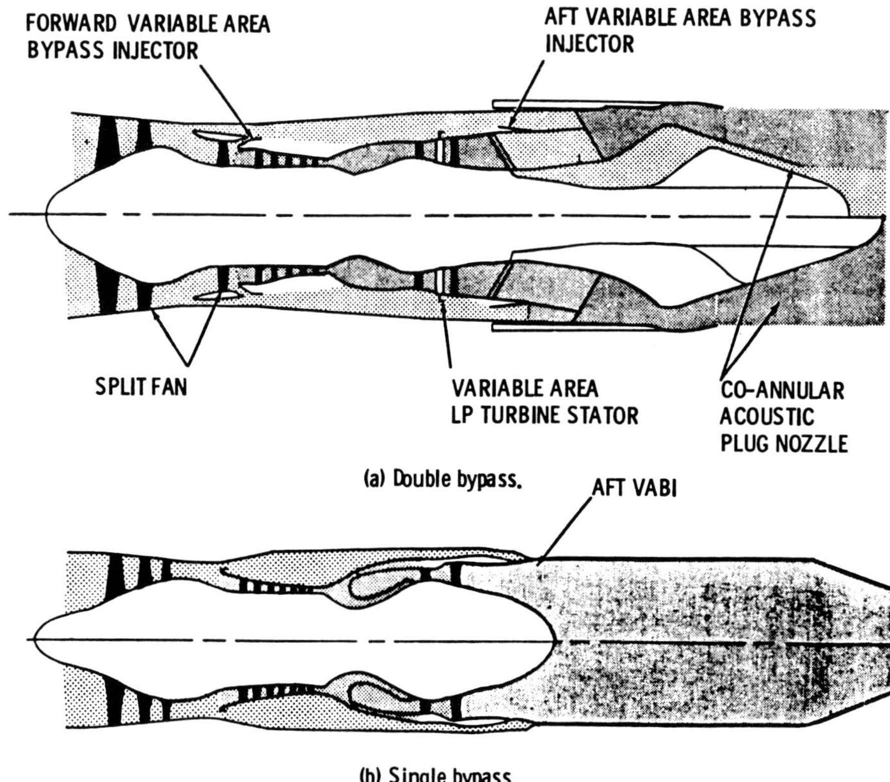



**Figure 10.5**   Operating modes of double bypass engine (Fishbach et al. 1982). *Sponsored by NASA. Copyright © AIAA 1982. Used with permission.*

A selector valve, placed at the entrance to the outer bypass duct, provides double-bypass operation when open and single-bypass when closed. Variable inlet guide vanes control the flow swings into the rear block occurring between double and single bypass operation. In the first the two bypass streams are merged into a single duct by a variable area bypass injector (VABI), reducing weight and simplifying the exhaust nozzle requirements. The VABI is a translating cylindrical sleeve which varies the discharge area of the inner bypass duct to match its static pressure to that of the outer bypass stream for efficient mixing over a range of double-bypass flow conditions. In the low-noise mode, bypass flow is then brought through cross-over struts to the inside of the plug nozzle, as shown in the view above the centerline in Figure 10.7. The aft portion of the plug centerbody is translated fore and aft to vary the exit area and thus control the flow of the cold fan stream. The hot turbine discharge gases flow around the nozzle support (crossover) struts and over the plug crown to surround the cold fan discharge stream to provide an inverted velocity profile for reduced jet noise. In the more conventional single-bypass operating mode, shown below the centerline in Figure 10.5, all the fan bypass flow goes through the inner bypass duct and is mixed downstream with the turbine discharge gases via the action of rear drop-chute VABIs located between the plug nozzle support struts. As with the forward VABI, the function of the rear VABI is to perform a static

pressure balance between the two streams for more efficient mixing. Mixing is desired at flight conditions where jet noise reduction is of no concern in order to provide a uniform exhaust velocity profile for greater propulsive efficiency. To stop the cold flow discharge from the inner plug, the aft portion of the plug centerbody must be translated fully aft in this operating mode.

The rear block fan is driven by the high-pressure (HP) turbine, as opposed to the conventional low-pressure (LP) turbine drive arrangement. This allows an otherwise underworked HP turbine to do more work and allows the enlarged front block fan to be driven by a single-stage LP turbine. Reduced turbine cooling also results from the arrangement because of the increased work extraction from the HP turbine stage, which reduces its average metal temperature, as well as that at the LP turbine inlet.

An intensive experimental program, using F404 engine hardware as the basis for the test vehicle, was carried out between 1976 and 1981. It started with component tests (single-bypass, double-bypass, forward VABI, acoustic nozzle) going through a core-driven fan stage test and a FADEC test to a full core-driven fan stage test bed engine which ran at Edwards AFB in 1981. All tests were successful, but in the last one a major engine failure occurred after two hours during conventional operation. This was caused by problems unrelated to the program, but prevented com-

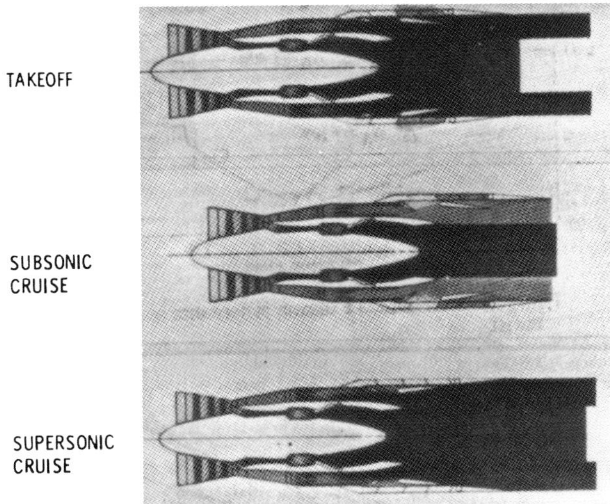



**Figure 10.6**   VSCE operation at various conditions (Fishbach et al. 1982). *Sponsored by NASA. Copyright © AIAA 1982. Used with permission.*

pletion of the planned tests, which should have given information about the overall engine noise. Nevertheless, the experimental program demonstrated that this type of engine performs according to specifications and should reduce the noise considerably, compared to the 1971 SST.

## 10.2.2. The Variable Stream Control Engine

Figure 10.4 is a schematic showing the basic arrangement of the major engine components of the P&W VSCE. The engine is a twin-spool configuration similar to a conventional turbofan but with the added feature of a burner in the fan duct. The VSCE derives its name from its ability to control independently the primary and bypass streams. The fan and compressor both have variable geometry components and are driven by advanced technology turbines. The main burner and duct burner both use low-emissions, high efficiency combustor concepts, based on NASA's Clean Combustor program. The coannular nozzle provides variable throat areas for the core and fan duct flow and also includes an ejector—thrust-reverser system.

The flexibility of this concept to meet the diverse requirements of low jet noise at takeoff and good fuel consumption at cruise can be illustrated by describing the operation at takeoff, subsonic, and supersonic cruise (Figure 10.6).

At takeoff, the primary stream is throttled to an intermediate power setting while the duct burner is operated at its maximum design temperature. The independent control of the two streams provides the unique inverted velocity profile that is needed to take advantage of the coannular nozzle noise benefit. The bypass jet velocity is about 60 to 70 percent higher than the primary jet velocity providing a significant reduction in takeoff jet noise.

At subsonic cruise, the main burner is throttled to a low temperature and the duct burner is turned off. Variable geometry features are used to "high flow" the engine to match the inlet airflow and thus reduce both the inlet spillage drag and the nozzle boattail drag. The velocity profile is nearly flat, and the engine approaches the performance level of a moderate bypass ratio turbofan engine designed strictly for subsonic operation.

At supersonic cruise, the primary burner temperature is increased relative to takeoff, and the duct burner is operated at partial power. The resulting velocity profile is nearly flat for good propulsive efficiency, and this concept provides a fuel consumption that approaches that of a turbojet cycle designed exclusively for supersonic cruise.

Two of the most critical technology components of the VSCE, the low emission duct burner and the low noise coannular nozzle were tested in the NASA Lewis component test-bed program; for this purpose they were added at the back of the F-100 engine, used as a gas generator.

An aero/acoustic design procedure for coannular nozzles was applied to several candidate exhaust systems to identify the most attractive nozzle for the VSCE. A schematic of the selected design in shown in Figure 10.7. At supersonic cruise, the nozzle is a conventional convergent-divergent configuration. Two internal clamshells are positioned to provide the initial portion of the expansion surface of the ejector shroud. Variable throat areas are provided for both the primary and fan flows. At low-speed conditions, the nozzle converts to an auxiliary inlet ejector. Actuated inlet flaps are opened to admit external airflow into the shroud. Panels located immediately downstream of the double-hinged doors are translated aft to provide additional area for the ejector, and the internal clamshells are aligned with the in-

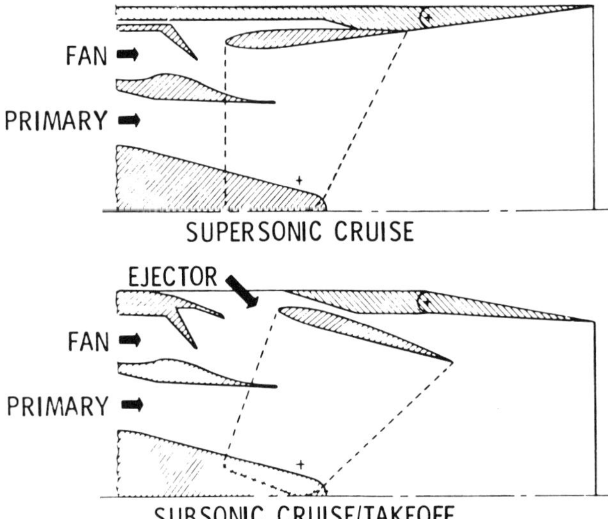



**Figure 10.7**   VSCE coannular nozzles (Fishbach et al. 1982). *Sponsored by NASA. Copyright © AIAA 1982. Used with permission.*

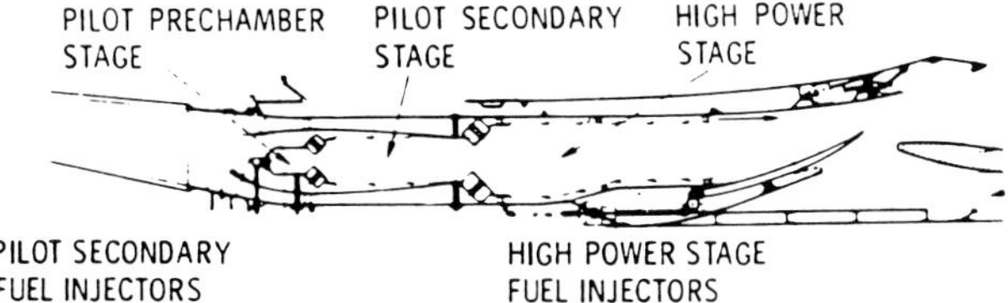

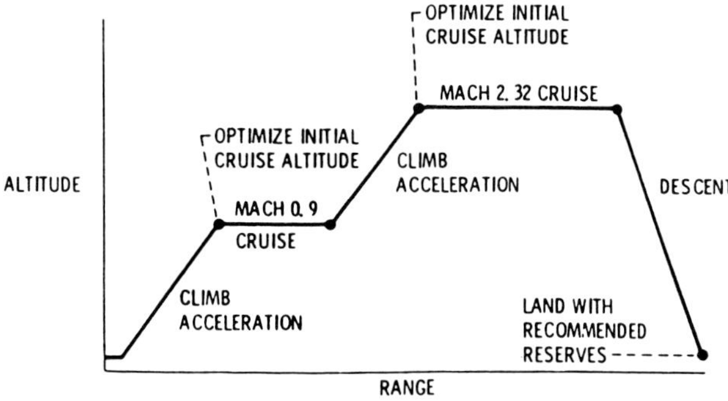

**Figure 10.8** Low emission duct burner (Fishbach et al. 1982). *Sponsored by NASA. Copyright © AIAA 1982. Used with permission.*

**Figure 10.9** Requirements for the turbine bypass engine (Franciscus 1981). *Sponsored by NASA. Copyright © AIAA 1981. Used with permission.*

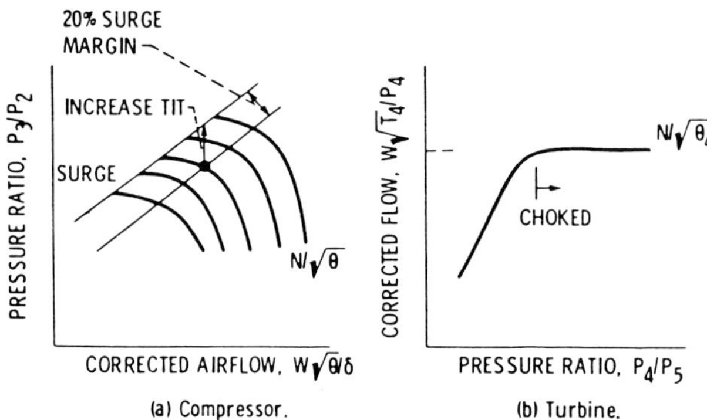

**Figure 10.10** Compressor-turbine matching (Franciscus 1981). *Sponsored by NASA. Copyright © AIAA 1981. Used with permission.*

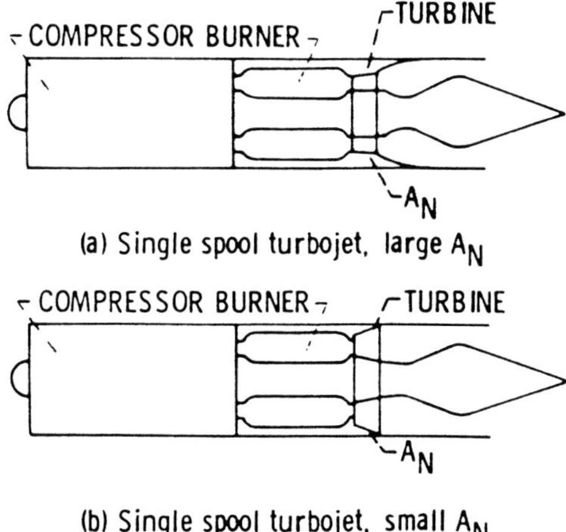

**Figure 10.11** Comparison of annulus area (Franciscus 1981). *Sponsored by NASA. Copyright © AIAA 1981. Used with permission.*

let flow. Floating tail feathers are aerodynamically positioned to provide the proper exit flow area. The clamshells are also used for thrust reversal by rotating them back to the nozzle centerline. The reversed flow is then expelled through the open inlet doors.

The initial experiments showed poor low-speed performance, so the nozzle was redesigned by changing the ejector inlet turning angles, minimizing internal overexpansion, static pressure mismatches, and the core/bypass flow impingement angle. The results obtained were satisfactory.

The low emission duct burner (Figure 10.8) is a three-stage concept: the first, the pilot preburner, is sized to allow stable operation at very low fuel-air ratios to provide a soft light (easy start) and to minimize disturbances to the fan operation. The combined first two stages (pilot prechamber and pilot secondary) operate at supersonic cruise, while the third, a high-power stage, is required for takeoff and transonic climb. Emission measurements performed both in a 2D segment test rig and in the F100 component test bed gave better performance and emission results than set forth in the program goals. CO and THC emissions were below design goal, while the $NO_x$ emission levels exceeded the requirements only slightly.

### 10.2.3. The Turbine Bypass Engine

Franciscus (1981) describes the single-spool turbine bypass engine (TBE), proposed by Boeing and further studied by P&W. The study was performed for Mach 2.32 supersonic cruise with a 556 km subsonic cruise. The conditions of a standard day (8°C) were assumed, and the mission range was fixed at 7408 km, including a 394 km descent (see Figure 10.9).

Figure 10.10 depicts the matching of a compressor and turbine for a single-spool turbojet. The turbine is choked for nearly all operating conditions indicated by the constant value of turbine corrected airflow $W_4 \sqrt{\theta_4}/\delta_4$. For variations in turbine inlet temperature, the compressor will operate at pressure ratios and airflows to satisfy the constant value of turbine corrected airflow. For a prescribed compressor airflow, the compressor operates at increasing pressure ratios with increasing turbine inlet temperatures. The compressor surge margin (usually about 20%) places a constraint on the upper limit of turbine inlet temperature. There are, of course, other constraints such as materials, cooling, etc. Decreasing the turbine inlet temperature at a fixed compressor airflow causes decreasing pressure ratios. Lower limits on the turbine inlet temperatures would have to be evaluated in terms of low compressor efficiencies or limits on nozzle area variations.

The study compares a large annulus area $(A_N)$ turbine to a small annulus turbine using the same compressor (see Figure 10.11); the large $A_N$ turbine can supply higher transonic and supersonic thrust, albeit at the expense of a higher SFC. On the other hand the small $A_N$ turbojet performs better at subsonic speed. The TBE was also compared to the VSCE, and it turned out that for the same mission the TBE has an 8% weight advantage, while for the same TOGW it has a 10% range advantage. On the other hand, one should remember that the VSCE would be inherently quieter than an unsuppressed TBE at takeoff, so that additional acoustic material could change the above relations.

## 10.3 Engines for Future Civil Transports

The Douglas Aircraft Co. was awarded a NASA contract to study high-speed civil transports in the framework of the HSCT program. They proposed two systems, one for a Mach 3.2 aircraft with a range of 11,700 km, which could start service around the year 2000 according to traffic forecasts, the other for a Mach 5 vehicle, which might find application after 2010. We shall describe briefly the engines envisioned in their report (Douglas Aircraft Co., 1989). Table 10.1 presents the initial results of this study, showing the possible engine cycles and those best suited to different speeds.

For the Mach 3.2 aircraft they chose as the baseline engine the P&W VSCE, described in Section 10.2.2, employing thermally stable jet fuel. The engine has a corrected airflow of 321 kg/s and a maximum SLS thrust rating of 275.35 kN augmented and 119.94 kN dry. The cycle and component design parameters for this engine are as follows: overall pressure ratio 14.3, fan pressure ratio 3.67, design by pass ratio 1.3, maximum compressor discharge temperature 1033 K, and TET = 2200 K (this assumes considerable advances in turbine design by the year 2000).

**Table 10.1** Douglas supersonic study engines (Douglas Aircraft Co. 1989).

*Phase One - High-Speed Propulsion Assessment (HSPA) Engines*
- Engine cycles
  - Turbojets
  - Turboramjets (Tandem, Turbofan, over/under)
  - Dual regenerator air turboramjet (ATR)

- Cruise Mach numbers - 3.5, 4.0, 4.5, 5.0, 6.0

- Fuels - kersene-based, endothermic (MCH), Liquid Methane (LNG), and liquid hydrogen

*Phase Two - Tailored engine cycles*
- Mach 2.2, Jet A fuel
  - Turbine bypass and variable cycle engine

- Mach 4.0, kerosene-based and methane fuels
  - Turbofan ramjets
  - Augmented and dry turbojets
  - Turbine bypass engine
  - Mixed flow turbofan
  - Duct burning (nonmixed flow) turbofan

- Mach 6.0, Liquid hydrogen fuel
  - Turbofan ramjet

The engines are individually mounted in nacelles located in the aft section of the wing where they are supported by wing-mounted pilons. A variable geometry nozzle using advanced noise suppression techniques is also proposed.

An alternative engine for this aircraft is the GE variable cycle engine (VCE) derived from the double-bypass engine discussed previously (Section 10.2.1). The corrected airflow is 334.5 kg/s and the maximum dry SLS thrust 289.88 kN. The engine is not augmented because of noise constraints at takeoff. In this, it differs from the double-bypass engine of the SCAR program; also, the fan has been sized to satisfy takeoff and subsonic requirements without augmentation. The cycle and design parameters for this engine are: overall

**Table 10.2** Comparison of some supersonic engines (Douglas Aircraft Co. 1989).

|  | Olympus 2 spool TJ | NK 144 3 spool TF | GE 4 single spool TJ | GE 21 2 spool TF |  | Douglas HSCT 2 spool TF | VCE 2 spool TF |
|---|---|---|---|---|---|---|---|
| M | 2.2 | 2.2 | 2.7 | 2.62 | 3 | 3.2 | 3.2 |
| Thrust (kN) | 167.7 | 196.1 (127.7 dry) | 311 | 159 | 272 | 275 (120 dry) | 290 |
| Passengers | 100 | 140 | 250–300 | 300 |  | 300 | 300 |
| Distance (km) | 5400 | 6500 | 9000 | 9000 | 12000 | 12000 | 12000 |
| TET (K) | 1455 | 1325 | 1560 | 1800 | 1945 | 2200 | 2200 |
| $\dot{m}$ (kg/s) | 186 | 250 | 300 |  | 184 | 321 | 335 |
| Efficiency (%) | 41 |  | 42 | 47 |  | 54 |  |
| Weight (kg) | 3386 | 2850 | 6012 | 2014 | 2382 |  |  |

pressure ratio 22, fan pressure ratio 4.8, bypass ratio 0.5, and TET 2200 K (to be achieved).

The baseline Mach 5 engine is the GE variable cycle turbofan/ramjet engine, using liquid natural gas (LNG) as fuel. Its airflow is 340 kg/s, the dry SLS thrust is 321.086 kN. Although the engine is augmented, it has been sized to takeoff partially dry, and the augmented SLS is not specified.

The turboramjet essentially phases out the turbomachinery during very high-speed operation. A ram air bypass duct is located around the basic engine, and a special stream control valve functioning in a manner similar to the VABI is employed to allow smooth transition from pure turbojet mode to pure ramjet mode as flight speed increases above Mach 3. If compressor windmilling cannot provide enough shaft power for power generation, a ram air turbine auxiliary power unit will supply airframe requirements.

The cycle and component design parameters for this engine are:

* Overall pressure ratio                 25
* Turbine rotor inlet temperature   2200 K (maximum)
* Bypass ratio                           1.50
* Fan pressure ratio                     5.5

For the Mach 5.0 baseline engine, a variable geometry two-dimensional inlet was selected. The inlet and nozzle are designed for Mach 5.0 cruise at an altitude of 25 km, which is the midpoint altitude between start and end of cruise. The inlet capture area is sized to satisfy engine, inlet bleed, and miscellaneous airflow requirements. The local flow conditions ahead of the inlet were determined by assuming a wing leading edge precompression resulting from a six degree flow deflection. One of the advantages of using LNG fuel is that it has a lower adiabatic flame temperature and therefore generates less $NO_x$.

A comparison of some of the supersonic engines discussed in this chapter is presented in Table 10.2.

# Chapter 11

# *Military Engines*

As already mentioned in Section 7.2 the main requirements for military engines are fast response and low vulnerability. This entails a low observable propulsion design, also known as stealth properties. The 1991 Gulf War has confirmed the importance of such an approach to combat engine design.

We shall first describe the requirements for military engines, which were spelled out very clearly by Lane in 1968. This will be followed by the description of some modern programs in which these principles were applied, the M88 built by SNECMA for the Rafale, and the Eurojet EJ200 under development for the Eurofighter by an European consortium. Then we shall treat in detail low observability requirements across the electromagnetic spectrum (in particular the radar cross-section and the infrared signature), giving, as examples, the American endeavours in this field. A review of contemporary Russian military engines concludes the section.

## *11.1 Requirements for Military Engines*

An advanced military strike-interceptor aircraft poses a number of propulsion challenges:

1. High augmented thrust at takeoff to confer a short takeoff length.
2. The capability to fly at transonic speed at low altitude for a considerable time with a heavy load of weapons; this demands high installed dry thrust.
3. Supersonic flight must be reached fast from the ground; therefore high augmented thrust is necessary over a large speed range, from sea level static to maximum speed.
4. This high augmented thrust should be available over a wide range of altitudes, giving the aircraft an extended flight envelope.
5. The engine should have low installed subsonic cruise SFC to allow for long range and considerable patrol time, and finally
6. High thrust for takeoff and acceleration with low cruise drag.

High altitude supersonic performance is represented typically by Mach 2.2 at 9 km and low altitude subsonic performance by Mach 0.9 at sea level.

To achieve the above goals low bypass ratios (up to 1) are employed, while the design pressure ratio is on the increase; Lane took 20:1 as a base, this has grown to over 30:1 in a modern version of the F-110 engine. The turbine entry temperature has increased from 1700 K, envisaged as a maximum by Lane (1968) and implemented for instance in the M88 demonstrator, to 1850 in the operational M88-2. As regards the engine architecture, the two-shaft configuration (sometimes with a fan-IPC arrangement) is preferred; only in the RB-199 we find three spools.

Reconnaissance aircraft have special requirements, such as Mach 3 and above with a ceiling of 20–25 km.

## *11.2 Some Modern Military Powerplants*

The French military authorities defined in 1978 the requirements for a multimission aircraft and its engine (Calmon 1989). On one hand it must allow low altitude penetration; on the other it should provide high altitude air superiority and interception. For the first mission low SFC, high compression ratio, high bypass ratio and a two-stream configuration are in order; the second one demands high specific thrust, low bypass ratio, high TET and reheating. Both require low infrared and radar signatures, low operational costs and easy maintenance. The conclusion reached was to strive for a bypass ratio between 0.3 and 0.5, a compression ratio of 25 and a TET above 1800 K. SNECMA responded to these specifications with the M88 program; as shown in Figure 11.1 in the first stage (1978–1984) a TET of 1700 K was achieved; a demonstrator was built between 1984 and 1988. In a second stage, started in 1984, the TET was raised to 1850 K and a 1996 date was fixed for operational deployment of the M88-2 in the Rafale C. This was preceded by flight tests starting in 1990 and series fabrication from 1992. The takeoff thrust of the engine is 75 kN, but growth to 100 kN with the same core is envisaged.

Figure 11.2 shows the architecture of the engine; the fan

**M 88, a family of engines for advanced fighters**

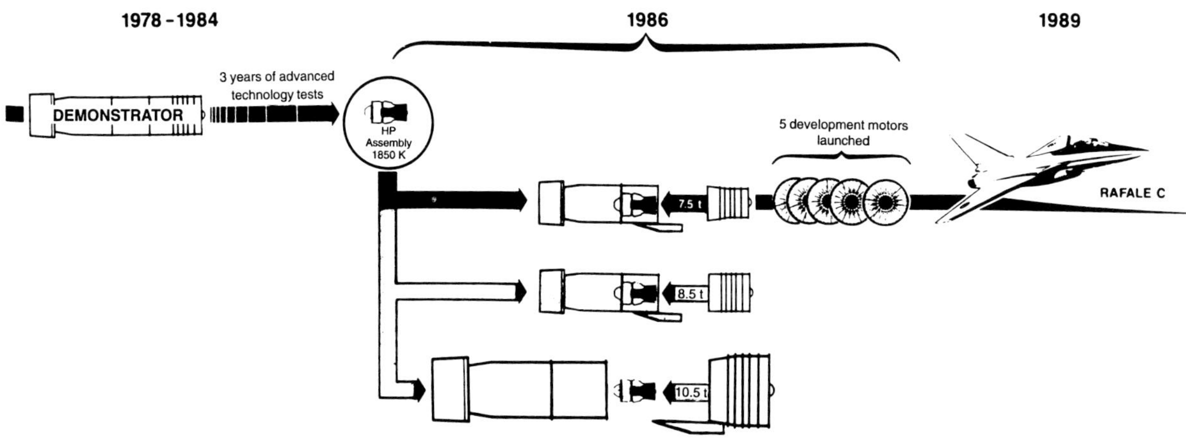


**Figure 11.1**   SNECMA M88 program (Calmon 1989). *Courtesy of L'Aéronautique et L'Astronautique.*

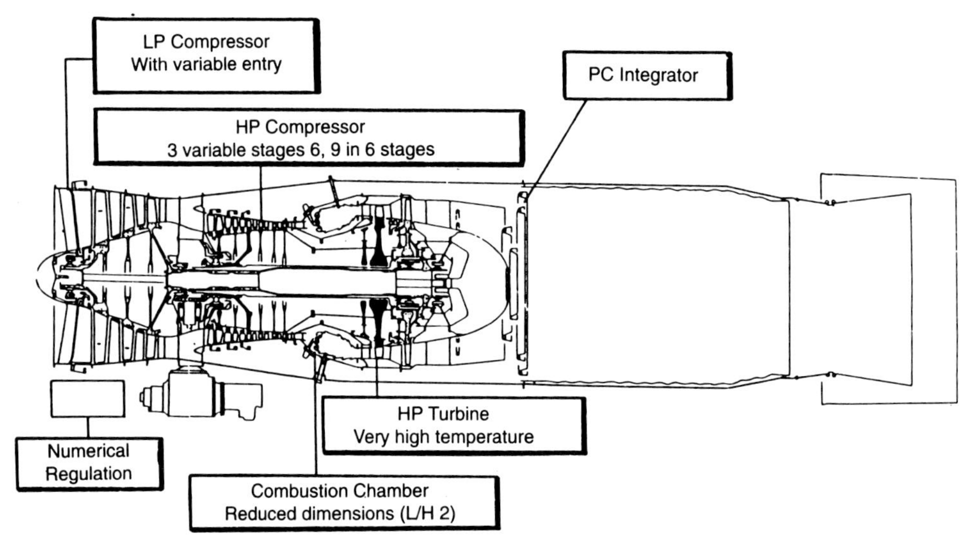


**Figure 11.2**   The SNECMA M88 engine (Calmon 1989). *Courtesy of L'Aéronautique et L'Astronautique.*

has three stages with a pressure ratio of 3.8, that can be raised in future to 4.5; the high-pressure six-stage compressor brings the total pressure ratio to 25. These results have been achieved by a combined effort of high technology experimentation, computational fluid dynamics, bench tests and finally flight tests. The height-length ratio of the combustor is 36% less than that of the M53, which in turn had improved by 30% on the Atar (these are SNECMA's earlier military engines). The fuel injection, shown in Figure 11.3, progressed from mechanical (Atar), through prevaporisation cans (M53, Olympus, Larzac) to aerodynamic; this ensures longer life and very low emissions. The TET is increasing in the last 20 years at a rate of 15 C/year vs. 10 earlier. This has been achieved by a combination of improved materials and better cooling (see Figure 11.4). Another advance is the use of monocrystalline alloys and the oriented solidification of the blades. The compressor discs are of Titanium alloys, while the final stages of the compressor are manufactured of Nickel type alloys, such as Inco 718 or Waspalloy. In the cooler parts fiberglass and kevlar, which can withstand 350 C are employed, while carbon-carbon and ceramic components appear in the afterburner and in the nozzle.

The EJ 200, shown in Figure 11.5, is an advanced turbofan for the Mach 2 Eurofighter, expected to enter service in 1997; it is built by the Eurojet Turbo consortium, consisting of FIAT Avio (21%), MTU (33%), RR (33%) and ITP, Industria de Turbopropulsores, from Spain (13%). This is a fully modular engine, allowing on-condition maintenance; health monitoring and test equipment are built-in. The total number of airfoils is about 60% of those used in the RB 199, the earlier European engine of the Tornado. The two-shaft augmented turbofan has a bypass ratio of 0.4; the LP compressor has three stages with 3D transonic blades, without inlet guide-vanes; its pressure ratio is over 4. The HP

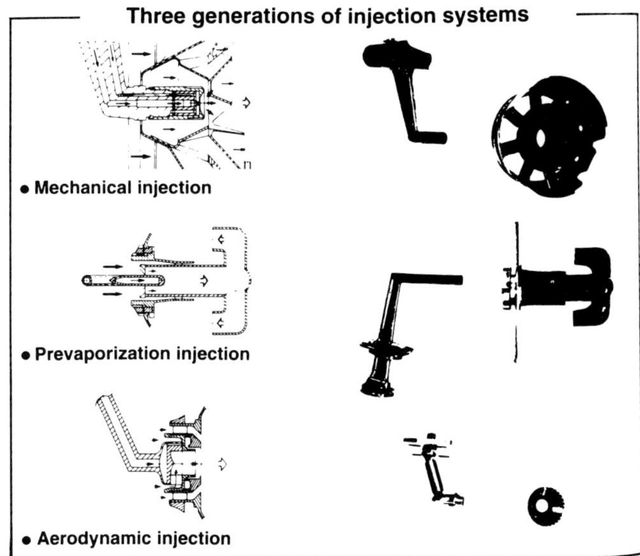


**Figure 11.3** Fuel injection system evolution (Calmon 1989). *Courtesy of L'Aéronautique et L'Astronautique.*

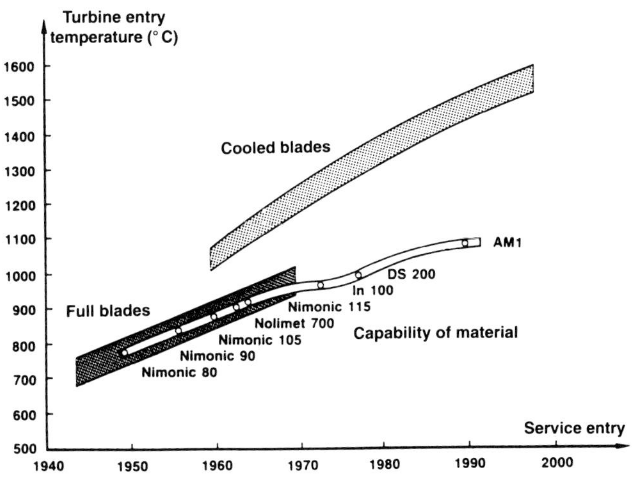


**Figure 11.4** TET evolution (Calmon 1989). *Courtesy of L'Aéronautique et L'Astronautique.*

compressor consists of five stages, with first stage variable inlet guide vanes; the overall pressure ratio is more than 25. The annular combustor has vaporizing burners; both turbines are single-stage units with powder metallurgy discs and low density air cooled single crystal blades. The nozzle is convergent-divergent and full authority electronic control (FADEC) is provided. The thrust is in the 90 kN class, with the mass around 1000 kg.

## 11.3 Low Observability Requirements

Although the Rafale and the Eurofighter demand low observable propulsion, not much has been published on the subject by their manufacturers. On the other hand American companies and in particular Lockheed have a number of papers on stealth aircraft. The problem of avoiding detection is very important for military aircraft and since World War II this was synonymous with a low radar cross-section.

The first aircraft designed taking this into account, was the Havilland Mosquito, in which wood was employed extensively, since it reflects less than metal. It became, however, apparent that the engines, avionics packages, electrical and hydraulic circuits reflect radar signals, often amplifying them. In the late fifties radar absorbing material was incorporated in the design—a typical example of the techniques used in this period is the Lockheed U-2 reconnaissance plane.

During the 1960s it became apparent that there are two ways to diminish detection; appropriate shaping of different components and the use of absorbing material. There was, however, not enough information to produce a completely balanced design and in most cases one component would dominate the reflection. This was the era of the SR-71. In the 1970s numerical methods allowing a quantitative assessment of the contributions from different parts of a body were developed. This allows the design of an aircraft with a balanced radar cross-section, minimizing the return from dominant scatterers. The F-117A and the B-2 are examples of contemporary design aircraft.

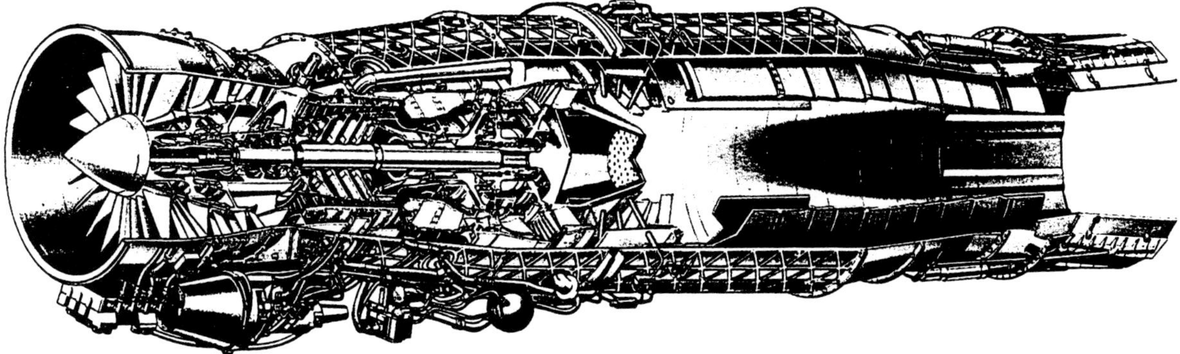

**Figure 11.5** The Eurojet EJ 200 (Rolls-Royce Magazine No. 49 1991). *Courtesy of Rolls-Royce plc.*

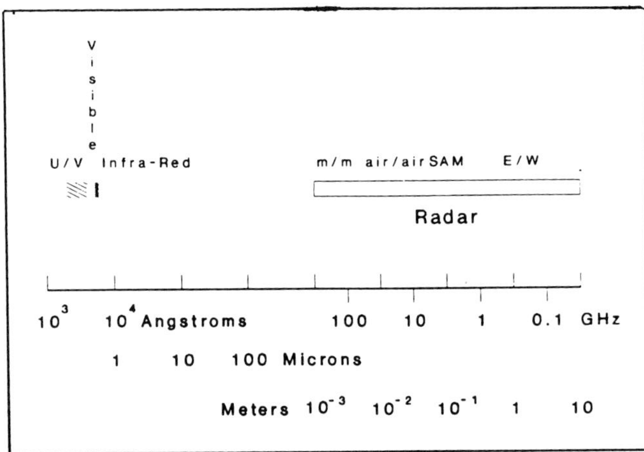


**Figure 11.6**  Relevant electromagnetic spectrum (Brown 1991). *Copyright © AIAA 1991. Used with permission.*

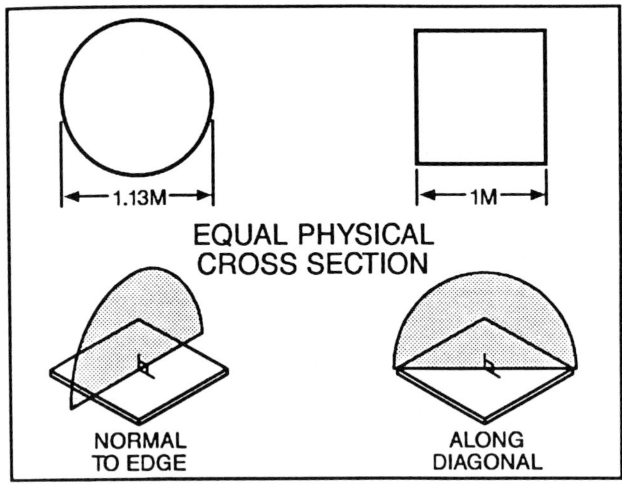


**Figure 11.7**  Return from different look angles (Brown 1991). *Copyright © AIAA 1991. Used with permission.*

### 11.3.1. The Radar Cross-Section

The profile of the electromagnetic spectrum relevant to low observability is shown in Figure 11.6. It covers eight decades from 0.1 micrometers to 10 m and comprises two important regions, one in the infrared, the other in the radar portion of the spectrum.

As mentioned earlier, shaping and coating are the two methods for diminishing the signature. There is a big advantage in positioning surfaces so that radar waves strike them at close to tangent angles and far from right angles to the edges. For a sphere, which has a diameter significantly larger than the radar wavelength, the radar cross-section is approximately equivalent to the geometric frontal area. In Figures 11.7 and 11.8 the return from a 1 m² sphere is compared to that from a 1 m² square plate at different look angles. The cross-sections are calculated for a wavelength, which is about 1/10 of the plate edge. At normal incidence the flat plate acts like a mirror and returns 1000 times (30 dB) the amount from the sphere. Rotating the plate about one edge, so that it is always normal to the incoming wave, one finds that the cross-section drops by a factor of 1000 becoming equal to that of the sphere when the look angle reaches 30° off normal to the sphere; increasing the angle farther the locus of maxima falls by another factor of about 50, giving a total change of 50,000 from the normal look angle. If instead the rotation is about a diagonal relative to the incoming wave, the cross-section drops by 30 dB when the plate is only 8° off normal and diminishes by another 40 dB by the time the plate is at a shallow angle to the incoming radar beam. This means a reduction of the radar cross-section by $10^7$.

Another way of attenuating the reflection is to employ multiple bounces from ducts, as shown in Figure 11.9. Coatings and absorbers having a 1/4 wavelength thickness are

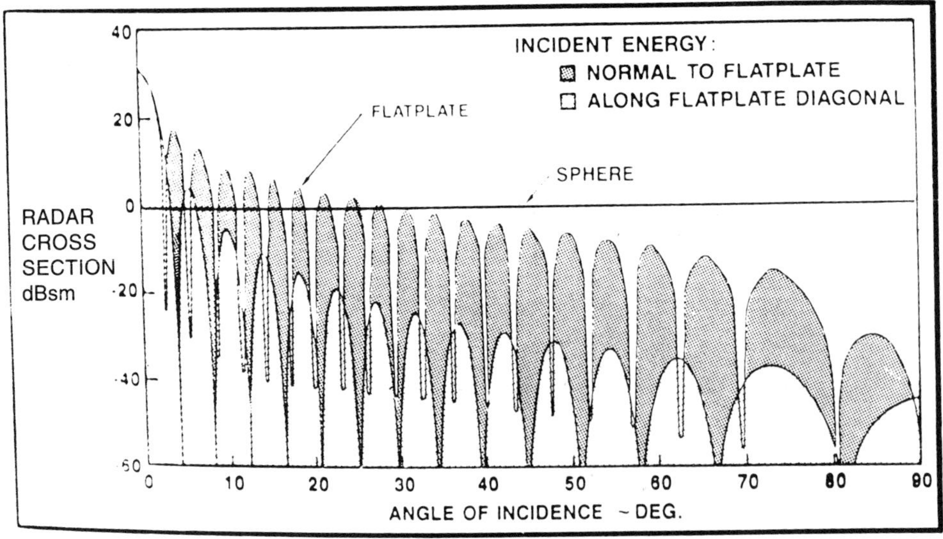


**Figure 11.8**  Return from different angles of incidence (Brown 1991). *Copyright © AIAA 1991. Used with permission.*

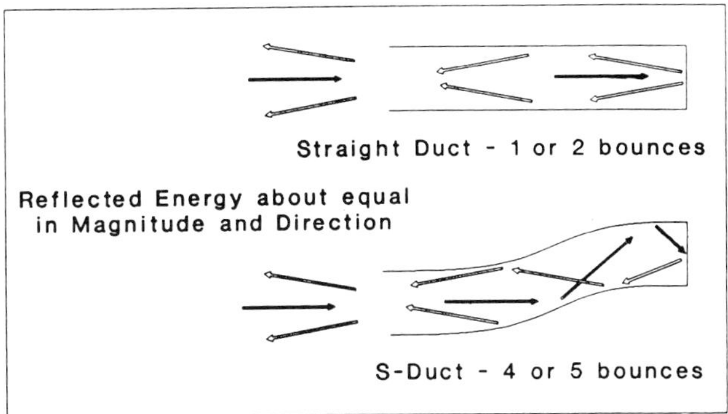

**Figure 11.9** Attenuation by multiple bounces (Brown 1991). *Copyright © AIAA 1991. Used with permission.*

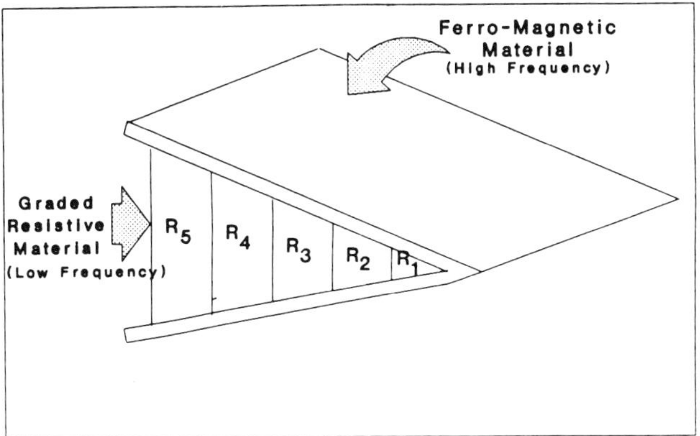

**Figure 11.10** Attenuation by graded coating (Brown 1991). *Copyright © AIAA 1991. Used with permission.*

additional means to diminish the radar cross-section; the key dimensions can vary in a range from 0.5 mm to 1 m. Graded coatings, like the one shown in Figure 11.10 were also found useful. When dealing with nozzles there is the additional problem of high temperature; for this component ceramic absorbers can be used.

### 11.3.2. Infrared Radiation

There are two main sources of IR radiation, hot parts and jet wakes. To reduce radiation both temperature and emis-

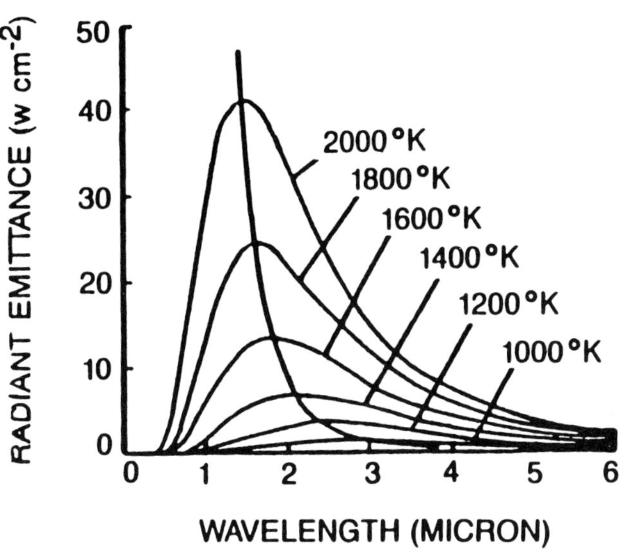

**Figure 11.11** Infared radiation emission (Brown 1991). *Copyright © AIAA 1991. Used with permission.*

sivity must be considered. One must try to diminish them, while in addition line of sight masking, employing thin multiple coatings, can be used; they are measured in angstroms (1 A = $10^{-4}$ micrometers) and are composed of metals or metal-oxides; the last are preferred because they are compatible with the radar cross-section.

Although the radiation emitted is proportional to the product of the emissivity by $T^4$, the frequency shift of radiation with temperature causes this relation to be closer to emissivity times $T^8$, as shown in Figure 11.14. Therefore diminishing the temperature of hot particles is very effective in attenuating the IR signature. The formation of soot must also be avoided, because of its high emissivity.

### 11.4 Examples

The U-2/TR-1 family, which first entered service in 1955, had its final aircraft delivered in October 1989. It is powered by a PW turbojet, the J75-P-13B which supplies 75.6 kN thrust. In 1989 a TR-1 began flight trials with a F101-GE-

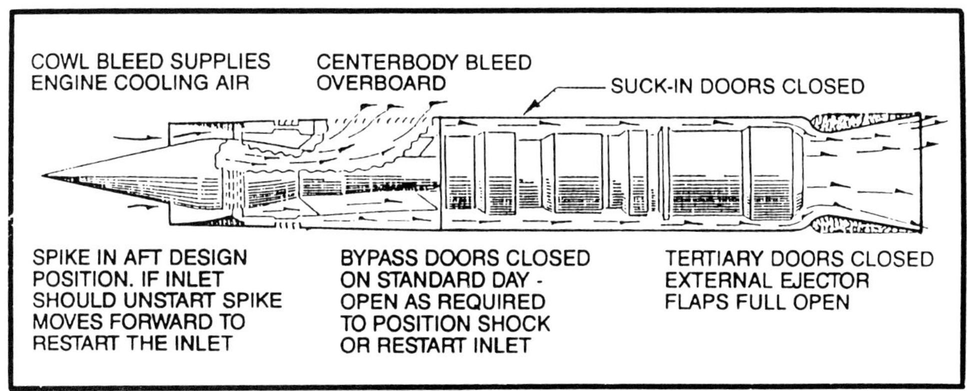

**Figure 11.12** SR71 attenuation design (Brown 1991). *Copyright © AIAA 1991. Used with permission.*

**Figure 11.13**   The JT 11D-20B turbojet of the SR 71 (Jane's 1970–71). *Reproduced with permission of Jane's Information Group.*

**Figure 11.14**   The F117A stealth aircraft (Jane's 1990–91). *Reproduced with permission of Jane's Information Group.*

F29 engine, derived from the F118 turbofan, which powers the B-2 stealth bomber.

The design of the SR-71 shown in Figure 11.12 stresses low observability. Its propulsion system is designed for high performance at Mach 3 at low angles of attack. The inlet is coated on the spike and inside the duct to attenuate incoming energy. The aircraft signature is low, because the very high flying altitude resulted in the IR and radar searching beams coming in at shallow angles. The engines are two PW 2JT11D-20B afterburning turbojets, each supplying 144.5 kN thrust (Figure 11.13).

The F117A was planned as a subsonic aircraft to diminish its observability; its shape (see Figure 11.14) is that of a multifaceted pyramid with the apex immediately behind the cockpit, which dissipates the radar signals by deflecting them in numerous directions away from the searching radar; a grid is used to attenuate the signature from the cavity-compressor combination. It is powered by two F404-GE-F1D2 nonafterburning turbojets, with a thrust of 48 kN.

The F22A, which won the ATF competition in 1991, is a supersonic fighter, with inlet ducts long enough to give 100% line-of-sight blockage of the engine face. Newly devel-

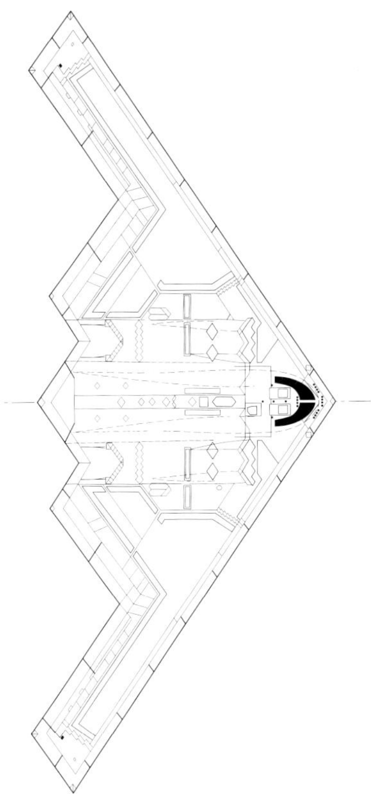

**Figure 11.15**   The B-2 bomber. *Courtesy of Northrop Grumman Corporation.*

oped high temperature absorbing coatings and closer cooperation with the engine manufacturers contribute to low observability.

The B-2 Northrop (see Figure 11.15) known as the stealth bomber, is the first all-composite aircraft, apart from the F118-GE-100 85 kN engines. Structurally the aircraft seems to consist of a pair of short, untapered outer wings mated to a trapezoidal 28 m-span torsion box, in whose middle lie the weapons bays, the engines and the main landing gear bays. The platform of the B-2 is dictated by the stealth requirements. There are two ways of using geometry to control the way a body scatters a radar wave. One, which we have seen in the F117A, is to make the shape flat or rectilinear, concentrating the reflection on one bearing, and to construct the body so that the reflection never goes towards the likely location of a receiver. The other is to scatter the wave over a carefully designed curve of constantly changing radius, so that each small part of the surface has its own tiny mainlobe reflection. The B-2 appears to combine both methods. The radius of action of the B-2 is about 8350 km without refueling, its cost is estimated around 500 million dollars.

The Lyulka AL 31F engine of the Su-27 Flanker is a twoshaft augmented turbofan with a bypass ratio of 0.4 (Figure 11.16). The fan has four stages with variable guide vanes; the compressor has also variable inlet guide vanes followed by a nine-stage HP spool with the first five stators variable; the total pressure ratio is 20. The annular combustor has 24 burners fed from an inner manifold. The single stage HP turbine has cooled blades, while the LP turbine is a twostage design. The thrust is 123.85 kN (dry 79.42). The latest Soviet entry is the type R, 245 kN turbofan, four of which power the Tu-160 Blackjack strategic bomber; they give the aircraft the capability of long range, long endurance with a large weapon load combined with the ability to penetrate enemy defenses at high subsonic speed. At high altitude (18.3 km) the aircraft can cruise at Mach 2. Its combat radius is 7300 km.

## 11.5 Thrust Vector Control

Thrust vectoring is the ability to point the thrust of an aircraft engine in a direction different from the flight path. The objective is to enhance aircraft agility and controllability and is applied to high performance military aircraft. It is particularly effective in situations where the aircraft control surfaces are less effective, say, at low airspeed and high angle of attack. Engine manufacturers are investigating the feasibility of incorporating thrust vectoring capability in the propulsion system of tactical aircraft. This is discussed by Herbst (1987), who distinguishes between four different experimental applications as shown in Figure 11.17. The first is the F-15 STOL experimental program, in which an F-15 aircraft was equipped with square nozzles which allows the jet to be deflected in the pitch plane. This improves takeoff and landing performance. An upward jet deflection enhances the aircraft rotation during takeoff, thus reducing the required rolling distance. The aircraft is also equipped with an additional control surface which interacts with a downward jet deflection allowing a lower landing speed.

A second American project, the experimental F-14, allows jet deflection in the yaw plane by means of jet spoilers. This reduces the safe approach speed on landing on an aircraft carrier. The lateral jet deflection improves also the controllability in the maximum lift range, allowing for an increase in flight maneuverability of up to 70° incidence.

In Germany the ND-102 Dornier project features square pitch nozzles with a design characterized by the absence of a horizontal tail. The deflection in the horizontal plane is used mainly to control and stabilize the longitudinal motion, compensating for the absence of the tail. The nozzle is similar to that of the F-15 STOL, allowing the integration of a thrust reversal mechanism. The aim was to use it also during flight instead of an aerodynamic brake.

The fourth project, the XD-31 A aircraft, is an international program involving DARPA-BMVg and MBB-Rockwell. This aircraft utilizes jet deflection in the pitch and yaw plane respectively and features a delta canard configuration, similar to the EFA. Here, jet deflection in a cone is

**Figure 11.16**   The Saturn (Lyulka) 2 shaft augmented turbofan (Jane's 1990–91).
*Reproduced with permission of Jane's Information Group.*

used to penetrate the flight region extending beyond the maximum lift, where the aerodynamic forces are not strong enough to allow the performance of controlled, tactically useful maneuvers; this type of operation is called superma-neuverability.

In a companion paper (Geidel 1987) the axisymmetric vectoring nozzle developed by MTU is discussed in detail and its advantages over the 2D C/D nozzle are highlighted.

This type of control was applied in an experimental program involving the deflection of the exhaust gases in the expansion nozzle of the GE F-110 engine. The thrust deflection is accomplished after the choked primary nozzle throat. The main control task is to change the position of the hy-

draulic actuators which accomplish the vectoring, and the goal is to integrate this control task with the airframe flight control and the engine control systems. Typically, a vector nozzle control system should be capable of an angular range of $\pm$ 20 degrees and a minimum vector rate of 40 degrees per second, with a frequency response similar to that of the aircraft flight control surfaces. Safety and reliability are very important requirements; that is, the system must be available when it is needed, and it must not induce inadvertent vectoring.

It was decided to actuate the vector control system hydraulically, since the primary exhaust nozzle is hydraulically powered. Different control systems were studied and a number of parameters was evaluated for each scheme.

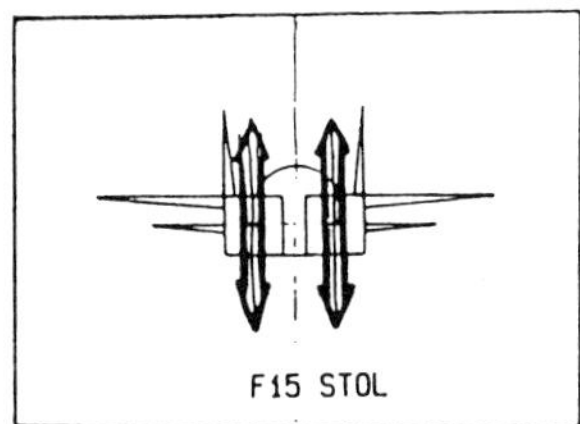


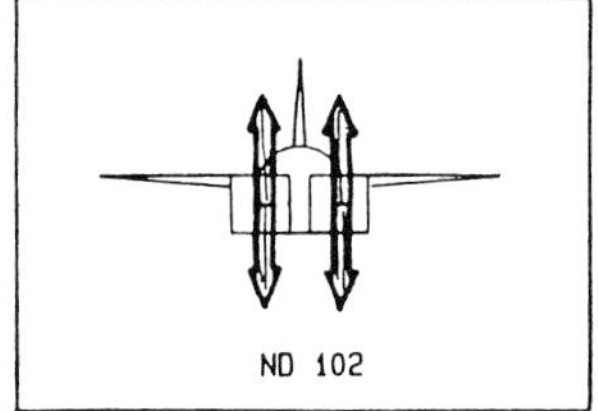


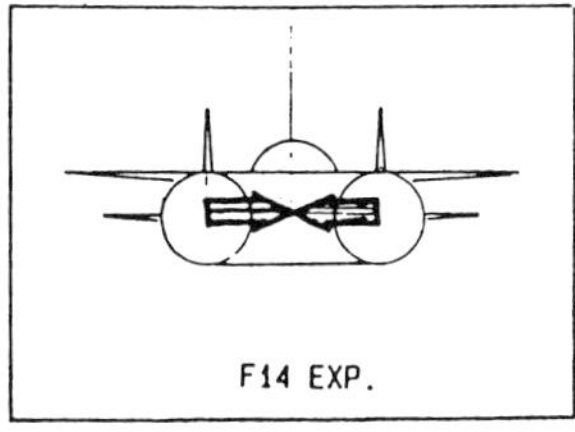


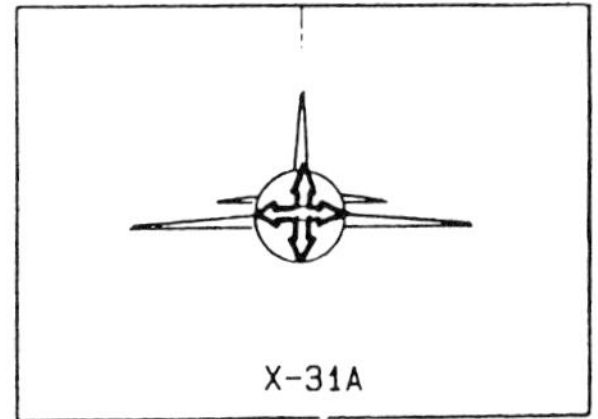


**Figure 11.17**  Different applications of thrust vectoring (Herbst 1987). *Copyright © AIAA 1987. Used with permission.*

Among these was the in-flight engine shutdown rate, which increased by only 1% and had similar values for the various systems; in-flight power loss also decreased by less than 1%. Loss of lateral thrust is alleviated by redundant vector control; loss of vector function can be caused by non-redundant hydromechanical components or by in-flight shut-down. Maintenance rate increases somewhat as does design complexity. Performance parameters evaluated were system response, accuracy and stability; they met the performance objectives in all cases tested. Development costs were similar for all options investigated, but redundancy made a difference in cost, both for hydraulics and engine control electronics. Integration of the engine and vector control resulted in lower cost, since this eliminated the need for one electronic box.

A comparison of the different approaches showed that a system with dual vector electronics and single hydraulics was the best, since it represents a reasonable compromise between safety, complexity, cost and weight. Addition of this system does not change the architecture of the engine control nor that of the engine hydraulics and requires only minimal changes in the engine and airframe systems.

The interested reader is referred to a recent monograph by Gal-Or (1989), which treats in depth vectored propulsion and supermaneuverability for aircraft, cruise missiles, RPVs and robot aircraft.

# Chapter 12
# V/STOL Propulsion Technology

An aircraft that can take off and land vertically while also operating in a normal forward flight mode with the speed, range, and efficiency of a fixed wing aircraft, is of great interest for both military and civil applications. In the former role such an aircraft eliminates the need for expensive landing strips, which can be easily detected, and permits operation close to the battle line or from a variety of ship sizes. For a commercial operator this capability would permit direct flights from one city center to another for passengers and high value cargo. While no fixed-wing civil aircraft have VTOL capability, it is interesting to note that the concept has been applied at subsonic speed in the military field for over twenty years (Harrier, AV 8, Yak 38 Forger). The Harrier was used with considerable success in the Falkland conflict and in the 1991 Gulf War.

There are a number of alternatives to achieve the goal: vectored thrust (Harrier), lift plus lift/cruise (Forger), ejector lift, remote augmented lift, a tandem fan, a shaft driven remote fan, and remote gas driven fan-in-wing. We shall start by discussing briefly the various possible options (Lewis, 1991), then we shall describe in some detail the Pegasus engine, including the experimental BS100 vectored thrust engine with augmentation provided by plenum chamber burning (PCB), concluding with the Forger powerplant.

## 12.1 The Options

The main characteristics of the vectored thrust engine are: (a) in the powered lift mode the full engine thrust can be vectored through a range of at least 90°, thus providing maximum STO performance with minimum complexity and cost; the position of the exhaust nozzles produces good handling characteristics, so the pilot does not need extensive automation; (b) the four poster arrangement gives low hot gas ingestion and good suck down characteristics in ground effect allowing small margins between available thrust and vertical landing weight; and (c) the jet footprint in the powered lift mode is within the limits that permit operation over a wide range of surfaces and conditions when coupled with small amounts of forward speed.

The lift plus lift/cruise option used in the Forger, has also been applied with some variations in a number of research aircraft (Shorts SC1, Mirage III V, VJ 101, and VAK 191B). In this concept (shown in Figure 12.1) part of the lift thrust is provided by one or more lift engines which are shut down for normal wing borne flight. The proportion of lift provided by these engines can be as high as 100%. An advantage of this system is that the total powerplant volume is small and the area distribution is good.

The principle of the ejector lift system (see Figure 12.2) is to direct some or all of the engine exhaust flow into nozzles which form the primary system of an ejector thus augmenting the engine thrust for lift. The experimental XV4 aircraft (Humming Bird), working on this principle, had great difficulty in the translation from jet borne to wing borne flight. In more recent tests the engine bypass air was used as the primary flow in the ejector and the LP turbine flow was exhausted separately through a vectoring nozzle, providing some acceleration force for the transition from jet borne to wing borne flight.

The remote augmented lift system is similar to the ejector concept, but the thrust augmentation is achieved by burning, rather than by increasing the airflow. The combustion takes place in the low temperature bypass air making possible a large increase in thrust with a modest tempera-

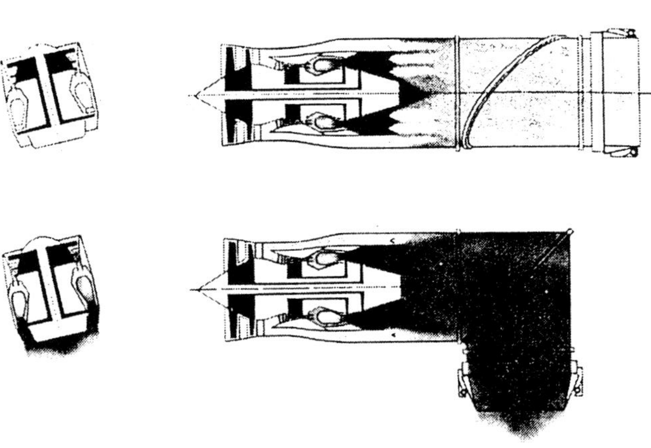

**Figure 12.1**  The lift plus lift-cruise option (Lewis 1991). *Published by kind permission of Rolls-Royce plc.*

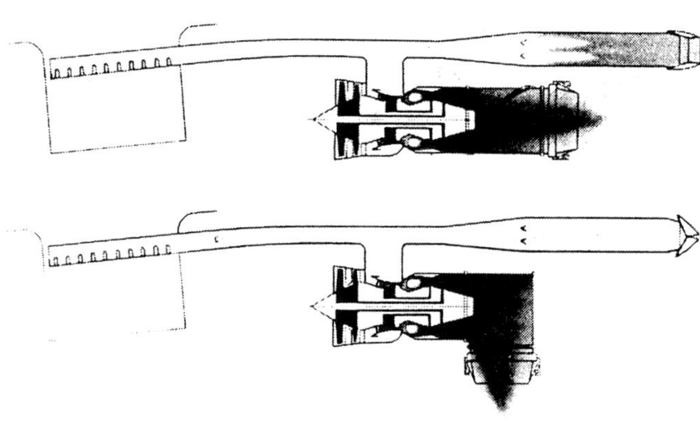

**Figure 12.2**  The ejector lift system (Lewis 1991). *Published by kind permission of Rolls-Royce plc.*

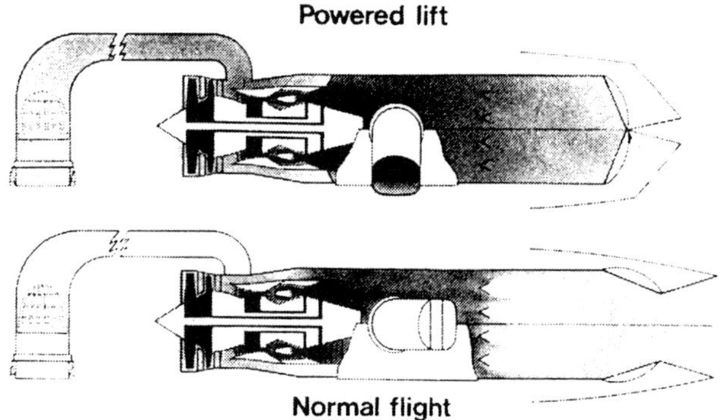



**Figure 12.3**  The remote augmented lift system (Lewis 1991). *Published by kind permission of Rolls-Royce plc.*

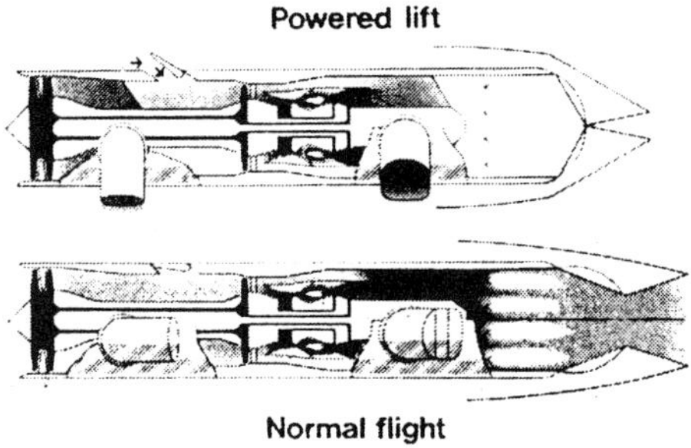



**Figure 12.4**  The tandem fan configuration (Lewis 1991). *Published by kind permission of Rolls-Royce plc.*

ture rise. This concept, shown in Figure 12.3, has not been tried in a flight vehicle.

In the tandem fan configuration (see Figure 12.4) a valve system between the stages of the fan of a conventional turbofan engine permits a large increase in airflow for hover conditions. Coupled with vectoring nozzles such a system could have operating characteristics similar to those of the Pegasus/Harrier system in the powered lift regime, going over to high performance turbofan for wing borne flight conditions. Ground tests of this engine gave promising results.

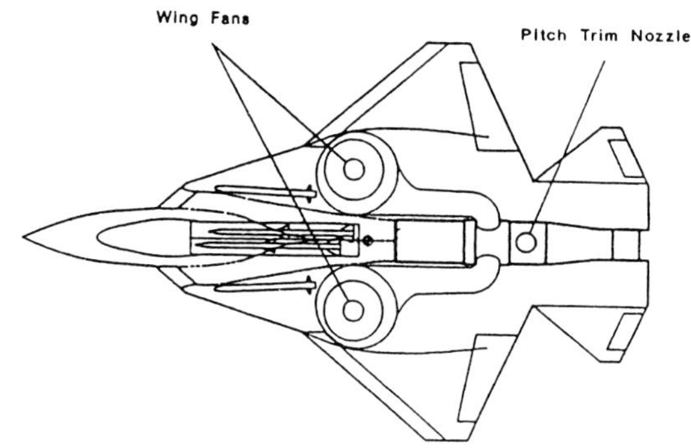



**Figure 12.5**  The gas driven fan (Lewis 1991). *Published by kind permission of Rolls-Royce plc.*

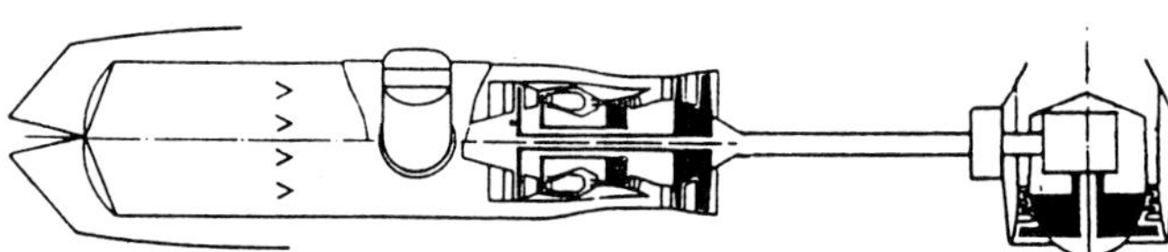

**Figure 12.6**  The shaft driven remote fan (Lewis 1991). *Published by kind permission of Rolls-Royce plc.*

In the gas driven fan, exhaust gases from the main engine are ducted to a turbine system which drives a high flow, low pressure ratio fan. The configuration, shown in Figure 12.5, has been flown in the Ryan XV5 A research aircraft, but unexpected difficulties, such as high levels of hot gas ingestion and the lack of full (90°) thrust vectoring were encountered.

Finally, the shaft driven remote fan, illustrated in Figure 12.6, is similar to the tandem fan, but the remote fan is declutched from the main engine for normal wing borne operation. This system has not been flown; its apparent advantages are the use of a smaller propulsion unit and a reduced cross-section. One must, however, remember that it requires the design of a light weight, compact, reliable gearbox, for transmitting 25,000 HP and the provision of a cooling system to reject the heat from this gearbox. The design of the clutch, the thrust control of the fan and the matching of this thrust with the engine thrust are additional areas requiring further study.

## 12.2 The Missions

English and American experts (Lewis and Simpkin 1981; Yackle 1983; Davenport et al. 1991) agree that the major advantages of V/STOL aircraft for military operations are that they offer greater speed and range than helicopters and better maneuverability and basing options than conventional aircraft. Missions having combined needs for high altitude, long range, low-speed maneuverability and flexible basing are prime candidates for V/STOL. Sortie rates of up to ten per day have been demonstrated. In the Gulf War in 1991 Harriers based on aircraft carriers performed many

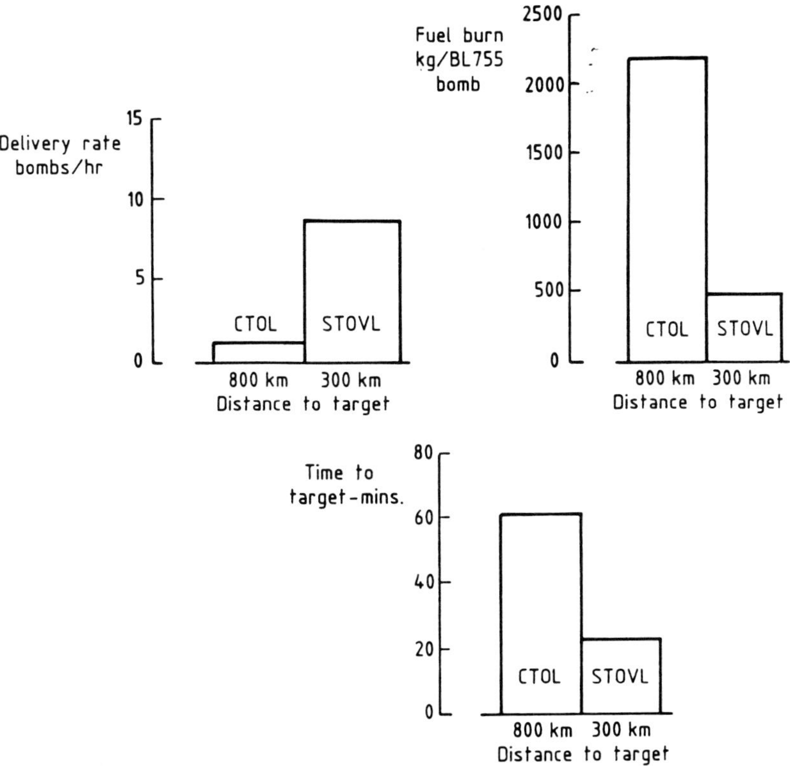


**Figure 12.7**  Weapon delivery rate comparison (Lewis and Simpkin 1981). *Copyright © AIAA 1981. Used with permission.*

close air support missions with considerable success. Other possible applications of V/STOL are fleet defense, personnel transport, search and rescue, and surveillance.

Lewis and Simpkin (1981) discuss advanced V/STOL aircraft that should have a supersonic capability and so be also suitable for air superiority missions. This could be achieved either by augmented vectored thrust as in the BS 100 engine or by the combination of lift and cruise engines, as in the Mirage III V, the only V/STOL aircraft that has exceeded Mach 2. They performed a detailed comparison between V/STOL and CTOL aircraft for supersonic air superiority and subsonic attack missions, based on equal operational effectiveness. In both cases the V/STOL aircraft were operating from bases significantly closer to the battle area than the CTOL aircraft. The flexibility of the V/STOL aircraft increases the options available in that operation from both airfields and dispersed sites is possible.

The subsonic aircraft is assumed to be used for strike missions and Figure 12.7 shows the weapon delivery rate (BL755 bombs/hour), the response time (time to target), and the fuel burn/bomb delivered for both the V/STOL and CTOL aircraft. The figure shows the enormous advantage that operation from bases close to the battle zone creates. The V/STOL aircraft can deliver two to four times as many bombs per hour and uses about one quarter of the fuel to deliver each bomb compared with the CTOL aircraft.

Though the currently accepted acquisition cost penalty of a V/STOL aircraft relative to a CTOL aircraft is about 25%, the above figures indicate such an overwhelming operational advantage for the V/STOL aircraft that the initial cost penalty is more than offset. On this basis, a V/STOL aircraft fleet would be approximately 35% cheaper to acquire than a CTOL aircraft fleet for the same overall effectiveness.

There is a similar advantage if supersonic air superiority aircraft are considered. The assumption here is that pairs of aircraft maintain a standing combat air patrol for one hour over the battle area and carry out a supersonic interception during the mission. Figure 12.8 shows the number of aircraft required to maintain this standing patrol along a 50 nm front assuming ten patrol stations. The number of V/STOL aircraft needed is half the number of CTOL aircraft needed to perform the same task. On the same unit cost ratio as above, the V/STOL fleet would be 25% cheaper to acquire than the CTOL fleet with equivalent effectiveness.

In order to fulfill this optimistic forecast the powerplant design must be optimized and the installation losses minimized. They discuss the basic engine parameters involved, thrust, weight, installation drag, and SFC, showing the effect of fan pressure ratio, thrust to weight ratio, and installation drag on the takeoff weight and explain under which conditions it may be advantageous to use afterburning.

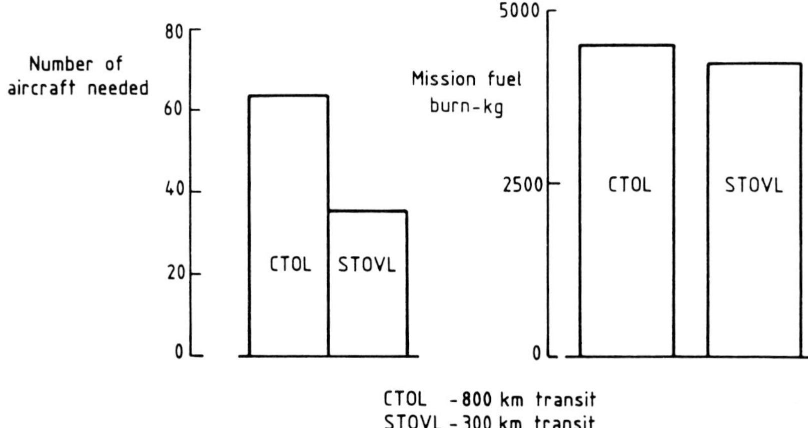



**Figure 12.8**   Air superiority comparison (Lewis and Simpkin 1981). *Copyright © AIAA 1981. Used with permission.*

Davenport and coworkers (1991) discuss the penalties and payoffs of V/STOL vs. CTOL for naval aviation. One of their proposals is to move the engine from the centre of the aircraft, using internal valving; the exhaust flow can then be switched as needed between rotating nozzles for vertical operation and a conventional tailpipe with afterburner for wing-borne flight (see Figure 12.9). High thrust levels (up to 220 kN corresponding to Mach 1.1) can be achieved; the engine can be located in the aft fuselage and not near the centre of the aircraft for better area distribution. Instead of mounting the forward nozzles directly on the engine, ducts carry the bypass airflow forward, allowing the nozzles to be remotely mounted wherever necessary to give the proper thrust-center of gravity balance.

They also point out that a V/STOL aircraft can be based on many sea-borne platforms not available to CTOL aircraft, since they do not need large-deck carriers, the catapult and arresting equipment. Any ship capable of operating a helicopter can carry a typical V/STOL aircraft. Special options include converted container ships, barges, and oil production facilities.

## 12.3 The Pegasus Engine

The Rolls-Royce Pegasus, shown in Figure 12.10, is a fan engine for STOVL applications consisting of two main rotating systems turning in opposite directions. Thrust vector control is achieved by four rotatable nozzles operating simultaneously on each side of the engine. The total thrust is divided between the four nozzles, with the resultant passing through a fixed point independent of the nozzle angle thus minimizing control problems. Bleed air from the high pressure compressor is used to stabilize the aircraft.

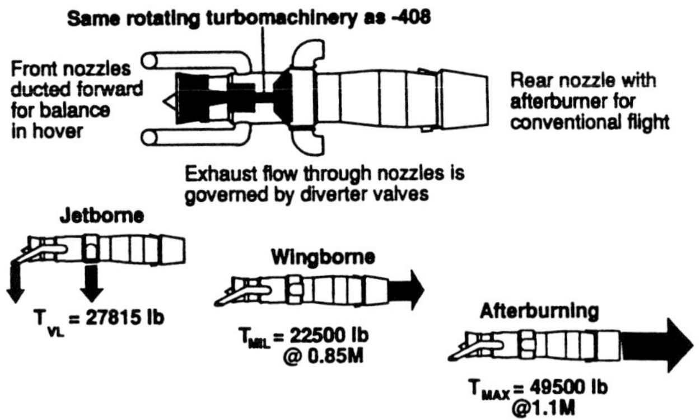

**Figure 12.9**   Proposed V/STOL engine with tail pipe exhaust (Davenport et al. 1991). *Copyright © AIAA 1991. Used with permission.*

**Figure 12.10**   Rolls-Royce Pegasus engine (Southeran 1971). *Published by kind permission of Rolls-Royce plc and Pergamon Press Ltd. © Rolls-Royce plc 1971.*

While the Pegasus engine dates back to August 1959, the initial operational version, the Pegasus 11 Mk 103, entered service on the Harrier V/STOL aircraft in 1969. This model was followed by the 11 Mk 104, the version for the Navy Sea-Harrier which has identical performance data but is 25kg heavier, due to minor changes in the casing, the coating, and the gearbox, to provide protection from the sea environment. The Pegasus 11-21 is an improved version for the Harrier 2 (or the MDD AV-8B), while additional improvements were introduced in the Pegasus 11-61. Those models are described in subsection 12.3.1; the studies performed for a supersonic version are covered in the next subsection.

### 12.3.1. Versions for Subsonic Aircraft

Chopping (1979) presents the early development of the Pegasus engine, from the first model, that developed a thrust of 40 kN in 1959, to the operational Pegasus 11 Mk 103 that entered service in 1969. This engine had a 95.64 kN thrust with an airflow of 196 kg/sec, a fan pressure ratio of 2.3, an overall pressure ratio of 14, and a bypass ratio of 1.4. It has a three-stage low pressure fan and an eight-stage HP compressor with Titanium rotor blades; the annular combustor has low pressure vaporizing burners, while the high pressure turbine has two air-cooled stages and the LP turbine is also a two-stage design. The most interesting aspect of this engine is the way the two exhaust streams are discharged; two pairs of exhaust nozzle-cascade sections replace the conventional nozzle. One pair exhausts the bypass air to either side of the engine via a plenum chamber situated at the fan exit, while the second pair exhausts the tur-

bine gases in a similar fashion at the rear of the engine. The nozzle-cascades are rotatable so that the gases can be exhausted rearward, downward, or forward to achieve full thrust vectoring. The cold nozzles are made of steel, while the hot ones are fabricated from nimonic. The activation is by duplicated air motors through shafts and chains under pilot control.

The following improvements appeared in the 11-21 Mk 105 version: better high pressure turbine cooling, a new shrouded LP turbine, and the introduction of FADEC control. As a result, the thrust was increased to 97.86 kN. Further improvements have been introduced in the 11-61 version, in which the fan pressure ratio has gone up to 2.6 with a thrust of 105.9 kN and the life cycle cost is lower.

### 12.3.2. Pegasus with Plenum Chamber Burning

In the sixties RR started studying a version of the Pegasus with combustion in the bypass duct (PCB) and later started developing the BS 100 PCB engine for the P 1154, the supersonic version of the Harrier. This was canceled in 1965, but much valuable knowledge was gained, as described by Southeran (1971). Since the late 1980s increased interest in a supersonic V/STOL has arisen, and RR is studying various options, some of which are mentioned in Section 12.1. We shall review here the pioneering effort of the BS 100.

A cut away view of the PCB system, seen from the front, is presented in Figure 12.11, while Figure 12.12 shows a rear view. One can distinguish the main combustor, the air and fuel manifolds, the plenum chamber in which the aug-

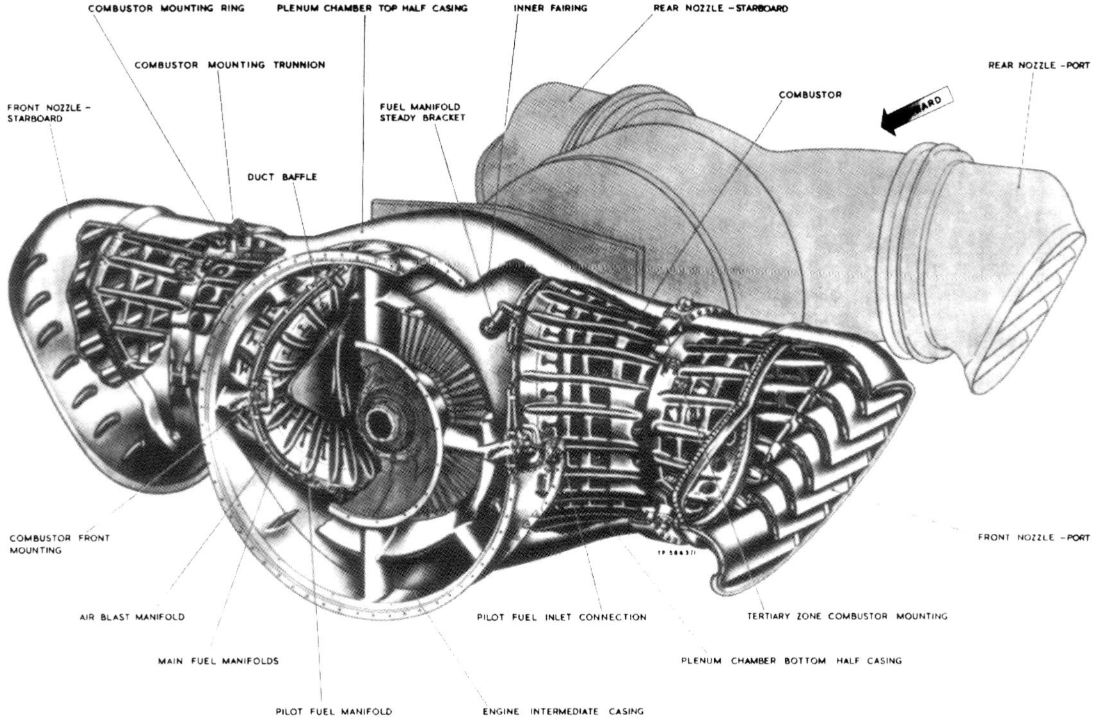


**Figure 12.11** Pegasus PCB system, front view (Southeran 1971). *Published by kind permission of Rolls-Royce plc and Pergamon Press Ltd. © Rolls-Royce plc 1971.*

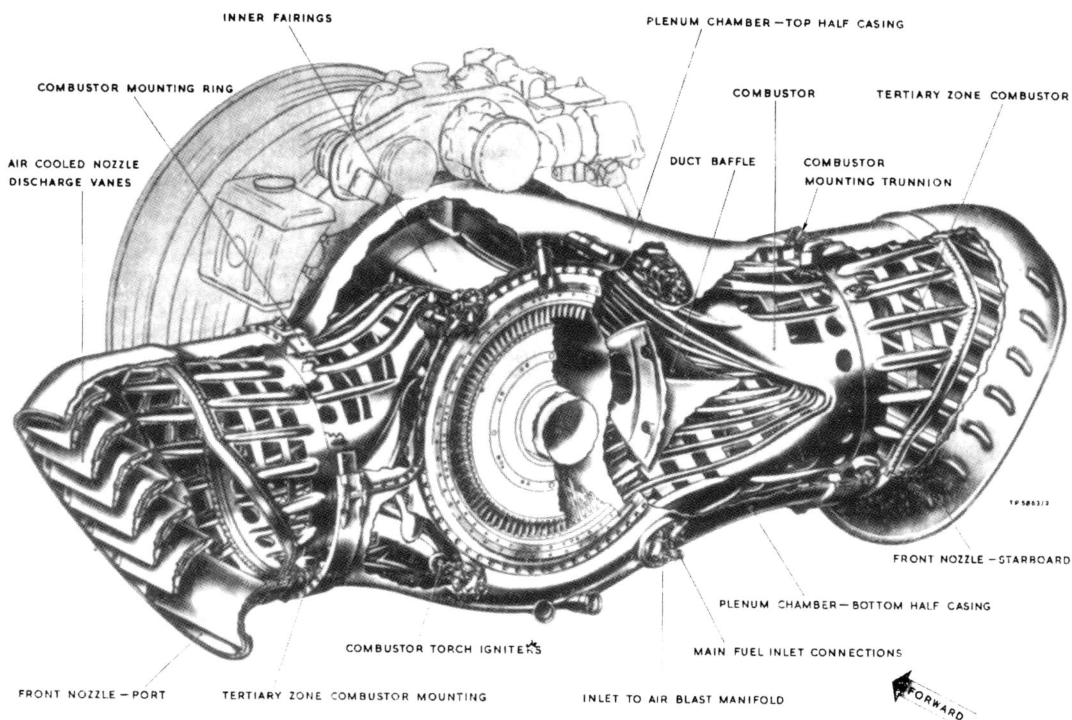



**Figure 12.12**  Pegasus PCB system, rear view (Southeran 1971). *Published by kind permission of Rolls-Royce plc and Pergamon Press Ltd. © Rolls-Royce plc 1971.*

**Figure 12.13**  Staggered gutter colander (Southeran 1971). *Published by kind permission of Rolls-Royce plc and Pergamon Press Ltd. © Rolls-Royce plc 1971.*

mented burning takes place in a colander, and the front nozzles through which these gases are exhausted. The staggered gutter colander, shown in Figure 12.13, is derived from a previously developed ramjet engine. Its advantage is that it performs well at low inlet temperatures and satisfies the low pressure loss and cooling requirements of PCB. It also appears to offer flow control possibilities. The figure shows its construction: the walls are highly perforated with an ordered array of rectangular holes separated by "longerons" which extend parallel to the gas flow and "crosstabs" set across the flow. A small depth V-gutter is attached to each cross-tab on its inner face and the longerons are also V-shaped. At the upstream end of the colander a high burning stability pilot zone is provided. The only flexibility in the choice of the colander shape was in the placement of the nose. After fixing the plenum chamber and determining the colander shape, the next step was to decide on the colander nose and align the longerons with the adjacent stream lines. The cross-tabs are normal to the longerons, and the flow enters through appropriate air feed holes. In the first design attempt the results were not satisfactory, so perspex models of the PCB system were built and the flow tested. This allowed the engineers to make changes in the nose and wall geometry and to develop an appropriate fuel injection system. The final version, with direct injection of the main fuel into the colander air entry ports, is shown in Figure 12.14. A pilot zone fuel was also necessary. Performance tests showed reasonable efficiency, as depicted in Figure 12.15, while pressure loss was not excessive.

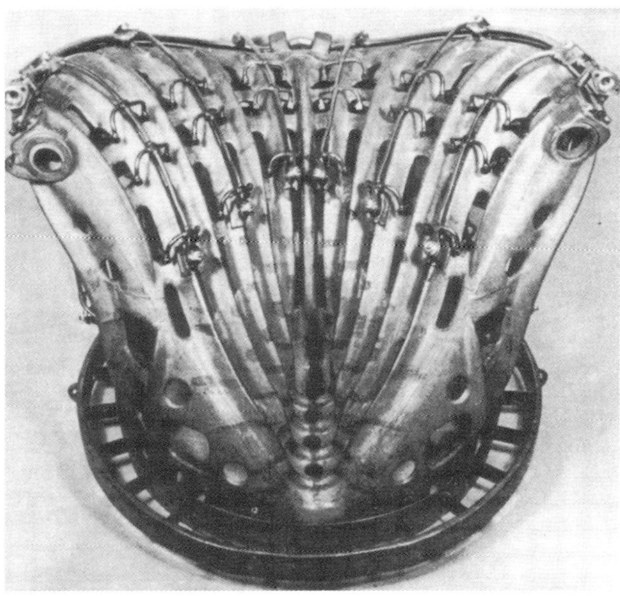

**Figure 12.14**  Direct injection of main fuel with colander (Southeran 1971). *Published by kind permission of Rolls-Royce plc and Pergamon Press Ltd. © Rolls-Royce plc 1971.*

## 12.4 Lift Engines

The Russian Yak 36 and Yak 38 use, as mentioned before, a combination of two lift jets, the Kolisov-Ribinsk RD-36-35 FVR and a Tumansky R-27-V300 turbojet. The lift-jets, each of which supplies 23 kN thrust, are installed in tandem immediately after the cockpit. They are inclined forward at 13° from the vertical exhausting their gases downwards and are also used for pitch and trim.

This single shaft turbojet has a five-stage compressor, partly constructed of composite materials, followed by an annular combustor with 12 burners; the power is supplied by a single-stage turbine, with air impingement starting. The Tumansky engine, which operates here without reheat, is a two-spool turbojet, the core consists of a five-stage com-

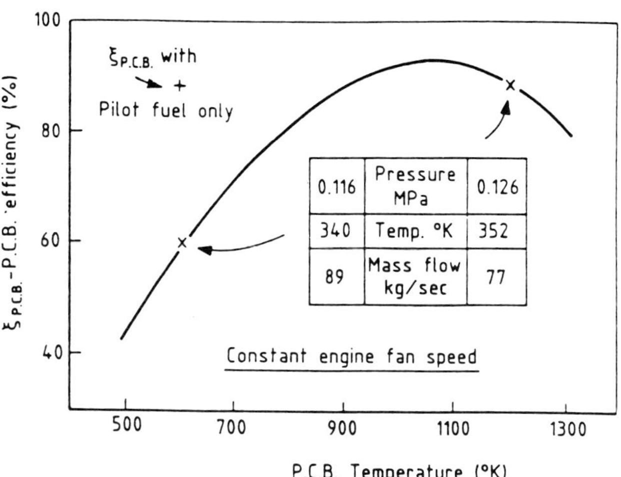


**Figure 12.15**  Efficiency of Pegasus PCB system (Southeran 1971). *Published by kind permission of Rolls-Royce plc and Pergamon Press Ltd. © Rolls-Royce plc 1971.*

pressor, an annular combustor and an air-cooled HP turbine; the low pressure assembly has a six-stage compressor and a single-stage turbine. The sea level thrust is 68 kN. The two nozzles are deflected to 90° for vertical takeoff.

The Russians were the first that took V/STOL to sea, putting them on their Kiev class light aircraft carriers, together with helicopters and have developed different techniques for using them. Yakovlev is developing a supersonic successor to the Forger, the Yak-141, which has been displayed in model form in the 1991 Paris Salon (Braybrook 1991), where they also showed videotapes of test flights. The aircraft, which can reach Mach 1.7 in level flight, has broken some time-to-height records previously held by the Harrier. The powerplant is similar to that of the Yak-38, but it has a more powerful lift-cruise engine, the R-79, comprising a Soyuz unit with a maximum horizontal afterburning thrust of 152 kN and two of the same RD-36-35 lift engines.

# Chapter 13

# Combustion Chambers

The task of the combustion chamber is to burn efficiently the fuel supplied by the burners with the hot air coming from the compressor, thus providing a uniformly heated, high temperature gas to the turbine. This requires the use of materials which can withstand extreme heat, together with a suitable distribution of the air between primary and secondary combustion and cooling air. Early engines had can combustors (an example is the RR Dart), later can-annular units were introduced (GE J-79), while today practically all engines use annular combustors.

Large combustors are situated geometrically in front of the HP compressor and followed by the HP turbine; in some smaller units reverse flow combustors are employed, as in the case of Garrett's ATF3 and F-109 engines (See Sections 9.2.3 and 9.3).

One of the most important requirements for modern combustors is that they should minimize pollution at both low and high altitudes. After describing these requirements we shall discuss pollutant formation, allowed pollution limits and the methods of achieving them, giving, as an example, the design of an advanced turbofan combustor. We shall then briefly treat measurements of combustion performance and durability characteristics. The next topic is special types of combustion, such as prevaporization and catalytic units. Then cooling methods will be discussed and the section will be concluded with a short description of combustion diagnostics.

## 13.1 Requirements for Combustion Chambers

The basic requirements common to all aircraft combustors are:

  a. high combustion efficiency, which means complete combustion of the fuel in the main chamber (in general over 99.9%);
  b. Good ignition quality, i.e., reliable and smooth ignition at all flight conditions, including high altitudes and low temperatures;
  c. wide stability limits over a wide range of pressures, velocities, and fuel-air ratios;
  d. uniform outlet temperature distribution, allowing high TET;
  e. low pressure losses;
  f. no combustion induced instabilities, such as pressure or velocity fluctuations;
  g. low emission of pollutants (carbon monoxide-CO, unburned hydrocarbons-UHC, nitroxen oxides—$NO_x$, and smoke)
  h. low cost, easy maintenance, and durability;
  i. capability of burning different fuels;
  j. small size and low weight.

Of course the importance of the different requirements will vary according to the type of aircraft in which the engine operates. For instance, size and weight will be very important for military aircraft, while low cost, ease of maintenance, and durability will be paramount for civil engines.

## 13.2 Pollutant Formation and Methods of Pollutant Reduction

In the last 30 years environmental issues have assumed increasing importance. During the 1960s air quality became an issue of public interest, first in the United States and then in Western Europe and Japan. Aircraft engine pollution is a primary concern at very high and very low altitudes. Below 1 km in the troposphere[1] emissions affect urban air quality close to airports. At high altitude cruise, which takes place in the stratosphere (extending up to 50 km), emissions affect the global environment (ozone formation, atmospheric heating).

The combustion characteristics that cause high pollutant emission levels are different at low power (including idle) and at high power. The main products formed at low power are CO, which is a toxic gas, and UHC, which produce odors and photochemical smog, while at high power the main pollutant is $NO_x$ (this usually includes NO and $N_2O_4$, although $N_2O$ has been shown lately to be also important); these

---

[1]This is the lower part of the atmosphere, extending up to a height between 9 and 17 km, varying with season and latitude.

**Table 13.1**  EPA engine classes (Petrash et al. 1979). *Courtesy of NASA.*

| Engine Class | Definition | Application |
|---|---|---|
| T1 | Turbofan/turbojet < 8000-lb thrust | General Aviation |
| T2 | Turbofan/turbojet > 8000-lb thrust | Wide body transports |
| T3 | JT3D models | B707, DC-8 |
| T4 | JT8D models | B707, B737, DC-9 |
| T5 | Supersonic cruise engine | Supersonic transports |
| P2 | Turboprops | General aviation |

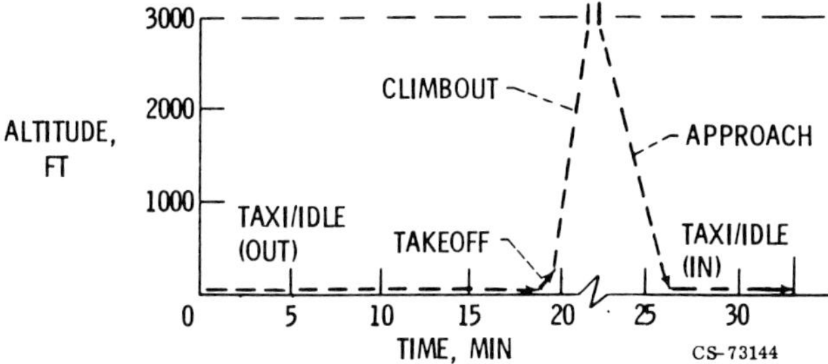


**Figure 13.1**  EPA takeoff and landing cycle (Grobman et al. 1975). *Courtesy of NASA.*

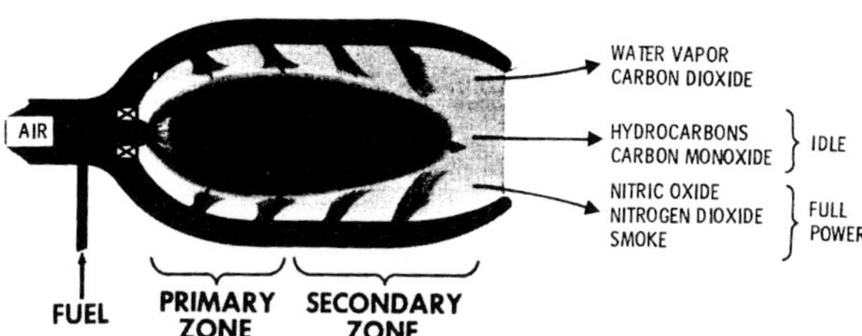


**Figure 13.2**  Schematic of early combustor (Petrash et al. 1979). *Courtesy of NASA.*

gases are toxic and contribute to photochemical smog and smoke, causing visible particle deposition.

The U.S. environmental agency (EPA) divided aeroengines into the six classes which are shown in Table 13.1 and established fixed gaseous emission standards. The EPA also defined a special parameter which is obtained by integrating the emissions of an aircraft over a specified takeoff-landing cycle, shown in Figure 13.1; more exactly, it is the mass of pollutant emitted during such a cycle divided by the takeoff thrust, in g/kN. This parameter, known as EPAP, has also been adopted by ICAO.

Figure 13.2 schematically illustrates a combustor used in the early large turbofans which had a single burning stage. Here the primary zone operates fuel-rich with large amounts of air bypassing it and coming in further downstream to cool and dilute the combustion products. These combustors

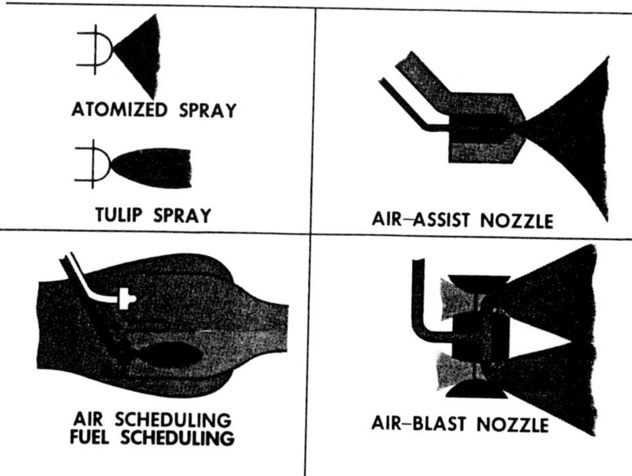


**Figure 13.3**  Low and high power emission causes (Petrash et al. 1979). *Courtesy of NASA.*

had a step louver and a film cooled liner, using a large portion of the total airflow.

Figure 13.3 shows that most techniques suggested early in order to reduce pollution conflict one with the other at different power levels, except the improvement of fuel atomization and distribution which is always beneficial. It was, therefore, necessary to find ways of designing an engine which could minimize pollution in all regimes without compromising performance. This problem was solved in the Energy Efficient Engine program, described in detail in Section 8.1, by introducing systems having a pilot and a main combustion stage. Combustor design features for further reducing the UHC and CO emission levels were developed in the early 1980s and incorporated into operational engines. An example for the GE CF6-80C2 engine is shown in Table 13.2. At cruise operating conditions $NO_x$ is the main pollutant and since combustor inlet air temperature and pressure are lower than at takeoff for subsonic aircraft engines, the $NO_x$ level goes down.

Recently, the effect of emissions throughout the flight cycle, particularly at high altitude, have been emphasized, due to the fact that the greenhouse effect is now one of the ma-

**Table 13.2**  Emission levels of CF6-80C2 engine model (Bahr 1992). *Reprinted by permission of Klewer Academic Publishers.*

| Emission | | Applicable ICAO Standard | | Status CF6-80C2/A3 |
|---|---|---|---|---|
| | | Regulatory Level | Compliance Test Limit* | (Rated Thrust: 262.2 kN) |
| Smoke | (SN) | 18.2 | 14.1 | 7.8 |
| HC | | 19.6 | 12.7 | 11.3 |
| CO | (g/kN) | 118.0 | 96.2 | 7 |
| $NO_x$ | | 103.3 | 89.1 | 49.9 |

*Based on 3 tests of a single engine

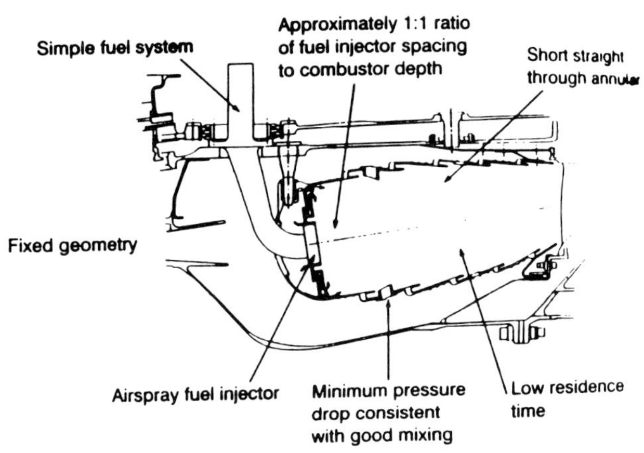



**Figure 13.4** State of the art combustor (Metcalfe et al. 1991). *Copyright © AAIA 1991. Used with permission.*

jor environmental issues. The natural greenhouse effect is due to absorption of infrared radiation from the earth's surface by atmospheric gases, mainly water vapor and $CO_2$, causing a temperature increase. There is now growing concern that increased emissions, due to human activities, may enhance this effect and cause major climatic changes. As regards the contribution of aircraft engines, their $NO_x$ emissions play an important role in the formation of ozone, which behaves as a greenhouse gas in the troposphere. According to calculations, the main contribution comes from cruising altitudes of 10–12 km (Metcalfe et al. 1991). Of even greater concern is the depletion of ozone in the stratosphere; that ozone layer shields the earth from ultraviolet radiation, which has harmful effects when it reaches its surface (increase in skin cancer and eye diseases, crop damage and lower timber yields). Supersonic aircraft cruising in the stratosphere appear to present a major risk of $NO_x$ emission, causing ozone depletion. Studies are now being conducted on this problem which may become even more important for future supersonic vehicles. Against this background of increasing environmental concerns the combustion technologist has attempted to improve all aspects of combustor performance.

The development of engine cycles for greater fuel efficiency has led to ever increasing turbine entry temperatures, with the consequent need for precise control of the turbine entry temperature distribution, and the requirement for advanced methods of combustor wall cooling. Although different designers have chosen different solutions for the many design problems encountered, a "state-of-the-art" combustor, for a turbofan engine, exhibits a number of fundamental characteristics, as shown in Figure 13.4 (Metcalfe et al. 1991).

Such a combustor will meet current emissions requirements and will have a service operating life of up to 20,000 hours, representing some 6 years of undisturbed service operation. All the requirements of the appropriate airworthiness authorities will have been met and this, along with service experience, has demonstrated the safety and reliability of the design. The challenge now is to reduce emissions further without detracting from other aspects of performance.

Until recently, future emissions control options have been perceived as being associated almost exclusively with combustion technology improvements aimed at reducing $NO_x$ emissions while maintaining current combustor performance. More recent concern over the effect of $CO_2$ and water vapor emissions, hitherto regarded as the ultimate aim of the combustion technologist, means that future airframe and engine cycle designers will have to consider these emissions in order to achieve an environmentally acceptable solution.

A design for an advanced combustor, along the lines mentioned previously, is described below.

### 13.2.1 Advanced Combustor Design

Sanborn and coworkers (1989) describe a combustion system design developed by Garrett for an advanced turbofan engine. They started by establishing a combustor envelope meeting the performance goals. The combustor size and its orientation with respect to the compressor and turbine have a direct impact on the engine geometry. A large diameter combustor increases the engine diameter, thus requiring a larger installation nacelle with higher aerodynamic drag. Minimization of the combustor length, resulting in a shorter engine, reduces the weight as well as critical speed shaft dynamic problems. The combustor performance influences the SFC through pressure drop and pattern factor characteristics. A 1% increase in pressure drop can increase the SFC by 0.5 to 1%, depending on the engine cycle. The pattern factor, which is defined as $(T_{\max} - T_{\text{avg}})/(T_{\text{avg}} - T_{\text{in}})$, influences the secondary cooling required for achieving acceptable durability of the turbine. The combustor operating requirements include the following: a temperature rise of 725 K, a pattern factor of 0.3, a 5.5% pressure drop, a lean blowout fuel/air ratio of 0.006, and a maximum metal temperature of 1150 K.

The combustion envelope resulting from engine size limitations and the combustor performance requirements is shown in Figure 13.5 and is remarkably similar to the configuration advocated by Metcalfe. The orientation of the compressor diffuser and deswirl vanes with respect to the combustor requires an analysis of the external flowfield around the outside of the combustor in order to optimize it. This was achieved with the aid of experiments performed on a test rig which was instrumented to measure pressure and velocities in cold flow. The internal flow field analysis was divided into two parts: definition of the fuel injection system with test evaluation of the fuel nozzle, determining the spray characteristics and detailed definition of the combustor internal flowfield. The initial and boundary condi-

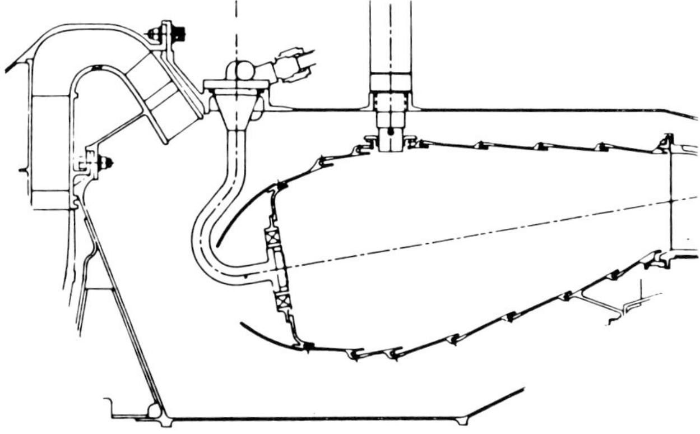

**Figure 13.5** Combustor envelope for advanced turbofan (Sanborn et al. 1989). *Copyright © AAIA 1989. Used with permission.*

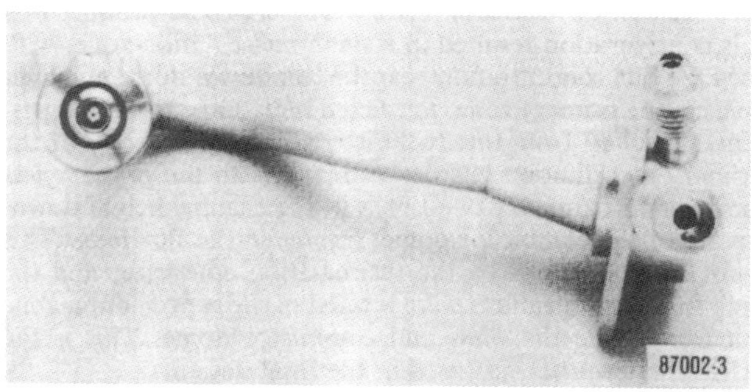



**Figure 13.6** Pressure atomizer piloted airblast fuel nozzle (Sanborn et al. 1989). *Copyright © AAIA 1989. Used with permission.*

tions for the internal flow analysis were the measured spray characteristics together with the local pressure drop and external velocities from the external flow analysis.

A pressure atomizer piloted airblast fuel nozzle, shown in Figure 13.6, with a single external flow divider, was chosen for the combustion system. At low fuel flows (ignition) all the fuel passes through the primary circuit pressure atomizer; after the flow has increased to a prescribed pressure, the flow divider valve opens, sending fuel to a prefilming airblast atomizer (secondary circuit). The nozzles were tested in a Malvern spray facility which determined the droplet size, the SMD (Sauter mean diameter), and the dis-

tribution. The flowfield was optimised by finding the best air entry locations.

After completing the combustor design and its fabrication, it was evaluated in a full scale annular test rig, simulating the engine flow path (see Figure 13.7). The parameters measured to determine combustor performance were ignition and blow-out fuel-air ratios, discharge temperature, and wall temperatures. The first two were significantly below design goals, while the combustor discharge temperature, measured with thermocouple rakes, was well within the maximum allowable envelope.

In conclusion, one can say that the application of advanced analysis and the use of specially built test rigs decreased the amount of rig test development time, achieving the required goal with only slight modifications. The final results are shown in Table 13.3. The system pressure drop was not reduced, since its impact on engine cycle performance was not significant.

GE and SNECMA developed an advanced combustor for the GE 36 engine of their propfan. The chamber is very short ($L/H = 1.75$) and gives a low SFC. The pollution levels measured are well below the EPA-ICAO requirements: CO, 34 g/kN vs. 118; UHC, 3 g/kN vs. 19.8; and finally, $NO_x$, 40 g/kN vs. 108. The low $NO_x$ level is obtained for a pressure ratio of 35.

## 13.3 Measurement of Combustor Performance and Characteristics

Tanrikut and coworkers (1981) describe the development of a combustor liner that can withstand high temperatures and stresses for a relatively long life. This work was performed at P&W as part of the Aircraft Energy Efficiency program. The first step was a screening study of different concepts, shown in Figure 13.8. The conventional film-cooled louver in which the coolant enters through discrete holes, mixes in the chamber formed by the lip of the preceding louver, and forms a film sheet at the slot exit, has low efficiency since it uses only some 5% of the total potential heat sink, which is at the compressor exit temperature. In the impingement-film technique, where the heat transfer on

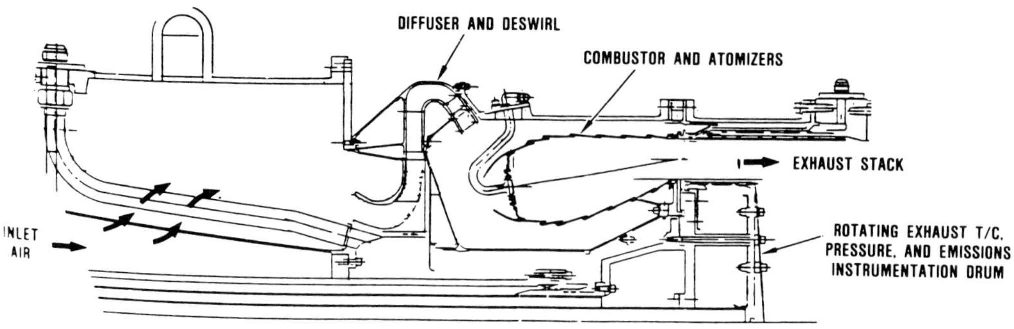



**Figure 13.7** Annular test rig (Sanborn et al. 1989). *Copyright © AAIA 1989. Used with permission.*

**Table 13.3** Comparison of combustor performance with goals (Sanborn et al. 1989). *Copyright © AIAA 1989. Used with permission.*

|                                      | Goal  | Measured |
|--------------------------------------|-------|----------|
| Combustor temperature rise, °F       | 1300  | 1300     |
| Pattern factor                       | 0.30  | 0.26     |
| Pressure drop, %                     | 5.5   | 5.7      |
| Lean blowout fuel/air ratio          | 0.006 | 0.005    |
| Maximum metal temperature, °F        | 1600  | 1500     |

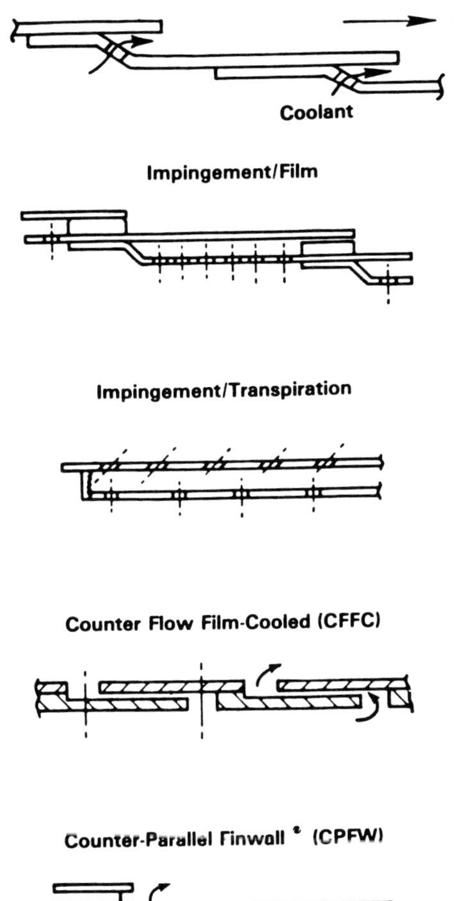


**Figure 13.8**  Different liner concepts (Tanrikut et al. 1981) *Copyright © AIAA 1981. Used with permission.*

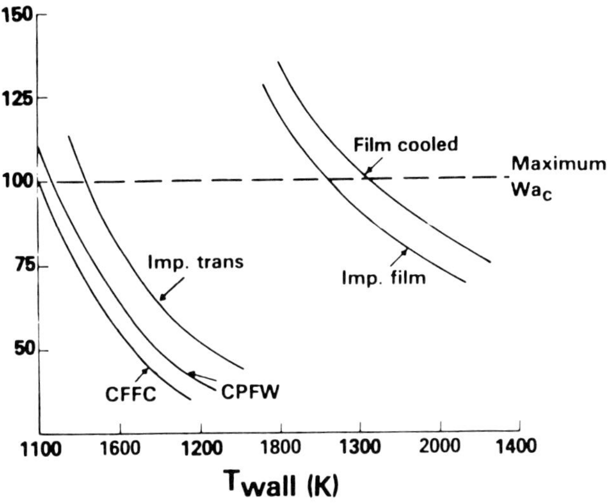


**Figure 13.9**  Counter-flow film cooling (Tanrikut et al. 1981). *Copyright © AIAA 1981. Used with permission.*

the back of the louver panel is augmented by an array of impinging jets, the improvement is marginal. Better results were obtained when using an impingement-transpiration technique. Here the liner hot wall contains many small inclined holes through which the coolant is injected, forming a film. An impingement plate on the back augments the back side heat transfer coefficient and reduces the pressure drop across the multiperforated heat sheet.

In the counter flow film cooled concept, the cooling air enters through holes in the cold sheet and flows the hot gases in channels, cooling the liner by convection, before injecting the fuel through holes in the hot wall and forming a film. Figure 13.9 shows that this method is somewhat superior to that described before. It was found, however, that the best technique is the counter-parallel finwall, in which the cooling air enters the cool side of the liner trough slots, then flows both forwards and aft through channels, cooling the liner convectively. The flows coming out at the aft and forward sections mix, providing film cooling on the hot side. One can extend the hot sheet of the aft section, providing a sheltered chamber that mixes better the two streams before filming.

In order to improve durability, a segmented liner construction, seen in Figure 13.10, was built, showing a 60% improvement over a loop configuration. This also made it possible to use cast turbine alloy material. Using B-1900, an investment cast nickel base alloy which has good strength in the 1000–1300 K range, the lifetime was 2.5 times longer than with Hastelloy X. A 90 degree rig was employed for testing during development; then complete liners were tested successfully up to 2.8 MPa inlet pressure and 810 K inlet temperature. This cooling scheme was applied to a 90 degree sector of the E$^3$ combusor developed by P&W (Greene et al., 1983) and an optimization program was run, first for the pilot zone, then for the main zone. For high power operation the optimum fuel split between the two zones was determined. The altitude relight characteristics were also evaluated, demonstrating ignition at 10,670 m height with a fuel flow as low as 21 kg/h.

## 13.4 Special Types of Combustion

In addition to the main type of combustors, such as those developed by GE, P&W, and RR for their large engines (see Section 9.1) there are a number of combustors that utilize

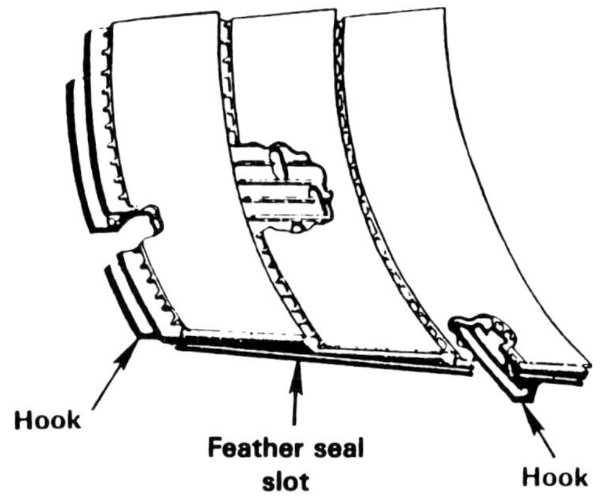



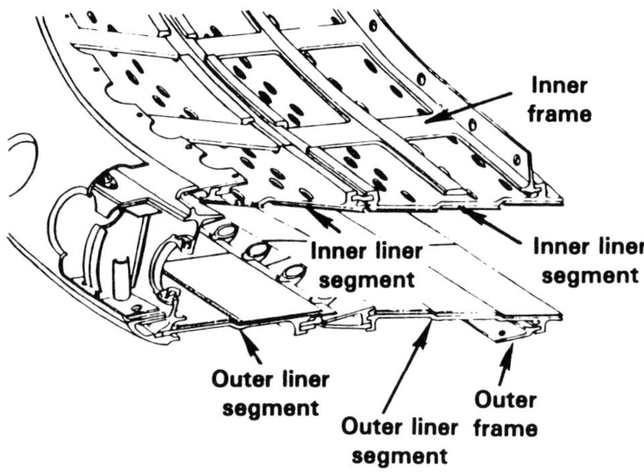



**Figure 13.10**   Segmented liner (Tanrikut et al. 1981) *Copyright © AIAA 1981. Used with permission.*

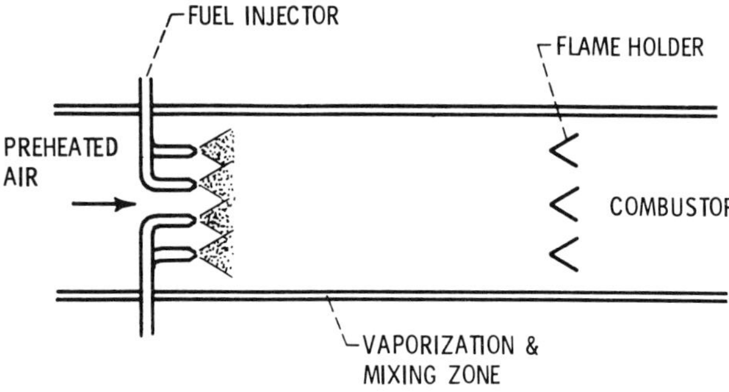



**Figure 13.11**   Premix flame tube combustor (Petrash et al. 1979). *Courtesy of NASA.*

special processes, like fuel prevaporization and catalytic, hot wall, and recuperative combustion. These will be described in this section.

### 13.4.1. Fuel Prevaporization

By evaporating the fuel completely before combustion and then mixing it with air, obtaining a lean primary zone, one achieves a low reaction temperature and eliminates hot spots thus drastically reducing $NO_x$ emissions. This is espe-

cially important for engines operating at supersonic speeds at high altitudes, where high combustor inlet temperature increases $NO_x$ production, with deleterious effects on the ozone layer. The essential features of such a combustor are shown in Figure 13.11. In order to use a lean premixed, prevaporised combustor, one must ensure a uniform fuel/air distribution, at the same time avoiding autoignition and flashback in the fuel/air mixing passages. Lean stability and altitude relight capability need special attention. Areas requiring additional research are shown in Figure 13.12.

Ferri (1972 and 1973) was one of the first to theoretically investigate this approach. He performed computations to determine the rate of $NO_x$ production in premixed flames of vaporised JP fuel in air, taking into account finite rate chemistry and turbulent transport processes. Figure 13.13 presents the dependence of the emission index on the flame temperature for an inlet temperature of 833 K. Replotting this as a function of equivalence ratio in Figure 13.14 shows that in order to achieve an emission index below 1, an equivalence ratio below 0.6 is required; Figure 13.14 illustrates the sensitivity of the $NO_x$ level to the equivalence ratio. A completely uniform mixture at an equivalence ratio of 0.6 produces 23 times the amount of $NO_x$ that would be created at an equivalence ratio of 0.4. Assuming that in a flow with an overall equivalence ratio of 0.4 there are 10% nonuniformities reaching an equivalence ratio of 0.6, this would increase the $NO_x$ production threefold over a completely uniform mixture. Experiments performed by Roffe and Ferri (1975) confirmed the validity of these calculations.

Early designs were used in the Mamba, Sapphire, and Viper engines in England and in the Wright J65 in the United States. Later RR, Lycoming, SNECMA, and MTU also used this approach. In general, vaporizer systems reduce soot formation, do not require high fuel-injection pressures, and are simpler and cheaper than mechanical atomizers. On the other hand, they have some drawbacks connected with the danger of autoignition and flashback.

### 13.4.2. The Catalytic Combustor

A catalytic combustor is a device in which chemical reactions initiated by a heterogeneous catalyst play an important role in the energy release process. The catalyst can be used either in the main combustor where segments of a monolithic support, usually a honeycomb, are located in the annulus or in the afterburner, where it may be used in the ignition or in the flameholder.

A model of a catalytic combustor, built from monolithic material such as a ceramic honeycomb, is shown in Figure 13.15; it acts as a bundle of tubular reactors with the energy release occurring as the reactants flow down each tube. This raises the temperature monotonically, with the maximum temperature occurring at the exit. Since there is no mass transfer between the tubes, one can assume, as a first ap-

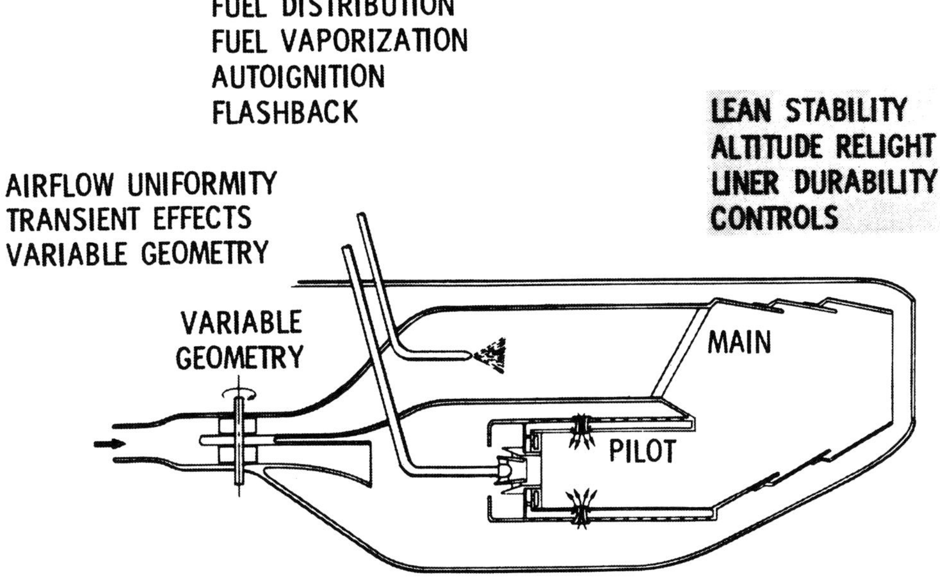


**Figure 13.12**   Lean premixed prevaporized combustor (Petrash et al. 1979). *Courtesy of NASA.*

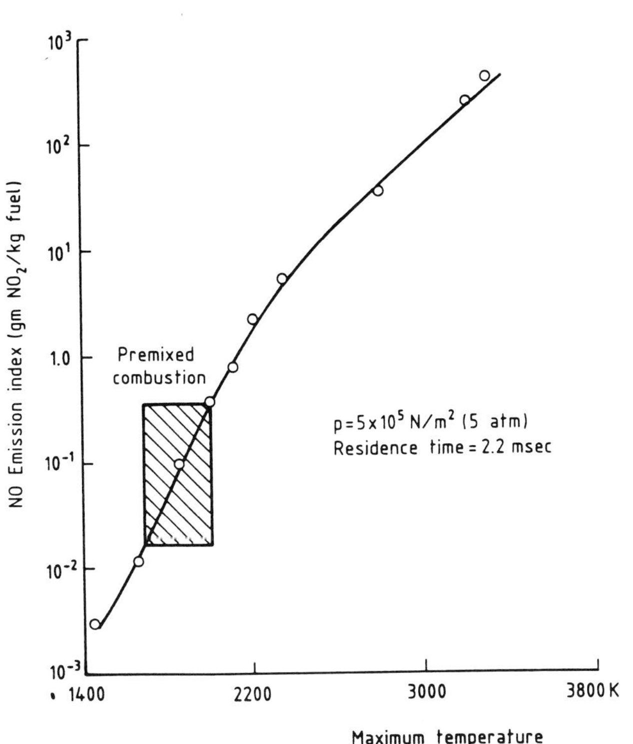


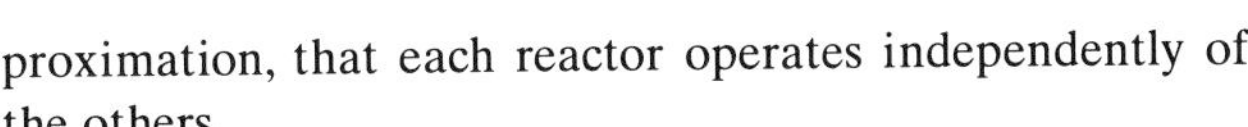

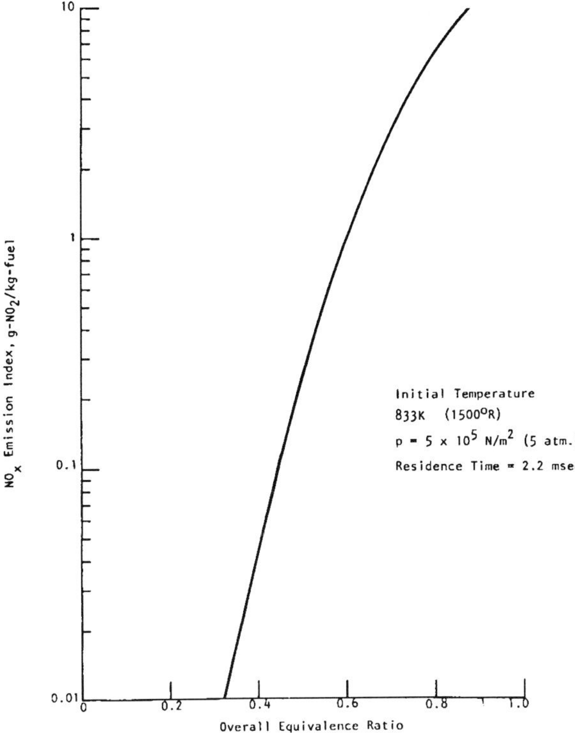


**Figure 13.13**   Equivalence ratio required for low emission index (Roffe and Ferri 1975). *Courtesy of NASA.*

**Figure 13.14**   NO$x$ level sensitivity to equivalence ratio (Roffe and Ferri 1975). *Courtesy of NASA.*

proximation, that each reactor operates independently of the others.

The processes taking place in the catalytic combustor are diffusion of the reactants to the catalyst, their absorption, chemical reaction, desorption of the products, and their diffusion into the bulk flow; homogeneous gas phase reactions also take place. The controlling energy release mechanism changes as the reactants flow along the tube, as shown in Figure 13.16. At the inlet there is no significant gas reaction,

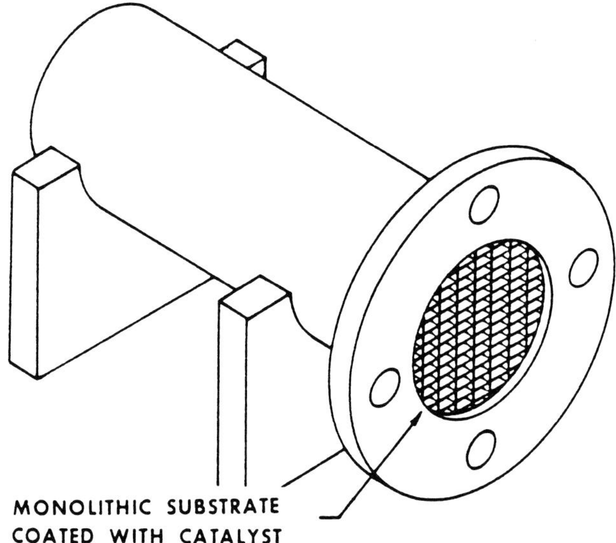



**Figure 13.15** Catalytic combustor test model (Rosfjord 1976). *Copyright © AIAA 1976. Used with permission.*

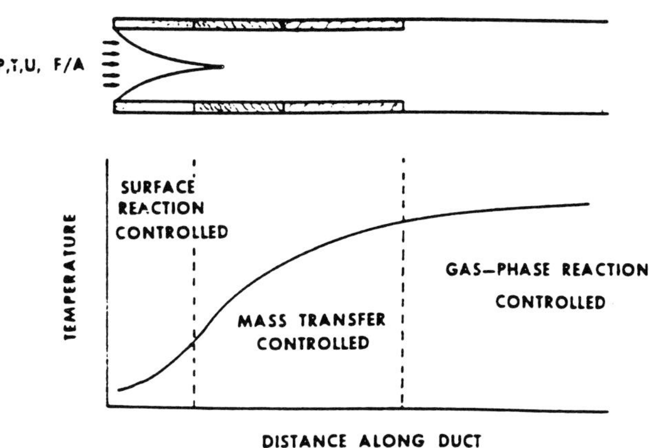



**Figure 13.16** Energy release mechanism (Rosfjord 1976). *Copyright © AIAA 1976. Used with permission.*

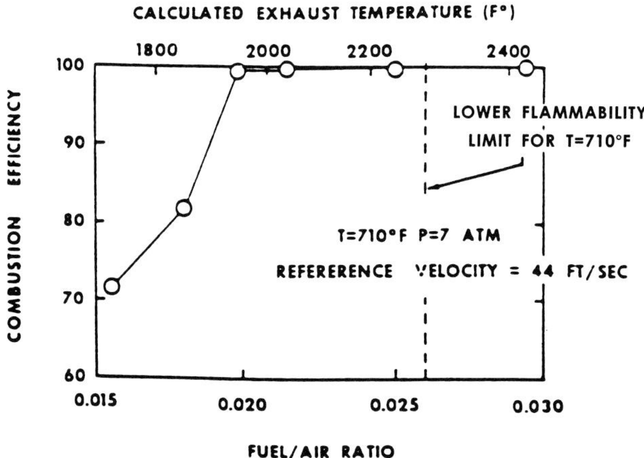



**Figure 13.17** Catalytic combustor feasibility study (Rosfjord 1976). *Copyright © AIAA 1976. Used with permission.*

due to the low gas temperature; there, the lower activation energy surface reactions predominate; with increasing temperature the heterogeneous reaction rates become fast enough, compared with the reactant diffusion rates; so mass transfer limits the rate of energy release over a large part of the duct. Finally, the temperature becomes high enough for gas-phase reactions to dominate. For most operating conditions, the transition from surface reaction to mass transfer control takes place within a distance of one tube diameter from the inlet. The change from reactant diffusion to gas phase control occurs at approximately 1250 K. Since the exit temperature of gas turbine combustors can reach 1700 K, part of the combustor does not need any catalyst, because gas-phase reactions will occur on their own.

The catalytic combustor has three components: the catalyst, the substrate, which supports it, and the fuel-air carburation system. Three classes of catalysts are of interest for this application: noble metals, such as Platinum, which offer the best low temperature capability but cannot sustain the higher temperatures near the combustor exit; transitional metal oxides, like Chromium or Cobalt oxides, which do not have the low temperature capability of the noble metals, but are good catalysts above 650 K; and the third group, the rare earth oxides (of Lantanum or Cerium) which have good high temperature capability (above 1100 K). As a consequence, the catalytic bed could be a sandwich of different wafers, with a low temperature catalyst near the inlet and a high temperature catalytic material near the exit.

As regards the substrate, the two leading candidates are cordierite, an alumina-magnesia-silica ceramic, and silicon carbide. The first is limited to a temperature of 1500 K, but is considerably cheaper; again using the sandwich approach, the cordierite can be employed close to the entrance and the silicon carbide near the exit. The fuel-air carburation system must provide a premixed, prevaporised inlet flow to the combustor. A feasibility study with JP-4 fuel showed that 100% combustion efficiency can be obtained if the fuel-air ratio is high enough, as illustrated in Figure 13.17. This can be achieved for mixtures below the normal lean flammability limit.

In a conventional combustor there are three zones of varying reactant mixtures. In the primary zone all the fuel is injected, but only part of the air, giving a local equivalence ratio close to unity; the flame is stable, but very hot (above 2500 K); the final stages of the hydrocarbon oxidation are completed in the secondary zone, while in the dilution zone the remaining air is added, reducing the temperature, as seen on the left side of Figure 13.18. The clean combustor, developed in the 1970s, operates in a more uniform fashion by working in stages; in these, the equivalence ratio is somewhat higher than the overall value of 1.00; however, the maximum temperature is lower and the pollutant level goes down. The catalytic combustor should, ideally, achieve stable combustion for uniform fuel-air mixtures and permit combustion to take place below an equivalence ratio of 0.5.

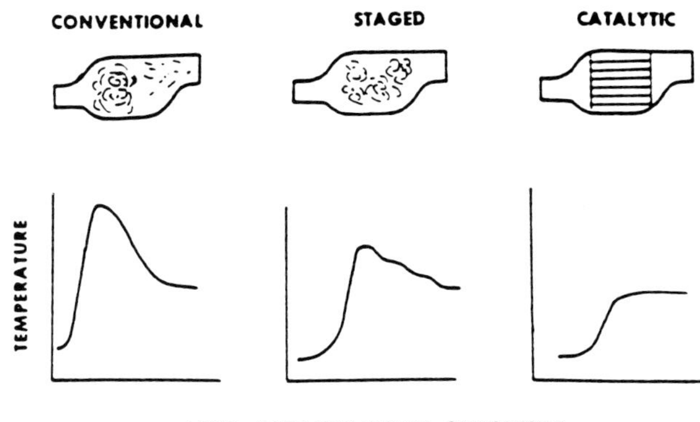



**Figure 13.18** Temperature profiles in conventional staged and catalytic combustors (Rosfjord 1976). *Copyright © AIAA 1976. Used with permission.*

In conclusion, catalytic combustors can lower the fuel consumption, increase the aircraft range, give higher thrust, and also enhance operational stability and reduce pollution emissions.

## 13.4.3. Other Low Emission Combustors

In a program conducted with the aim of reducing low power emissions, GE developed three concepts capable of diminishing the amount of CO and UHC pollutants to 10 grams and 1 gram per kilogram of fuel, respectively, while keeping the $NO_x$ emissions below 4 g/kg. This would result in a combustion efficiency of 99.7% at idle. These combustor concepts were designed and tested at the conditions given in Table 13.4. They include the hot wall combustor, the recuperative cooling combustor, and the catalytic converter combustor, each of which was discussed in detail in the preceding section.

The hot wall combustor, shown in Figure 13.19a, uses a thermal barrier coating along the inside combustor liner. It employs a 1–3 mm thick thermally sprayed yttria-stabilized zirconia layer, which allows the combustion gases near the wall to be at a higher temperature, minimizing wall quenching of the combustion reactions. The liners, shown in Figure 13.21, are of double wall construction, having

**Table 13.4** Concept emissions comparison (Mularz et al. 1979). *Copyright © AIAA 1979. Used with permission.*

| | At design point (422 K, 304 kPa, 23 m/s, f/a = 0.0105) | | |
| --- | --- | --- | --- |
| | HC g/kg fuel | CO g/gk fuel | $NO_x$ g NO/kg fuel |
| Hot-wall (H4) | 0.05 | 1.3 | 2.6 |
| Recuperative (R7) | 0.5 | 9.0 | 5.4 |
| Catalytic (C7) | 0.3 | 1.3 | 4.0 |
| Goal | 1.0 | 10.0 | 4.0 |

[a]$NO_x$ corrected to standard humidity, 6.3 g/kg.

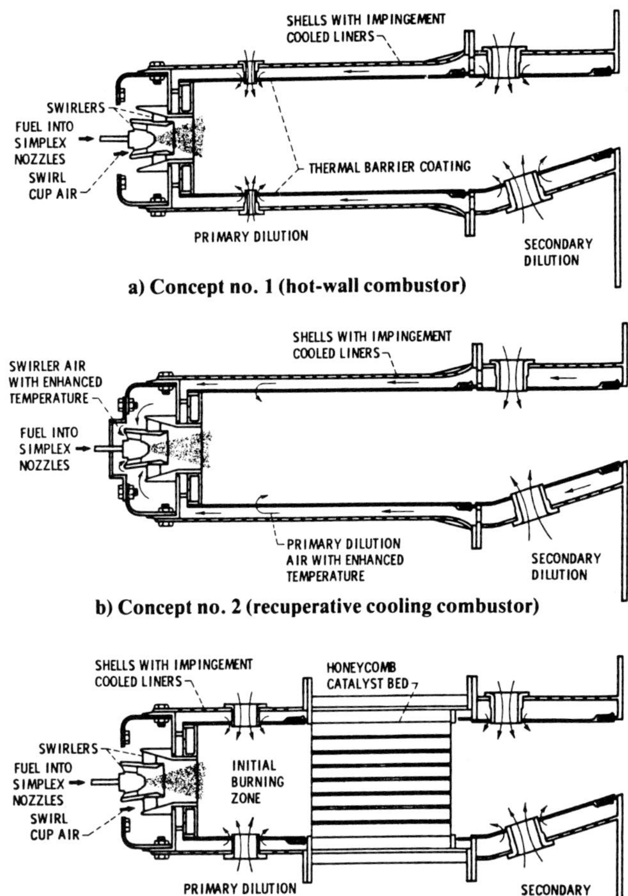



**a) Concept no. 1 (hot-wall combustor)**



**b) Concept no. 2 (recuperative cooling combustor)**



**c) Concept no. 3 (catalytic converter combustor)**

**Figure 13.19** Schematic of three combustor concepts (Mularz et al. 1979). *Copyright © AIAA. Used with permission.*

equally spaced holes drilled in the outer wall for high velocity air impingement cooling. After impinging off the back of the inner liner, the cooling air flows between the liner walls, until it reaches the dilution air thimbles, then passing into the combustor as a coannular dilution jet.

The recuperative cooling combustor (Figure 13.19b) sends all the primary combustion air through the annular passage of the combustion liners before it is admitted into the combustor either through the air swirlers of the fuel injectors or through the primary dilution holes. In this way the air temperature goes up by 50 to 100 K, according to the operating conditions; this reduces pollutant emissions by increasing the reaction rates. The secondary air dilution thimbles were sized to prevent cooling air from passing into the combustor as a coannular jet. A cover attached to the cowls of the combustor dome prevents the air from flowing directly into the air swirlers from the diffuser.

The catalytic converter combustor (Figure 13.19c) consists of a standard primary zone followed by a ceramic honeycomb catalyst bed. The fuel is first burned in front of the catalyst bed with about half of the airflow, resulting in

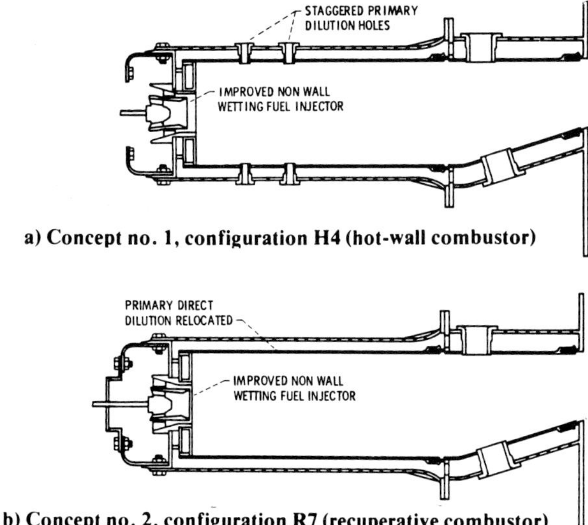



**a) Concept no. 1, configuration H4 (hot-wall combustor)**

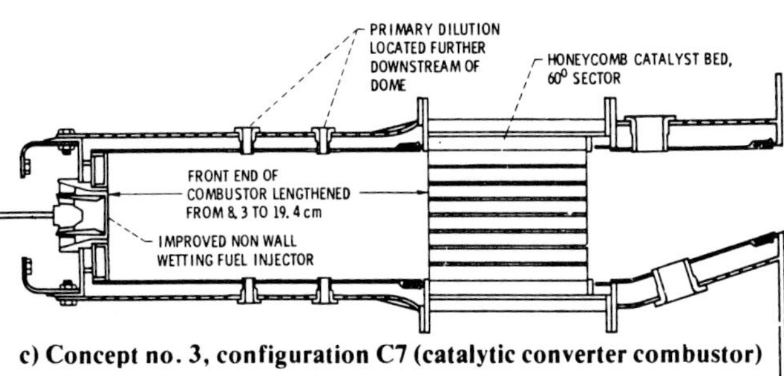



**b) Concept no. 2, configuration R7 (recuperative combustor)**



**c) Concept no. 3, configuration C7 (catalytic converter combustor)**

**Figure 13.20**   Three most promising configurations (Mularz et al. 1979). *Copyright © AIAA 1979. Used with permission.*

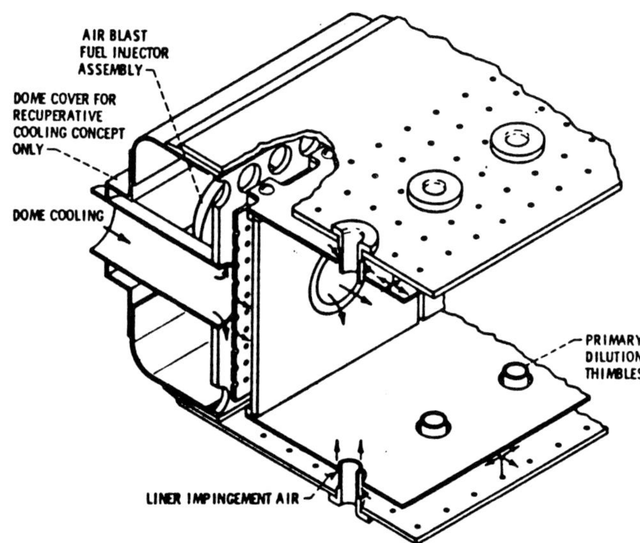



**Figure 13.21**   Construction of combustion liners (Mularz et al. 1979). *Copyright © AIAA 1979. Used with permission.*

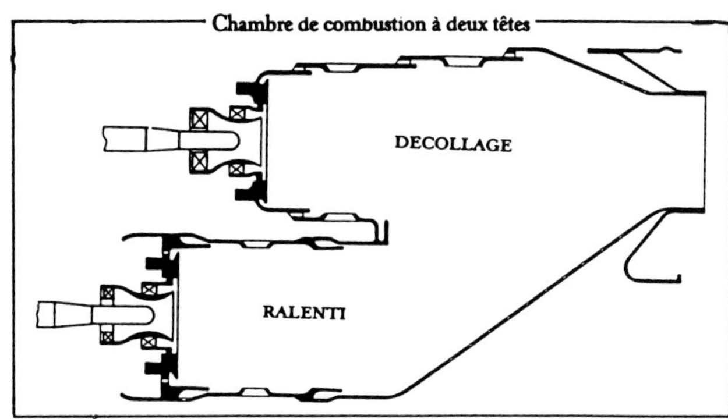



**Figure 13.22**   Double-head combustor (Bayle-Labouré 1991). *Courtesy L'Aéronautique et L'Astronautique.*

an average equivalence ratio of 0.3 at idle on the face of the catalyst bed. The catalyst acts as a cleanup reactor for the UHC and CO in the combustion gas. The lean equivalence ratio protects the catalyst bed from high temperatures and lowers $NO_x$ formation rates.

The three designs were fabricated and tested in a 60° sector combustor rig in a facility which provides nonvitiated preheated air at the required pressures. Seven configurations of each combustor concept were tested. The three most promising are shown in Figure 13.20. In the hot wall combustor the design of the air-blast fuel nozzles was changed, eliminating the fuel wetting of the barrel wall and lowering the CO and UHC emissions to levels that were below program goals. The recuperative cooling combustor was also modified by using an improved set of air blast fuel nozzles similar to those of Concept 1 (see Figure 13.19a). This configuration met the emission goals for CO and UHC, but exceeded that for $NO_x$. The catalytic combustor front end was lengthened to reduce peak temperatures in front of the catalyst bed (see Figure 13.20c). A set of non wall-wetting fuel injectors, similar to those of Concept 2, was used and two rows of dilution holes were located further downstream from the dome than the single row in the original design. CO and UHC emissions were very low, while the pressure drop was also below design value.

The three concepts are compared at the design point in Table 13.4. In all cases the UHC and CO emission goals were achieved, while the recuperative combustor had a somewhat high $NO_x$ level. The hot wall combustor, which is the simplest aerodynamically, gave the best results and appears to have the best characteristics.

Finally, the double-head combustion chamber, studied by SNECMA (Bayle-Labouré 1991), should be mentioned. In this concept, shown in Figure 13.22, the low power head has a large volume and, therefore, a long residence time and receives a relatively small amount of air, allowing the stoichiometry of the primary zone to be optimized for these conditions. The high power head, however, has a small volume and is fed by the air coming from the compressor, creating a sub-stoichiometric primary zone. After leaving it, the flow is diluted by secondary dilution holes and by the air coming from the low power head. Experiments have shown that a 30% reduction in $NO_x$ emission is thereby obtained.

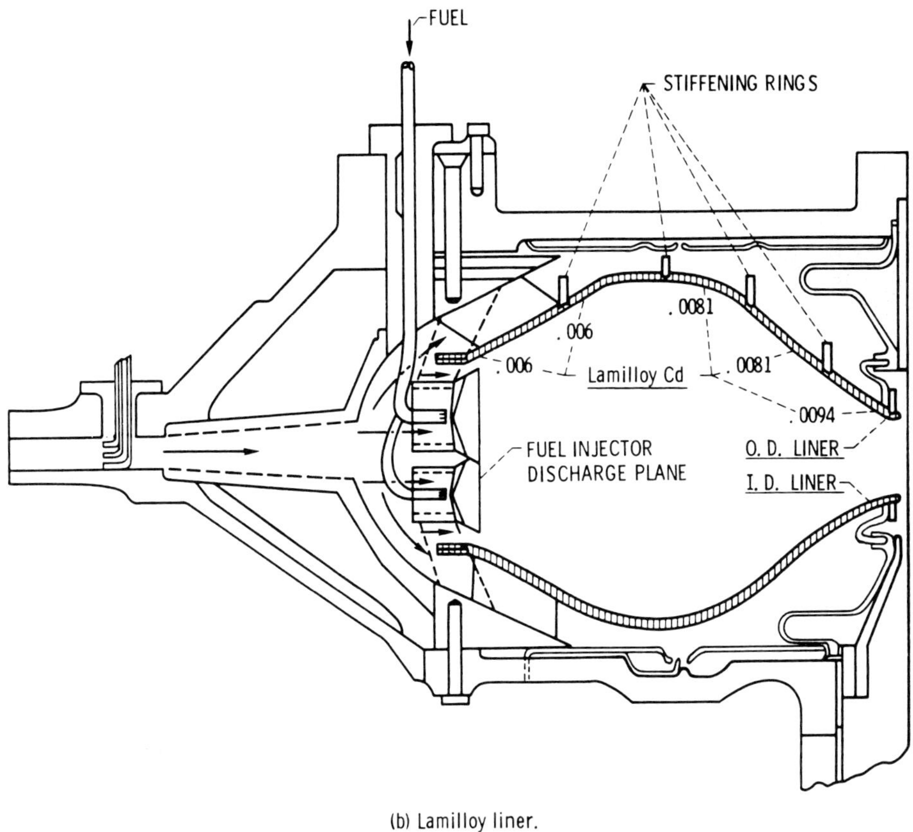



(b) Lamilloy liner.

**Figure 13.23**   Cross-section of test section (Essman et al. 1983). *Copyright © AIAA 1983. Used with permission.*

## 13.5 Combustion Chamber Cooling

In Section 13.3 various types of combustor cooling were discussed. It was shown that counter-parallel finwall cooling had many advantages over conventional film and impingement cooling techniques. Another advanced method employs a semitranspiration cooling material. This will be described next to complete the treatment of cooling techniques.

### 13.5.1 Lamilloy Liner for a High-pressure, High Temperature Combustor

Lamilloy, a semitranspiration cooling material developed by General Motors, was used for a combustor liner (Wear et al. 1978) and later tested in an Allison TF41 engine (Essman et al. 1983). Lamilloy is a multilaminate porous structure, fabricated from diffusion bounded, photo-etched metal sheets. In a typical geometric arrangement for a wall structure, air enters from one side through regularly spaced holes and flows through small etched passages to holes in the next layer of metal, continuing the process until it goes out through regularly spaced holes on the flame side of the liner. This liner is designed to provide higher durability than a conventional step louver liner; Figure 13.23 shows a cross-section of the high pressure combustor test section with the Lamilloy liner.

The material selected for the liner was Haynes 188, which is preferred to Hastelloy X because it has better high temperature strength and is preferred to TD-Nickel-Chrome because fabrication and welding are easier. The liner is built as a series of Lamilloy sheets welded together; the ring portions were hydroformed to the proper shape and then welded to form the complete unit. A photograph of the liner is shown in Figure 13.24. The design requirements are presented in Table 13.5.

Initial experiments were conducted with a temperature-indicating paint; later, thermocouples were employed for temperature measurements. The tests showed that the lamilloy liners performed according to expectations.

Lamilloy was also used in the dome and the barrel of the combustor of the TF41 64.5 kN turbofan engine (see Figure 13.25), a can-annular design with ten combustion liners. An accelerated mission test, simulating approximately 1,000 hours of flight service, including flight, start, and ground test cycles, validated the durability of the Lamilloy, showing that it reduces significantly the cooling in comparison to a combustor using conventional materials. A pseudo-transpiration material similar to Lamilloy was developed by

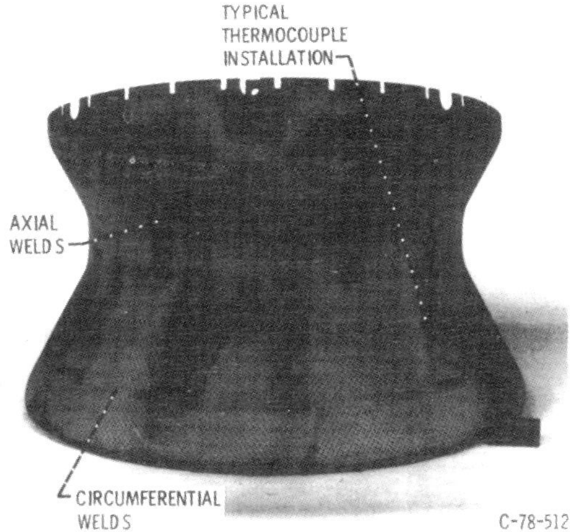


(a) Inner liner

(b) Outer liner

**Figure 13.24**  Photograph of liner (Wear et al. 1978). *Courtesy of NASA.*

Rolls-Royce, which gave it the name Transply (Bangher et al. 1984). The need for it arose from the unsatisfactory performance of their Spey Mk 555 engine in the Fokker F28. They built a new combustor using this material and succeeded in reducing the metal temperature as well as the pollutant levels.

## 13.6 Combustion Diagnostics

The most important measurements in a combustor are those of temperature, pressure, flow velocity, and product composition. One must distinguish between probe methods and nonintrusive techniques. The former have the advantage that they can provide values at exact locations, but they always interfere with the process. For both types it is necessary to provide access to the interior of the combustor.

For photography, cinematography, and other nonintru-

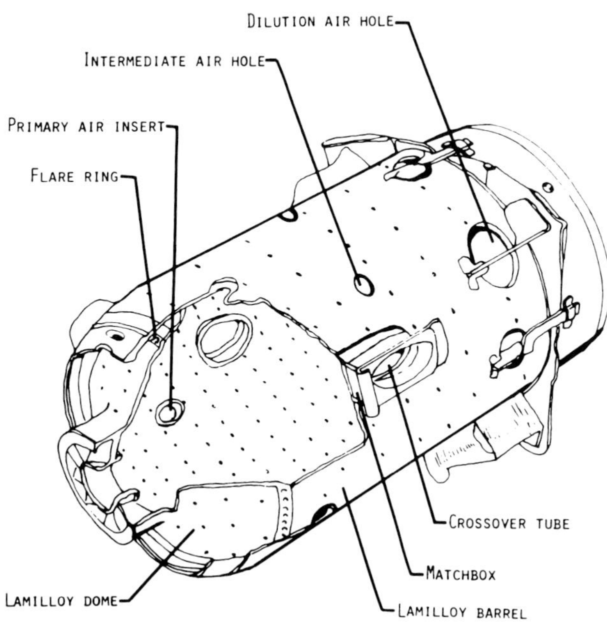


**Figure 13.25**  TF41 combustor configuration (Essman et al. 1983). *Copyright © AIAA 1983. Used with permission.*

sive optical methods, windows in the wall (generally made of quartz) are necessary.

Pressure can be monitored by mechanical devices, mostly slow, or by faster strain gages, which give an accuracy of about $\pm 0.5\%$. For temperature measurements Pt-Pt/Rh is used; employing 10% or 13% Rh the maximum temperature is 2030 K, which is sufficient for most combustors. One must provide probes sturdy enough to withstand the impact of the gases. The principal sources of error, when using bare thermocouples, are heat losses due to radiation and conduction along the leads, heat gain caused by catalytic reactions

**Table 13.5**  Low emission combustor design concepts (Mularz et al. 1979). *Copyright © AIAA 1979. Used with permission.*

1) Hot-wall concept:
   Refractory coated surfaces
   Minimized wall quenching

2) Recuperative cooling concept:
   Preheated primary air
   Increased combustion reaction rates

3) Catalytic converter concept
   Precombustion and catalytic cleanup
   Rapid residual CO and HC combustion
   Catalyst bed defined and fabricated under subcontract with
       Engelhard M&C Corp.

4) All concepts:
   Impingement-cooled liners, no film cooling
   Air-blast fuel injectors
   Near stoichiometric primary zone equivalence ratio
   Dilution air admitted far down combustor length
   Common dome assembly
   Common aft dilution-transition assembly

on the thermocouple surface and thermal inertia. This can be compensated by suitable electrical circuits and signal processing (Shephard and Warshawsky 1952; Lockwood and Odidi 1975). Nonintrusive methods include the line reversal technique which is reliable and accurate if the hot gases are in equilibrium (Gaydon and Wolfhard 1979); they estimate that at temperatures below 1800 K an overall accuracy of 10°C can be achieved. Rayligh thermometry has also been used successfully (Dibble and Hollenbach 1981). Laser Induced Fluorescence (LIF) is another method which gives good results (Logan 1987); this technique can also provide measurements of species concentration (Cattolica and Vosen 1986). Flow velocity can be obtained by LDV (Levy and Lockwood 1981) or by L2F techniques (Schoedl 1980).

Concentration measurements can be performed by probe sampling followed by different analytical techniques. Fristrom and Westenberg (1965) describe this method in detail. The sample is withdrawn, quenched and analysed; the disturbance to the flame can be minimized by using a very fine microprobe. After withdrawing the sample various analytical techniques, such as mass spectroscopy, gas chromatography, and absorption spectroscopy, can be employed. They discuss in some detail the interpretation of a measurement, showing the effects of the disturbance on the aerodynamic flow, of concentration gradients, and heat losses, as well as the influence of possible catalytic effects; they also deal with the proper way of quenching chemical reactions.

Nonintrusive techniques include spectroscopy, discussed in detail by Gaydon (1974), the above mentioned LIF technique and another modern method, photodeflection spectroscopy, which allows simultaneous measurement of concentration and temperature (Rose and Gupta 1984).

# Chapter 14

# *Noise Problems*

In this chapter we shall treat first the fundamental properties of noise in aeronautics, giving the appropriate definitions and relations. Then we shall treat in more detail aeroengines, describe noise regulations and the means to comply with them using attenuation, inverted velocity profiles, and other techniques.

## *14.1 Fundamentals of Aeronautical Noise*

The problems of noise and acoustic fatigue in aeronautics are dealt with in depth by Richards and Meads in their book (1968); here the pertinent definitions and results will be recalled.

The most efficient means of sound generation is by forcing the mass within a fixed region to fluctuate; this corresponds to an acoustic pole (sometimes called a monopole) or source. It can be visualized like a small balloon, undergoing successive inflation and deflation, so that the mass of fluid within a fixed region surrounding the balloon fluctuates; mathematically, $\int \rho \, dV$ fluctuates, where $\rho$ is the density and $V$ the volume. Another, less efficient, method is when the mass flux of the fluid across the boundary varies; that is, $\int \rho \, v_i \, dV$ varies (with $v_i$ the velocity), although the mass of fluid contained in a fixed region is constant. One can visualize the mechanism as a positive source, where the mass enters the region, and a sink where it leaves; together they constitute an acoustic dipole. This mechanism is relevant when solid bodies are present. If we consider a fluctuating fluid flow free of solids, there is no physical mechanism by which the mass can vary in a fixed region. However, the momentum flux can vary. This is the rate at which $\rho v_i$, the momentum in the $x_i$ direction, is convected with velocity $v_j$ in the $x_j$ direction. That is, a change in momentum flux means $\int v_i v_j \, dV$ will change. Since the momentum $\rho v_i$ entering the fixed region must be balanced by momentum leaving the region across a different part of the boundary, the sound generated is caused by two almost cancelling acoustic dipoles, i.e., by an acoustic quadrupole.

As aircraft noise is heard by the population, it is important to assess it from a subjective point of view. The human ear detects not only the difference between sounds of discrete frequencies, but is also sensitive to different qualities of sounds (that is, it has been found that people are more sensitive to high frequency noise than to high frequency pure tones). The main stimuli to the brain, resulting from the response of the human ear, vary approximately as the logarithm of the pressure fluctuations. Therefore the sound pressure level (SPL) is defined as

$$SPL = 20 \log_{10} \overline{p}/p_o$$

where $\overline{p}$ is the r.m.s. pressure fluctuation and $p_o$ is a reference value, usually taken as 0.0002 dynes/cm$^2$. The sound pressure level as defined above is measured in decibels (dB).

One should also note that sound at one frequency may mask a lower amplitude sound at another frequency. In general, low frequency sound is more effective in masking high frequency sound than the other way around. The duration of the signal reaching the brain is also important; for short duration impulses, such as sonic booms, the ear integrates the energy over a short period, thus forming a judgment. It is also important to remember that excessive exposure to intense sound or loud noise can cause damage. The normal ear drum will rupture at levels of about 185 dB, while pain is felt deep in the ear at 130 dB; above levels of 90 dB there can be cumulative and permanent damage to hearing.

As a consequence of the ear's response, one must weigh the noise energy in different frequency ranges. This can be done in several ways—(1) by using an electronically weighted sound level meter which accentuates the energy differently in different ranges, by establishing a series of equal loudness curves, indicated by the reaction of listeners or (2) by a physiological study of the response of the ear, from which suitable response curves are established.

The first of the above methods is used extensively in industry. Figure 14.1 presents three scales—the A scale is most suitable for low sound levels, such as motor car and machinery noise, the B scale is best in the 70 dB range, and the C scale for loud noises. Normal equal loudness contours for pure tones are presented in Figure 14.2. Taking into account

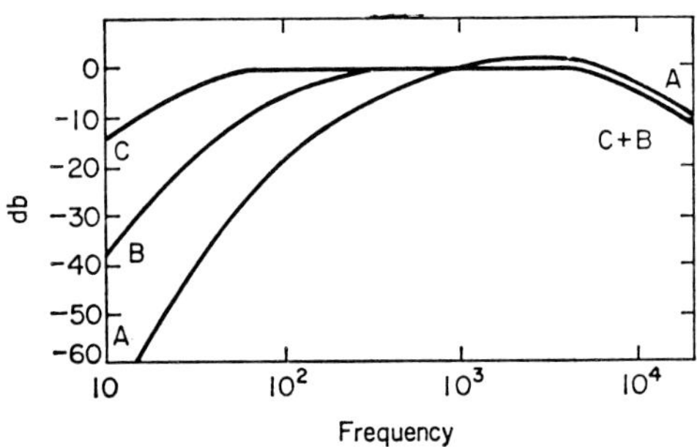

**Figure 14.1** Different noise scales (Richards and Mead 1968). *Reprinted by permission of John Wiley and Sons, Ltd.*

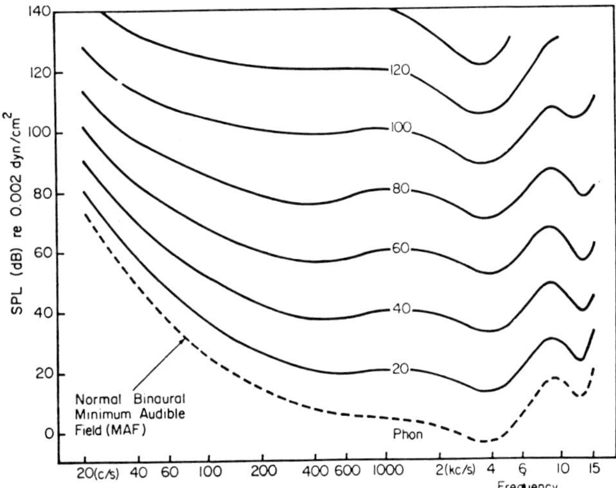


**Figure 14.2** Equal loudness contours for pure tones (Richards and Mead 1968). *Reprinted by permission of John Wiley and Sons, Ltd.*

the factors mentioned above an $N$ scale has been developed for aircraft noise measurements.

Finally, one should note that the dominating sources of jet sound, which is the most important close to airports, are found in a high shear region as shown in Figure 14.3. If a fluid element moves downstream from $A$, where the velocity is high, to $B$, where it is lower because of the gradient $\partial U/\partial y$, the deceleration sets up forces, which radiate sound. They are proportional to the gradient $\partial U/\partial y$ and to the random turbulent velocity in the transverse direction $v$. As a consequence severe noise will arise when a region of high velocity shear is associated with a high turbulence region.

## 14.2 Engine Noise and Its Attenuation

The main sources of noise from aero engines are

a) the propulsion jet, where the turbulent mixing process generates quadrupole type noise, which is mainly influenced by the velocity and is proportional to the eighth power of the jet velocity;

b) the compressor and the turbine, where the interaction of the blades with turbulence and wakes generates dipole type noise, due to the presence of solid boundaries; it is proportional to the sixth power of the relative velocity; and

c) combustion noise, which is monopole in character and is proportional to the fourth power of the velocity.

Often the noise coming from compressor, combustor, and turbine is treated together as engine core noise. Figure 14.3 presents the sources of noise in a turbofan engine. In general, the main disturbance is felt close to the airport and the American FAA has a fixed number of measurement points as shown in Figure 14.4. These are the approach points, under the aircraft flight path, 1 nautical mile (1852 m) before touchdown, the takeoff point, also under the flight path, 3.5 n.m. (6482 m) from brake release and the sideline point on a line parallel to the runaway at a distance of 0.35 n.m. (648.2 m) for a four engine aircraft and 0.25 n.m. (463 m) for an aircraft with less than four engines. The amount of noise allowed is given in FAR part 36 (it has also been adopted by ICAO) and goes from an effective perceived noise level (EPN) of 92 to 110 EPN dB for the takeoff and

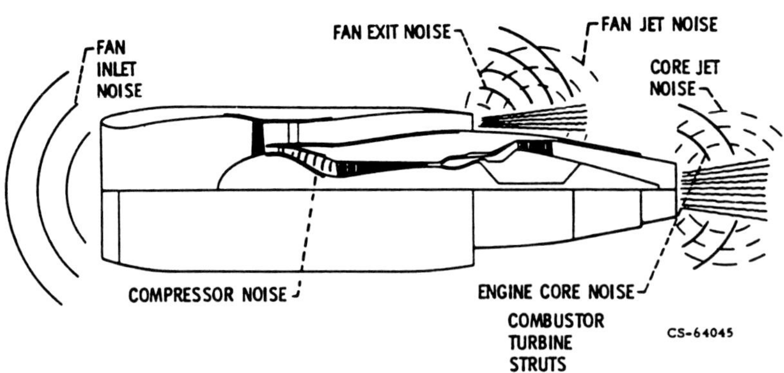


**Figure 14.3** Noise sources in turbofan engines (Feiler et al. 1975). *Courtesy of NASA.*

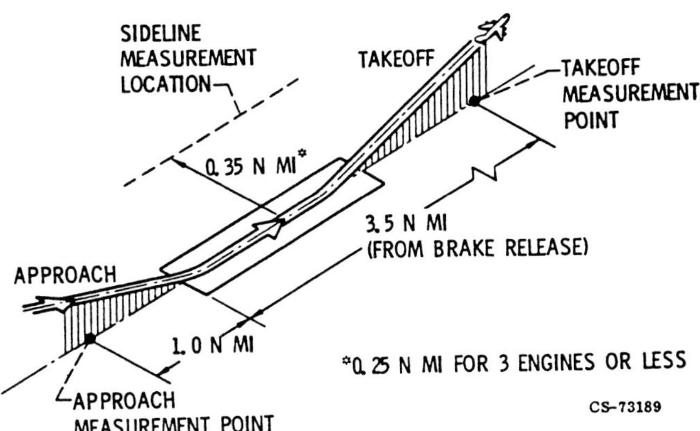



**Figure 14.4** FAA noise measurement locations (Feiler et al. 1975). *Courtesy of NASA.*

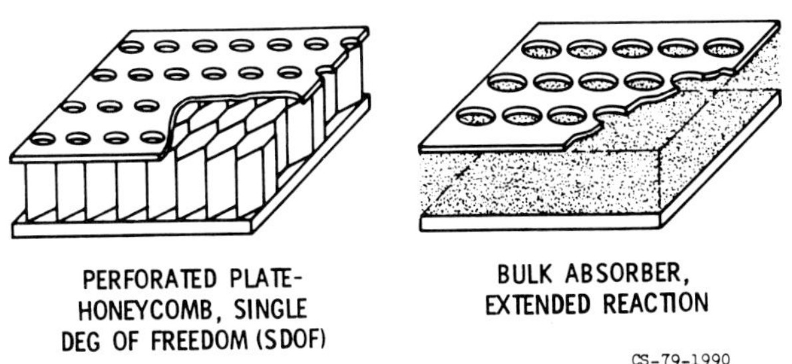



**Figure 14.5** Lining materials and construction (Feiler et al. 1979). *Courtesy of NASA.*

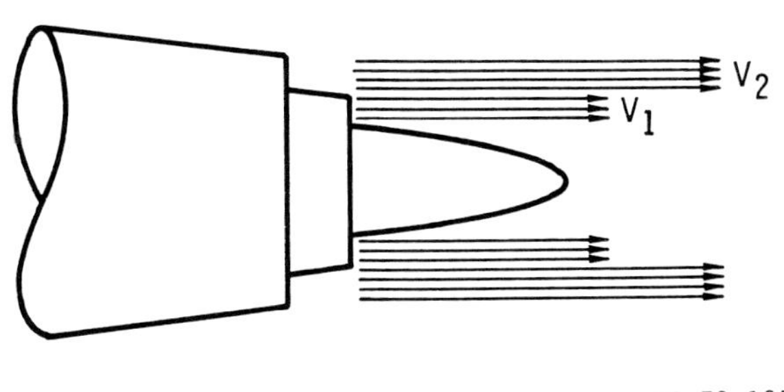



**Figure 14.6** Inverted prolife coannular jets (Feiler et al. 1979). *Courtesy of NASA.*

from 102 to 110 EPN dB for the approach noise, according to the gross takeoff weight.

The fan is a dominant noise source in high bypass ratio turbofan engines; the principal mechanisms producing this noise are

a) rotor-stator interactions, in the form of rotor wakes and vortices, which impinge on the stators and are particularly important at subsonic tip speeds during landing approach;

b) rotor alone noise caused by nonuniformities in shock

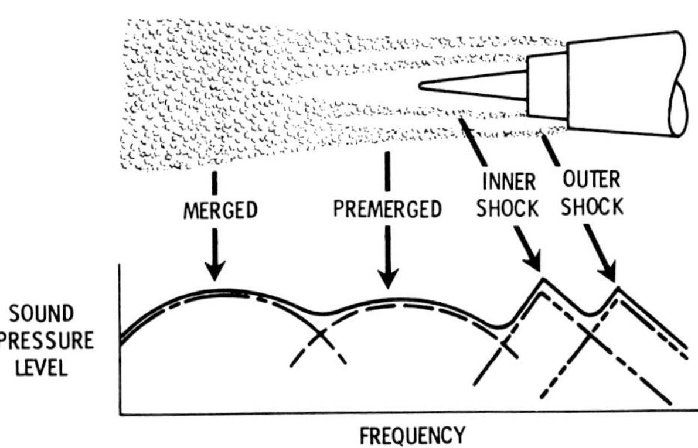



**Figure 14.7** Inverted velocity profile jet noise sources (Feiler et al. 1979). *Courtesy of NASA.*

## ACOUSTIC LINING CONCEPTS

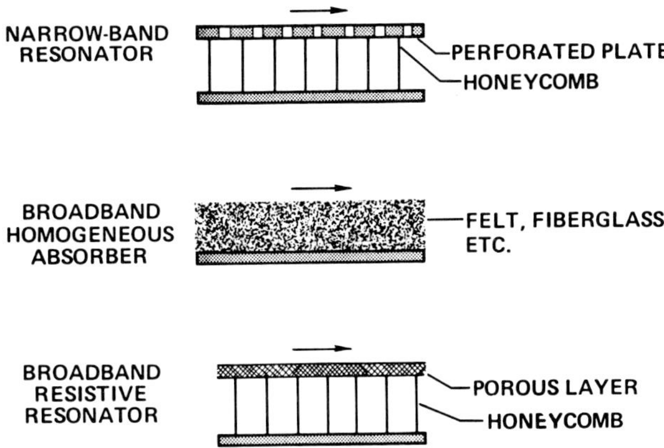



**Figure 14.8** Characteristics of duct lining materials (Mangiarotty et al. 1968). *Courtesy of NASA Langley Research Center, Hampton, VA, U.S.A.*

wave patterns forming at the leading edge at supersonic tip speeds, occurring mainly at takeoff; and

c) rotor interaction with inflow disturbances.

In order to suppress the noise coming from the fan and the high velocity jet two methods are used—introduction of attenuating material or inversion of the velocity profile. The first can be of the perforated plate honeycomb or of the bulk absorber type (see Figure 14.5); the second noise suppression technique is shown schematically in Figure 14.6. Here the high velocity flow is exhausted through an annulus, with the lower velocity flow in the middle. This can be achieved, for example, by crossducting the fan and core streams or by combustion in the fan duct. Figure 14.7 illustrates the complex processes occurring: four noise generating regions must be considered, the low frequency noise generated downstream, the higher frequency premerged mixing noise near the nozzle exit; when either or both streams are supersonic, turbulent eddies passing through shock waves of both inner and outer stream type must also be considered.

## LINING MECHANISMS

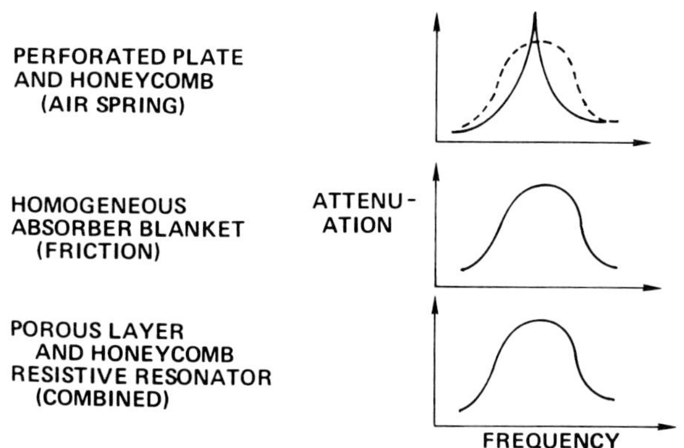



**Figure 14.9**  Attenuation mechanisms (Mangiarotty et al. 1968). *Courtesy of NASA Langley Research Center, Hampton, VA, U.S.A.*

## EFFECT OF DUCT HEIGHT

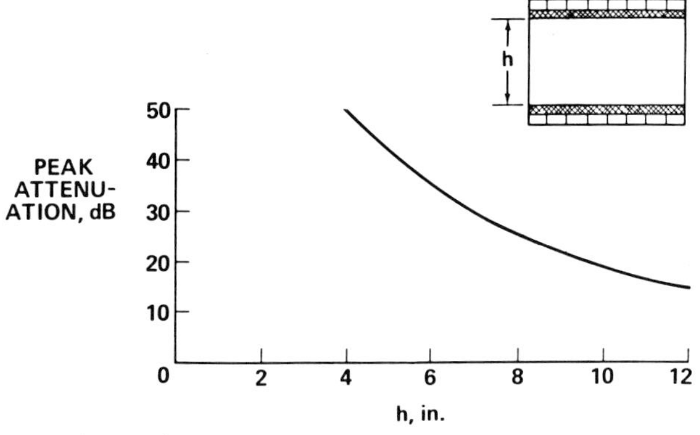



## RELATION BETWEEN LINING DEPTH AND DUCT HEIGHT

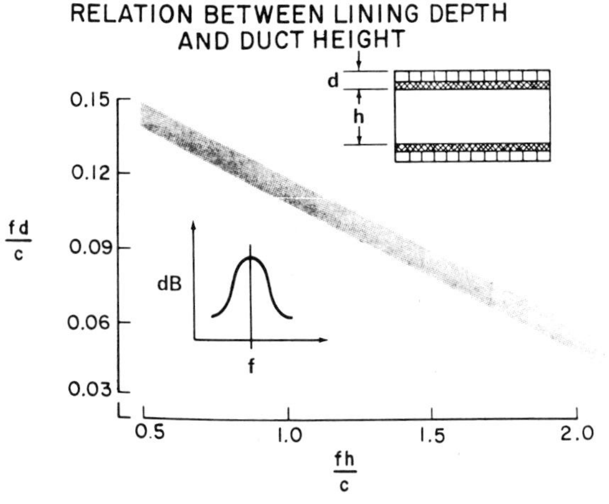



**Figure 14.10**  Effect of duct geometry (Mangiarotty et al. 1968). *Courtesy of NASA Langley Research Center, Hampton, VA, U.S.A.*

## WALLS TREATED
### AREA EFFECT

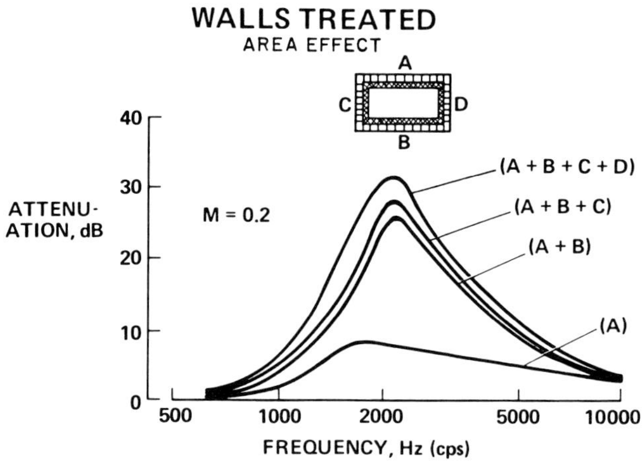



## EFFECT OF TREATMENT LENGTH
### M = 0.28

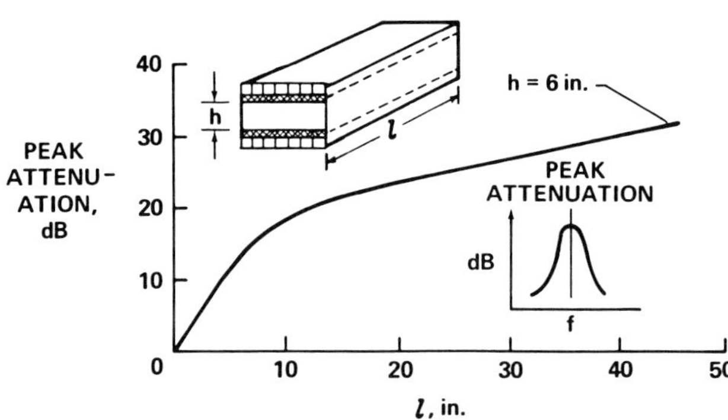



**Figure 14.11**  Effect of area treated and absorber length (Mangiarotty et al. 1968). *Courtesy of NASA Langley Research Center, Hampton, VA, U.S.A.*

Mangiarotty and coworkers (1968) studied the characteristics of various duct lining materials and proposed a number of concepts—the narrow band resonator, the broad band homogeneous absorber (such as felt or fiberglass), and the broad band resistive resonator, shown in Figure 14.8. They succeeded in explaining the mechanism operating in the different cases, as shown in Figure 14.9. The effect of the duct geometry, the area treated, and the length of the absorber are presented in Figures 14.10 and 14.11. Their conclusions were that fiber-metal materials are desirable where contamination is not serious, while resin-coated fiberglass cloth is preferable when contamination is a problem.

A comprehensive computer program, the aircraft noise prediction program (ANOPP), was developed by NASA taking into account the aircraft trajectory and the various noise sources, as shown in Figure 14.12. Its results are in good agreement with experimental data for both subsonic and supersonic aircraft, as shown in Figure 14.13.

Finally, one should note the renewed interest in supersonic jet noise, stemming from the studies for a modern SST. Preisser and coworkers (1990) at Langley utilized the

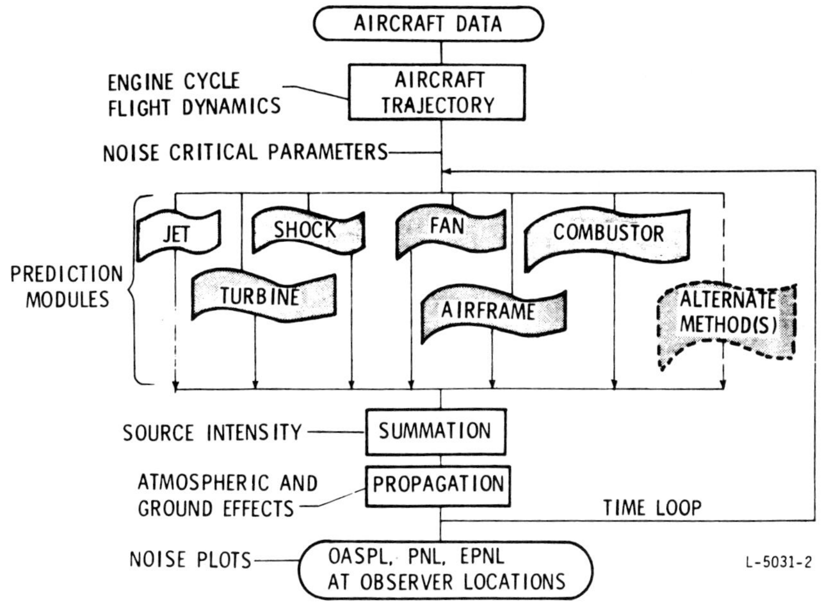


**Figure 14.12** ANOPP CTOL noise prediction (Feiler et al. 1979). *Courtesy of NASA.*

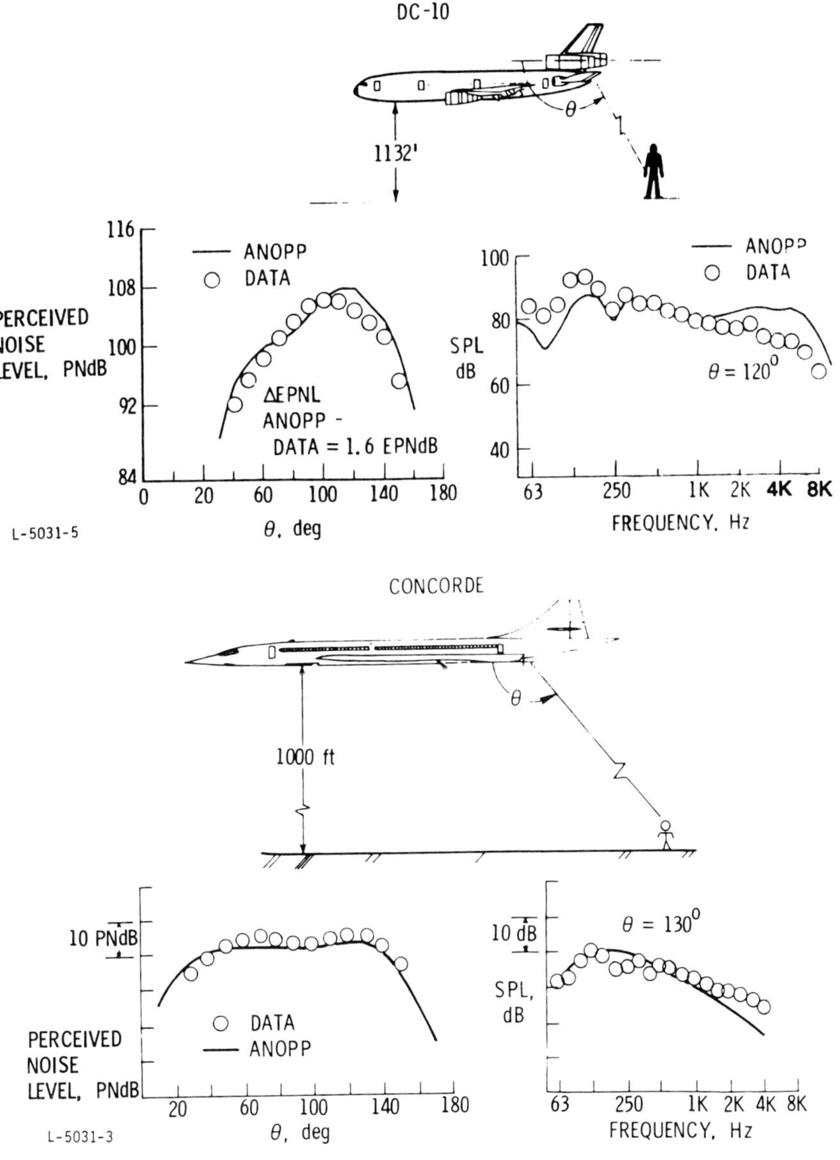


**Figure 14.13** ANOPP flyover noise validation for Concorde and DC10 (Feiler et al. 1979). *Courtesy of NASA.*

ANOPP code to assess the noise for the conceptual AST-205 supersonic aircraft, powered by the proposed GE21/J11-B14 engines (see Section 10.1). They concluded that noise suppression must be directed to all noise source components and recommended additional research avenues. In a 1989 Boeing study which was performed in the framework of the high speed commercial transport program, it was found that the sideline noise limits are the most difficult to meet. Techniques to be used to meet the FAR36 Stage 3 noise requirements, such as oversizing the engines and employing a programmed lapse rate that reduces takeoff thrust after liftoff, are proposed.

# Chapter 15

# *Components and Materials*

This chapter first treats new materials, in general, then components and materials for turbine engines, and finally those required for scramjet and combined propulsion.

Initially, new materials are used on aerospace vehicles in less critical areas until their performance can be confirmed under operational conditions. Such materials then are used in more critical applications and, if warranted, they are used in the most critical components (Ruhmann et al., 1992a, b).

Most commonly used materials for aircraft structural members are conventional cast, wrought, and heat-treated aluminum alloys. Recent developments in aluminum alloys are centered on ingot alloys, in the form of aluminum-lithium for low-density structures, and powder alloys processed via rapid solidification and mechanical alloying for complex forms with high integrity. Powder systems offer the potential of improved strength, toughness, corrosion resistance, and high temperature tolerance.

New aluminum alloys incorporating iron or other alloying additions have been identified as being attractive for certain high-temperature (higher than conventional aluminum alloys) aerospace applications. These alloys, processed by rapid solidification, demonstrate higher temperature tolerance than conventional aluminum alloys and, along with the new aluminum-lithium alloys, could replace some of the titanium alloys in applications where weight reduction is a critical design requirement.

Advanced composite materials are used widely in aerospace structures. Composites are hybrid materials, consisting of reinforcing fibers embedded in a matrix. The high-performance resin matrix composites used in today's airframes are composed primarily of graphite and, to a much lesser extent, aramid fiber-reinforced plastics. Boron filament, carbon fiber, aramid, and glass filament are incorporated to provide additional strength and stiffness. Thermosetting epoxy is the matrix material most commonly used for high-performance applications. Matrix resins with improved toughness are very desirable for hostile operating environments.

Toughness can be improved by incorporating microscopic dispersions of elastomeric polymers as a discrete phase. This might, however, reduce tolerance to elevated temperatures. Thermoplastics offer improved performance in applications where toughness is a key requirement. Unlike thermosets, thermoplastic materials can be consolidated when exposed to initial manufacturing temperatures. Consequently, they can be reformed from bulk material into final shapes. Another advantage is that thermoplastics do not require an autoclave; therefore, cycle times are comparatively short.

Thermoplastics are well-suited to forming operations such as injection molding and stamping. Polyetheretherketone (PEEK) is one thermoplastic material which has begun to see broader use in aerospace applications. Composites based on this matrix display a high degree of toughness, as well as stiffness and solvent resistance. However, thermoplastics generally exhibit lower thermal strength and corrosion resistance than thermosets. Current thinking is that thermoplastics with crystalline domains and the capability of cross-linking in the amorphous regions will minimize some of the shortcomings of conventional thermoplastics.

Metal matrix composites (MMC) consist of reinforcement fibers or ribbons coupled with a metal matrix. The reinforcement typically exhibits high strength and stiffness, with little or no ductility, whereas the metal matrix has significantly lower strength but much higher ductility.

MMCs are also attractive candidates for high-temperature components in hypersonic aircraft, including skin panels and other hot sections of the airframe. Recent developments in materials for hot structures suggest that glass and glass-ceramic fiber composites may be viable candidates to replace superalloys. Although they exhibit the desired levels of strength, stiffness, and toughness, the softening temperature of glass matrices is relatively low. However, researchers are confident that the use of more refractory glasses or recrystallized glasses will make it feasible to develop glass fiber composites with temperature tolerances suitable for hot structure applications.

Aggressive development of high-strength, temperature-tolerant materials is also the key to the enhancement of next-generation aerospace propulsion systems. Many of the forecasted improvements in engine performance will evolve from the increased use of composite and composite-reinforced structures, structural ceramics and ceramic composites, and ceramic coatings on turbine airfoils. This will allow tomorrow's aircraft engines to operate at higher rotor speeds, temperatures, and operating pressure ratios. This will result in lighter-weight, fuel-efficient commercial aircraft engines and dramatically improved thrust-to-weight ratios for military aircraft engines.

Composites may also be in the form of laminates which can be designed to have higher specific strengths, improved fatigue resistance, and lower density than those obtainable with current monolithic aluminum sheet material.

MMCs can be classified as continuous or discontinuous, depending on whether the reinforcing fibers are continuous, short fibers, or particles relative to the dimensions of the composite components. Continuous MMCs are often, but not always, in sheet form, and they contain reinforcement in only one direction. These sheets can be layed up in different orientations and bonded with matrix foils to form thicker sheets with more isotropic properties. Discontinuous MMCs incorporate short fibers, whiskers, platelets, or particulates in a metal matrix, consolidated via powder or solidification processing techniques.

Research in continuous MMCs will provide an increased capability for tailoring material properties (such as higher temperature tolerance) and improving fabrication techniques that yield higher-quality, less expensive parts. Significant work is also being done in discontinuous metal matrix composites, and researchers are confident that combinations of high strength, stiffness, and modulus of elasticity, with acceptable levels of ductility can be achieved.

Carbon/carbon composites will be used in high-temperature applications where high strength and toughness are also required; however, there are several barriers which must be overcome before they are more widely used. Current research is aimed at improving their oxidation resistance at higher temperatures, increasing their operating temperatures, extending their life, and reducing their processing costs.

Intensive research in advanced composites over the past decade has produced high-quality starting materials (prepregs). Mathematical models are evolving which will assist in material and process selection. For example, it may be possible to monitor the dielectric and acoustic properties of parts being cured in an autoclave in order to define the chemical and physical state of the material as it is being processed. The twofold objective of this computerized process control technique is to achieve reproductibility and to reduce processing costs.

Processing is a critical factor in determining the final properties of any of the new materials being considered for use in aerospace structures. For such materials, processing technology and controls have not yet reached the point where the final material structure and properties can be reproduced on a consistent basis. In some cases, characterization of the product, in terms of microstructure and final chemistry, remains a difficult challenge. For instance, in monolithic inorganic materials, identification and location of major and minor phases is necessary; however, it is often an ardeous and time-consuming task.

For metallic materials, trends in the industry are toward using methods such as isothermal forming, hot isostatic pressing, metal injection molding, rapid solidification rate processing or superplastic forming to achieve complex, neat-net shape parts. This will minimize the machining, joining, and other processes that tend to consume excessive energy and produce waste while, at the same time, producing components with superior mechanical properties. The design of such processes often relies on computer modeling techniques that incorporate appropriate descriptions of material behavior, interactions between tooling and the workpiece material, and equipment characteristics. Through these efforts, build-and-try methods of process design can be avoided, resulting in reduced cost and developmental lead time.

## 15.1 Turbine Engines

The key elements for an efficient engine are the core compressor and the high pressure turbine; they have a major effect on engine performance, fuel consumption and maintenance costs. Increasing cycle efficiency requires total pressure ratios of 40 and more with turbine entry temperatures up to 2000 K. For the core compressor the trend is to use a smaller number of stages, while improving the aerodynamic blading design technology, reducing wear and erosion. In modern compressors variable angles are used in the guide vanes and in the initial stator stages.

In the high pressure turbine high temperatures are attained by using advanced materials and modern fabrication techniques, such as single crystal and directionally solidified blades as well as improving the cooling, so that the blades can withstand temperatures that are above the melting point of the material. The future introduction of ceramic components will allow further advances. Different methods of cooling or insulating blade walls are shown schematically in Figure 15.1. The simplest way is to use convective cooling with augmented surface area. Better results can be obtained with impingement cooling, where air from one or more rows of small holes coming from an insert in the blade impacts on the wall, reducing its temperature; adding local film cooling gives additional temperature control. Full coverage film cooling, utilizing a multitude of small holes, is even more efficient. Another advanced method which reduces the amount of air required is transpiration cooling, while the use of thermal barrier coatings on cooled turbine

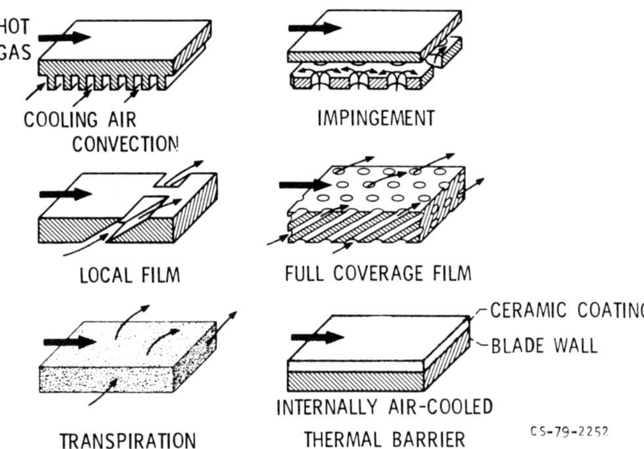



**Figure 15.1**  Turbine blade wall cooling techniques (Hauser et al. 1979). *Courtesy of NASA. Copyright © AIAA 1979. Used with permission.*

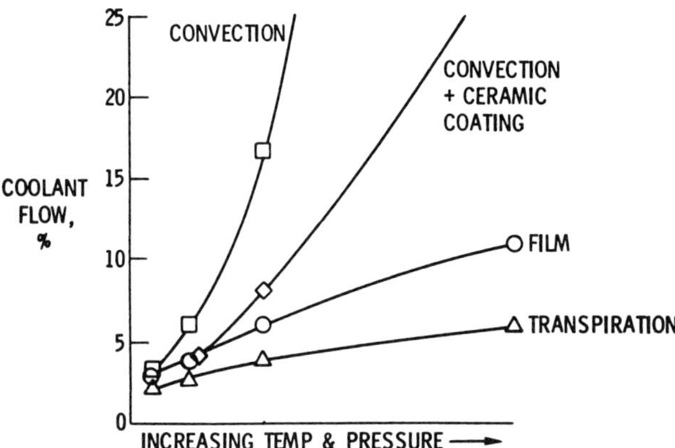



**Figure 15.2**  Comparison of cooling methods (Hauser et al. 1979). *Courtesy of NASA. Copyright © AIAA 1979. Used with permission.*

blades offers an additional possibility of diminishing the amount of cooling air.

Experiments carried out with different cooling techniques (Esgar et al. 1970, Houser et al. 1979) determined how the various methods could be improved and pointed out their limitations. Figure 15.2 compares the effectiveness of the different methods. A recent paper by Haas and co-workers (1988) dealing with film cooling of turbine blades by injection from a row of full coverage holes demonstrates that this simple method is still under consideration with the appropriate geometry and velocity distribution.

Another component that has considerable influence on performance, particularly for large, civil aviation engines, is the fan. The fan blade weight can be reduced by introducing advanced materials, such as modern composites (Stoltze and Graff 1981); these can be employed also in the discs, shafts, bearings, and engine mounts with a cascading effect.

Vogelsang and Gunnink (1985) evaluate the following three types of materials, which show promise for applica-

tion in aircraft structures: aluminum alloys, composites, and aramid aluminum laminates. Aluminum-lithium and aluminum-boron give a sizable weight advantage and are increasingly being used. The aramid aluminum laminates combine high strength, low density, a high modulus of elasticity, good corrosion resistance and fatigue properties. Their main weakness is low fracture toughness and, therefore, their use should be mainly in secondary structures. Many composites such as graphite-epoxy, graphite-polymide, composite aluminum, and boron-epoxy have been used for blades.

Smith (1990) presents a projection of advanced materials that could be used in the year 2015 in an ultrahigh bypass engine; accordingly, the new materials should be available by the year 2007. Figure 15.3 features polymatrix composites (PMC) with a 650 K capability for the fan blades and frame, the nacelle and the thrust reverser, ceramic matrix composites (CMC) for the combustor and the HP turbine vanes and blades, with intermetallic and metal matrix composites for high temperature discs and single crystal nickel aluminides for LP turbine blades and vanes. A comparison of the density and temperature of advanced technology engine materials with those currently in use is presented in Figure 15.4. The author sounds a note of caution about the high price of the proposed materials and tries to determine when their usage is justified.

Picard (1988) has a similar classification of modern materials: composites with high strength and intermediate modulus carbon fibers, aluminum-lithium alloys, and improved titanium alloys, describing also their applications in civil and military aircraft.

High performance composites are lighter than metals and may have better characteristics. Their remarkable qualities are based on the mechanical properties of fibers (carbon, boron, or organic), in particular, their outstanding tensile strength and stiffness when embedded in a metallic or organic matrix. The role of the matrix is to bond the fibers, keep them aligned, and transfer the loads applied to the different elements. The resins selected for the matrix are mainly of the epoxy type, with temperature limits corresponding to the requirements of aircraft speeds up to Mach 2 or slightly above. Industrial fibers available include boron fibers, high strength carbon fibers, and high and intermediate modulus carbon fibers.

The unit layer of a composite is unidirectional, so its strength in the fiber direction is of the order of 1500 MPa, but in the perpendicular direction it may be twenty times less. Therefore, crossed multi-plies or layers are used to ensure the required strength in different directions. One of the main advantages of composites is their outstanding fatigue strength, so that a weight saving of 20% to 30% can be achieved. On the other hand, epoxy resins suffer from aging effects and this must be taken into account. Early successful uses of these materials in French aircraft include a boron

## Advanced Materials Application GD 20/100

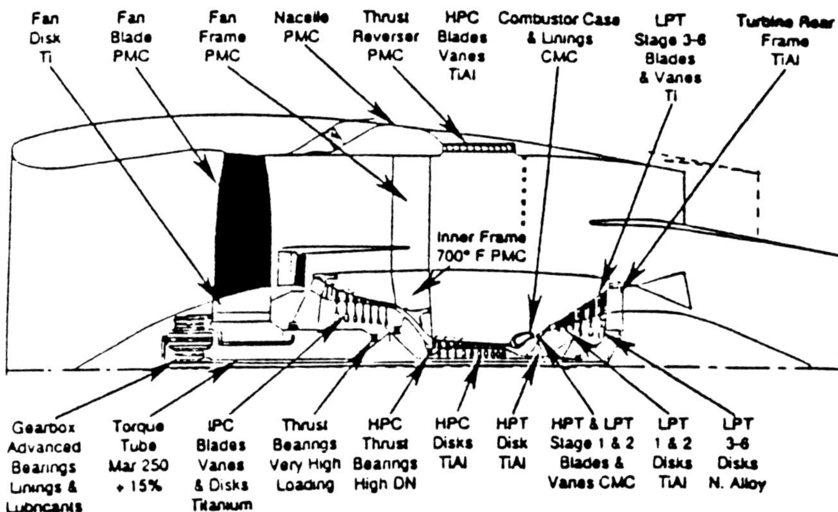


**Figure 15.3**   Advanced material for 2015 airplane: polymatrix composites (Smith 1990). *Copyright © AIAA 1990. Used with permission.*

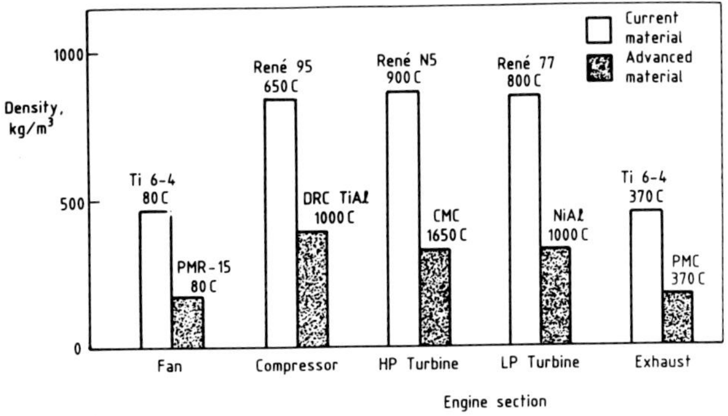

**Figure 15.4**   Comparison of materials (Smith 1990). *Copyright © AIAA 1990. Used with permission.*

composite horizontal stabilizer for the Mirage F1, a carbon-epoxy rudder for the Mirage III, and carbon-epoxy ailerons in the Falcon 50. In the Mirage 2000 a number of parts, such as elevons, various doors, and panels, for a total of 342 Kg, have been manufactured in carbon epoxy or hybrid boron-carbon-epoxy components yielding a weight reduction of 82 Kg.

In the Rafale program, which started with a demonstrator for the new generation of fighters, it was decided to introduce composites in primary structures, such as wings, fins, canard, elevons, and fuselage skins. Figure 15.5 shows the materials used; these include high-strength epoxy, intermediate modulus composites, aramid, aluminum-lithium, titanium, and aluminum. As a result 24% of the structural weight is composite laminate. There are also advantages in using aluminum-lithium alloys which could provide a 10–15% weight saving for various aircraft structural parts. Finally

Picard treats titanium alloys and describes two manufacturing methods: diffusion bonding and superplastic forming, which can be combined together. This technique has been used to manufacture the leading edges of the Rafale wing.

## 15.2 Hypersonic Engines

Ronald (1990) describes the materials used in the NASP program. Early in the development the five prime contractors took each the leading role in the development of a class of materials. General Dynamics heads the refractory composites area, involving carbon-carbon composites and ceramic-matrix composites. Rockwell leads the titanium-aluminide development and scale-up effort, based on the $Ti_3Al$ and TiAl classes of materials. McDonnell Douglas manages the effort on titanium metal-matrix composites, consisting of fiber-reinforced titanium alloys and $Ti_3Al$ intermetallics. Rocketdyne is responsible for high conductivity materials, comprising copper-matrix composites and beryllium alloys. Pratt and Whitney has undertaken the high creep strength materials activity, involving materials and structures intended for hot, actively cooled engine components. This latter area centers on monolithic and reinforced TiAl, using potentially compatible fibers such as titanium diboride.

Almost all of these materials are potentially important for hot, load-bearing structures in both the airframe as well as the engines of the vehicle. On the airframe they would be used as lightweight skin panels of honeycomb-core, truss-core, or integrally stiffened thin sheet configuration. Where necessary they could be cooled with hydrogen fuel by incorporating appropriate coolant passages into the structures. In the engines they would be used as the sidewall

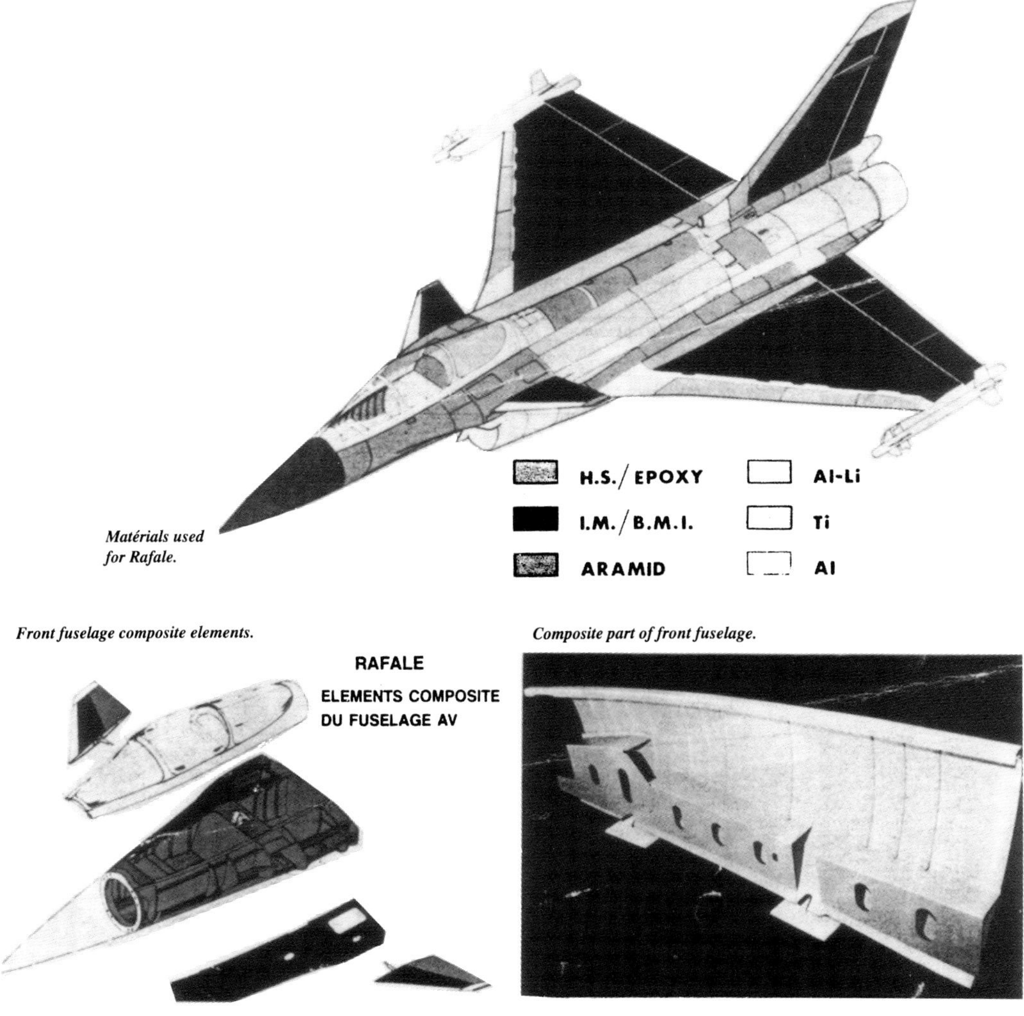


**Figure 15.5** Materials for Rafale primary structure (Picard 1988). *Courtesy L'Aéronautique et L'Astronautique.*

panels in the hot gas path of the ramjet/scramjet. The engine application represents a particularly severe thermal, acoustic, and mechanical loading environment. The structures will have to be actively cooled; the materials will be in contact with hot hydrogen from the fuel, hot oxygen from the incoming air, and the gaseous products of combustion.

Titanium-aluminide intermetallics are candidates for structures in both the engines and the airframe. They have essentially the same density as titanium but lead to the possibility of much higher use temperatures. The two intermetallic systems that are of primary interest are based on the $Ti_3Al$ and $TiAl$ compositions and have potential temperature capabilities of about 815°C and 980°C respectively. The principal drawback of the aluminides is their limited ductility and toughness properties at temperatures less than a few hundred degrees. Coupled with the requirement for very high fabrication temperatures, this makes their processing difficult. New processing methods are used to modify aluminide compositions and microstructures to yield structurally useful materials. The goal is to achieve an appropri-

ate balance of strength with toughness or ductility while retaining the low density and high temperature characteristics that make the aluminides so attractive. One successful approach employs rapid-solidification powder methods to develop improved alloys, and much of the NASP-related work in this area uses a rotary atomization process to produce rapidly solidified powder. The powder is consolidated into fully dense billets that are then processed into appropriate product forms. Advanced thermal-mechanical processing methods have been developed for both the powder-produced intermetallics and those made using conventional cast and wrought methods. These processes have improved the characteristics of the $Ti_3Al$-based alloys significantly and good quality sheet products have been produced.

Metal-matrix composites based on titanium-aluminides offer significant improvements in stiffness and strength over their monolithic counterparts, making them attractive for the thin-gauge skin structures required for NASP. The basic technical challenge in making these composites is to incorporate reinforcement fibers into the matrix without creating adverse reactions in the fiber/matrix interface. A rapid-solidification plasma-deposition process is used to fabricate titanium-aluminide composites. The matrix material starts as a powder that is fed through a plasma arc to convert it into molten droplets. These are deposited onto reinforcing fibers that are spiral-wrapped on a large diameter drum, where on impact they are rapidly quenched to a solid state. Rotation and translation of the drum allows the build-up of a layer of matrix material on and between the fibers. This solidified deposit of matrix material, containing a single layer of fibers, can be subsequently slit and stripped off the drum and several of these layers can be stacked together and hot pressed to make a multilayer composite. The process has been demonstrated successfully with SiC reinforcements in $Ti_3Al$ matrix materials, obtaining good mechanical properties.

XD composites refer to a process that can be applied to a variety of materials to form fine, close-spaced distributions of reinforcing second-phase particles. The term XD refers to the proprietary technique that is used to create the reinforcements. In essence, it is a process that results in the fabrication of discontinuously reinforced metal matrix composites where the matrix can be any one of a number of alloys and the reinforcing particles can be varied in terms of composition, size, shape and distribution. A unique feature of the process is that the reinforcements are formed and grown in situ within the matrix, as distinct from being mechanically mixed as a separate additive; as a result, the particle/matrix interfaces are clean and well bonded, thereby enhancing the effectiveness of the reinforcements. The process can be tailored to produce a variety of second phase particle distributions, where the particle shape can vary from spherical to needle-like. Mixtures of different reinforcements also can be formed that include coexisting sizes, shapes and types of particles. The XD process has been used to make titanium aluminide composites where the reinforcing phase is titanium diboride. The microstructures resulting from the XD process are attractive principally because they lead to significantly improved strength levels over the useful temperature ranges of the aluminides. The properties are also essentially isotropic, making the materials useful for complex-shaped structures that would be difficult to fabricate from continuous fiber reinforced materials.

Carbon-carbon composites have the potential for use as lightweight structures that could be exposed to temperatures in excess of 1400°C without the need for active cooling. Because of this capability, they are candidates for the airframe, where they could be used as large, integrally stiffened skin panels on the hotter parts of the vehicle. They may be useful also for engine structures. Carbon-carbon composites can be regarded as mature materials. There is a large base of knowledge available regarding their fabrication and practical use, and they have been used in a variety of applications. Several companies specialize in the manufacturing of components, and there are several basic methods available for making structural shapes.

Ceramic-matrix composites have the potential for use at temperatures in excess of 1300°C, with the added advantage of a much higher degree of inherent oxidation resistance. Unlike the carbon-carbon materials, they are not as mature as a class of structural materials and do not enjoy the same broad base of manufacturing experience. There are two general classes of ceramic-matrix materials that may be important for NASP: glass-ceramic-matrix composites, useful up to temperatures of about 815°C, and advanced ceramic-matrix composites, potentially applicable at much higher temperatures. Glass-ceramic-matrix composites are relatively well characterized and can be fabricated into product forms such as honeycomb-core panels, truss-core panels, and other complex shapes. Advanced ceramic-matrix materials—such as silicon carbide fiber reinforced silicon carbide (SiC/SiC) —are not as mature, principally because of the lack of a fiber that has sufficient stability in the matrix when exposed for long times above 1000°C. Improvements are being made in these materials, and large, complex-shaped demonstration components have been manufactured by specialist companies. A particular interest in the ceramic-matrix materials for NASP stems from their inherent resistance to hot hydrogen and they would be useful for actively cooled engine components.

Beryllium based alloys are of interest since Beryllium is a commercially available material that possesses the advantages of low density, high elastic modulus, and very good thermal conductivity. Its disadvantages include poor toughness characteristics, crystallographic-texture-sensitive properties, a limited use temperature of about 540°C, and environmental concerns associated with the toxic nature of the oxide. In spite of its problems, beryllium is successfully used in a wide variety of applications, and there is a considerable body of experience concerning its fabrication and handling. In addition to the benefits of light weight, its thermal conductivity characteristics make it particularly useful for structural components that are designed to transfer heat efficiently from one location to another. For NASP, the beryllium would be used primarily in heat exchangers or actively cooled engine panels, where it would be in sheet form in honeycomb-core or truss-core panel structures, that would contain integral cooling passages within the structure.

Since NASP will make extensive use of actively cooled structure, there is a particular interest in high thermal conductivity materials, including copper-matrix composites. Copper itself has a good thermal conductivity but is heavy and its upper use temperature is limited by its low mechanical properties. Pitch-based, high modulus graphite fibers have excellent thermal conductivity—better than the copper itself in the direction of the fiber—and the addition of

**Figure 15.6**  Power turbine based on ceramic matrix composites (Boury et al. 1990). *Courtesy of Advisory Group for Aerospace Research & Development, NATO.*

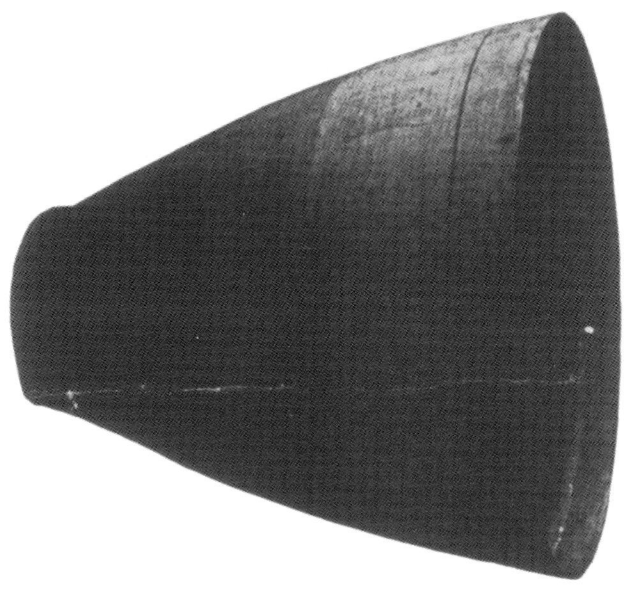

**Figure 15.7**  Ariane third stage advanced material nozzle (Sepcarbinox Novoltex) (Boury et al. 1990). *Courtesy of Advisory Group for Aerospace Research & Development, NATO.*

these fibers to copper reduces density, increases stiffness, raises the use temperature, and significantly improves thermal conductivity in the direction of the fibers. One approach to the fabrication of these composites starts with a process that places a layer of copper around each fiber in a graphite fiber tow. The coated fibers are subsequently packed together and hot pressed into a fully dense material containing a high volume percent of fibers. In practice, crossplied lay-ups will be used to tailor the thermal conductivity and compensate for the directional effects of the fiber.

Boury and coworkers (1990) in France have reviewed materials and systems for combined cycle propulsion. Their evaluation of materials is similar to that of Ronald. As regards the various components, they propose a metallic structure with thermal protection for the air inlets. Two technologies are considered for the compressor, which must withstand 1625 K: active cooling, similar to that used in the turbine blades of engines, or the use of a composite material

based on a ceramic matrix (Sep Carbinox). This has to be checked for impact, (e.g., of birds). The power turbine could be realized as a monobloc of composite material with a ceramic matrix—as shown in Figure 15.6. This is a small unit; a larger size may present problems. The combustion chamber must be actively cooled, either by air or by hydrogen. Boury and coworkers favor for it a composite material with a glass matrix. Finally the nozzle could be regeneratively cooled by air or by the fuel, or it could be manufactured from a composite with ceramic matrix, like the smaller nozzle for the third stage of Ariane, built by SEP and shown in Figure 15.7.

It is of interest to note that a recent paper by Jacobson (1992), which deals with materials for HSCT combustion, that must withstand similar temperatures (up to 2000 K), but somewhat higher pressures, recommends the same type of compounds, with the addition of reinforcements, such as alumina-fiber reinforced alumina.

# Chapter 16

# *Maintenance and Reliability*

Maintenance and reliability of aeroengines are part of the overall requirements for the aircraft, which are of primary importance in ensuring their airworthiness and will be treated in this context. In a seminar held in 1989 (Aircraft Eng., 61,3,1989) the views of regulatory authorities, manufacturers and operators were presented, giving a complete picture of all aspects of the process. It is necessary to provide a system safety assessment, based on requirements for replacement of parts, inspection, servicing, and system monitoring. The last category is the most recent and requires a reliability program for aircraft systems and components. For this purpose one must measure reliability levels using in-service data and take appropriate action when the measured values approach levels showing this is necessary by certification analysis. In-service data sources, data reduction, and analysis together with a method for comparing in-service results to predictions are required.

Fickeisen (1991) gives a detailed example, treating the extended range operation of two-engine airplanes (ETOPS), applied to the 767 aircraft. His conclusions are that one must compare predicted to actual reliability levels and review individual events that have taken place during operation. The results of the process verify that required reliability levels have been achieved and define needed or desired system improvements. Judicious employment of the continuing airworthiness process can contribute significantly to better commercial aviation reliability and safety.

Cumming (1991) gives an interesting review of the maintenance standards of British Airways, based on 30 years' experience with 12 airframe and 15 engine types. The goal is to ensure maximum passenger safety, together with the best financial returns. For this purpose British Airways strives to ensure high standards of safety, high utilization with consequent high revenues, robustness of operation through dispatch reliability, and structural and system integrity.

Cumming treats in particular new aircraft; nowadays these are designed taking into account maintenance standards. This is achieved by cooperation with the manufac-

turer; an example is shown in Figure 16.1, presenting the maintenance program development for the B747-400. This results in a maintenance planning document, to which are added specific and cost effective company requirements, such as engine health monitoring and hydraulic fuel sampling programs. After deciding on standards of maintenance for the cabin, the document is submitted to the CAA and when approved it becomes the approved aircraft maintenance schedule.

The philosophy adopted by all airplane users nowadays is on condition monitoring and maintenance, for which they have special control programs. Figure 16.2 shows how BA's condition monitored maintenance and reliability programs are structured. They are managed by cross department committees, responsible for reviewing unscheduled work arising during aircraft operation and maintenance, issuing the approved maintenance schedules, ensuring that component overhaul and repairs meet the required airworthiness, reli-

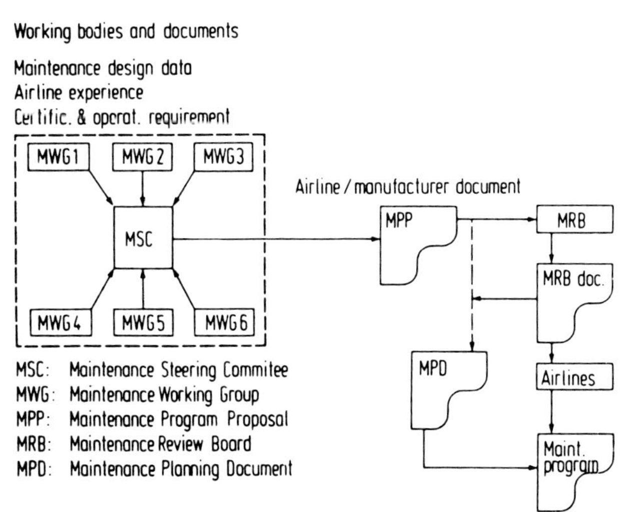


**Figure 16.1** Maintenance program for Boeing 747-400 (Cumming 1991). *Reprinted with permission of Z. Flugwiss. Weltraumforsch.*

FORMAL CONDITION
MONITORING SYSTEM

*STRUCTURE*

M C P Alert Control System

Aircraft Technical Control Committee (ATCC)

Aircraft Performance Review Committee (APRC)

Engine Maintenance Control Committee (EMCC)

Component Reliability and Review Committee (CRRC)

Engineering Control Review Board (ECRB)

CAA Monitoring

*GENERAL AIRCRAFT C. M. SYSTEMS*

Repetitive and Acceptable Deferred Defects control

Reaction to significant aircraft defects

Technical Delay control

Incident/Accident reporting and reaction

Technical Log and hangar defect data collection

Hangar and workshops routine AMS task defect data
collection

Zone sampling system

Reaction to problems reported by external organisations

**Figure 16.2** British Airways formal condition monitoring system (Cumming 1991). *Reprinted with permission of Z. Flugwiss. Weltraumforsch.*

ability and cost standards and reviewing engine removal rate. An important part of the process is the air safety report, shown in Figure 16.3, which deals with all incidents occurring during flight and ground operations. This may be followed by recommendations to modify the approved maintenance schedule for different reasons as shown in Figure 16.4.

It is interesting to notice that aircraft manufacturers, operators, and regulatory agencies have become aware of problems presented by aging aircraft as a consequence of a number of incidents, and have set up an aging aircraft working group, which made the following recommendations: to identify the necessary structure modifications in order to operate beyond a specified lifetime; to develop specific corrosion control programs to be incorporated in the maintenance practice; and to reassess repair standards, in order to enable to extend the airworthiness of the aircraft.

An interesting example of maintenance work employing modern technology is the use of high-speed waterjets for removing jet engine coatings. In this application a modified version of waterjets developed for cutting operations is used; it was necessary to raise the maximum pressure from 240 to 380 MPa to ensure positively that all coating material is removed. Figure 16.5 shows a set of modified waterjet components using rotary nozzles, which have been employed successfully; such a unit is usually installed on a robot arm. An additional advantage of the technique is that it can incorporate a filter unit.

The schematic of a typical coating removal system is presented in Figure 16.6. One sees that a coating removal cell is required; the reasons for this are the high noise of a waterjet running in free air, the high humidity levels that would be created otherwise in the work environment, and

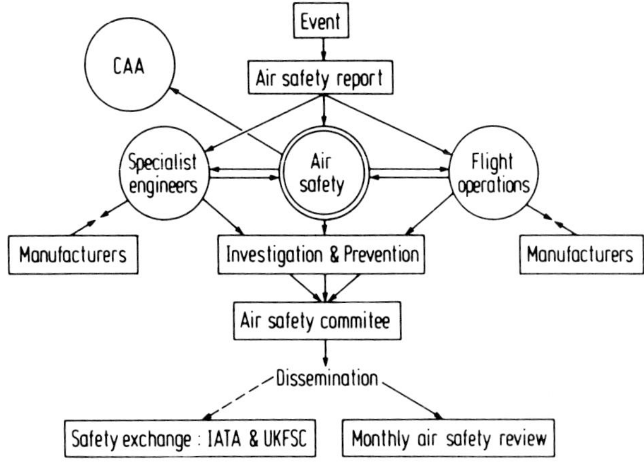


**Figure 16.3** Action following on air safety report (Cumming 1991). *Reprinted with permission of Z. Flugwiss. Weltraumforsch.*

THE NEED TO AMEND THE AMS
CAN RESULT FROM:

1. (a) CAA and FAA Mandatory requirements
   (b) CAA requirements
   (c) Manufacturers' Bulletins

2. Action required by C. M. Systems

3. Design changes to the aircraft thus changing the original Logic Analysis

4. Operational problems

5. Maintenance/Production and Supplies Department problems on existing programme implementation and costs

6. Economic considerations

**Figure 16.4** Causes for modification of maintence schedule (Cumming 1991). *Reprinted with permission of Z. Flugwiss. Weltraumforsch.*

**Figure 16.5** Rotary nozzles used in waterjet removal of jet engine coatings (Watson and Sisson 1992). *Reprinted with permission © 1992 Aerospace Engineering Magazine, Society of Automotive Engineers, Inc.*

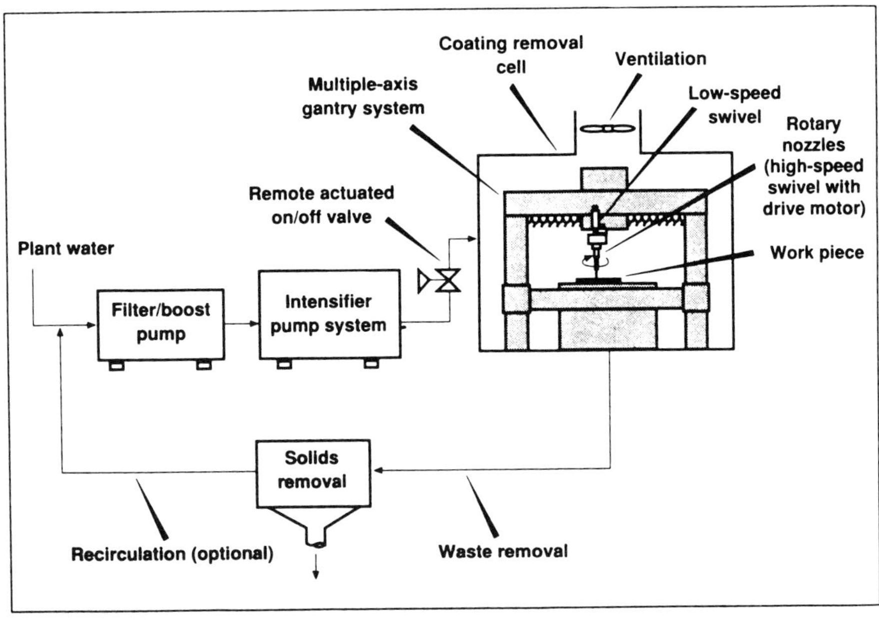


**Figure 16.6** Typical coating removal system schematic (Watson and Sisson 1992). *Reprinted with permission © 1992 Aerospace Engineering Magazine, Society of Automotive Engineers, Inc.*

the need to eliminate the large mess that would be created by the water spray and the removed coating material in the absence of such a device. Another advantage of the coating removal system described is the elimination of hazardous chemicals and the more effective control of waste coating material.

Such a removal system is operated by a programmable controller, which monitors the filter stage, pump conditions, and cleaning cell status. These inputs allow a dedicated processor to monitor and control water pressure, nozzle rotation, and robot position. P&W has already approved the use of waterjets by Northwest Airlines to remove coatings from various components of the PWJT8D engine, such as first stage stators, diffuser housings, intermediate case, rear compressor case, and burner cases. Figure 16.7 is an example of the results obtained with a JT8D burner can by this technique.

**Figure 16.7** Results of coating removed from JT8D burner case by the waterjet process (Watson and Sisson 1992). *Reprinted with permission © 1992 Aerospace Engineering Magazine, Society of Automotive Engineers, Inc.*

# Questions

## Chapter 1

1. Describe the main groups of modern airbreathing engines and discuss the difference between them.

2. What are the distinguishing characteristics of airbreathing jet engines used for (i) civil and (ii) military applications?

## Chapter 2

1. A ramjet performs ideally in all components, except that combustion occurs at a fixed Mach number $M_3$. Show that the specific thrust is given by

$$F/m =$$
$$a_0 M_0 \left[ \left\{ (T_{04}/T_{03}) \left( 1 - \frac{2}{(\gamma - 1)M_0^2} \left[ (T_{04}/T_{03})^{(\gamma-1)M_3^2/2} - 1 \right] \right) \right\}^{1/2} - 1 \right]$$

2. It is known that in order to swallow the shock-wave standing before a supersonic diffuser with fixed geometry, it is necessary to accelerate the engine to a speed above the design Mach number $M_D$ (see sketch). Even if it was possible to accelerate the diffuser to $M = \infty$, the value of $M_D$ would remain finite. What is the value of this $M_D$? Assume one-dimensional flow and $\gamma = 1.4$

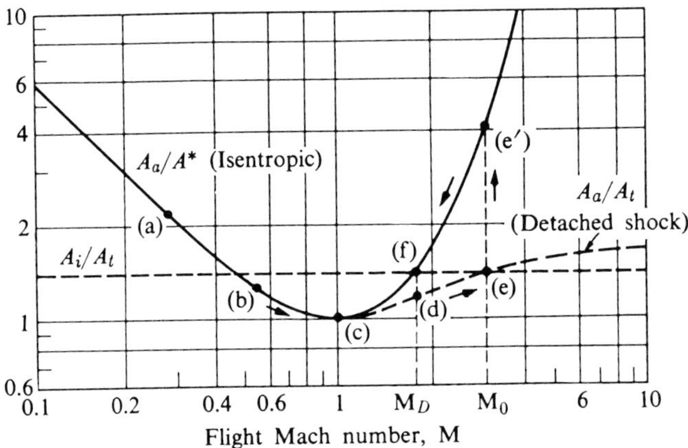


Acceleration and overspeeding of a one-dimensional supersonic inlet (Hill and Peterson 1992). *Reprinted by permission of Addison-Wesley Publishing Co. Inc., Reading, MA.*

3. An inlet is designed for flight at $M = 1.7$. The minimal area, which determines the mass flow, is 0.05 m$^2$. For flight at 10,000 m ($T_a = 233$ K, $P_a = 0.026$ MPa, and $\gamma = 1.4$) it is proposed to use a two dimensional inlet with a control wedge, having a half-angle of 15°. Calculate for the design condition and for $M = 2.5$ the following:

a) The stagnation pressure at the end of the inlet.

b) If the inlet lip is 0.4 m from its symmetry plane, by how much does the wedge forward end stand out of the entry plane?

c) What is the minimum Mach number that can be achieved without detachment of the oblique shock from the wedge?

d) What wedge angle will give maximum stagnation pressure at the exit from the diffuser?

Remember that the flow goes through an oblique and a normal shock.

4. An ideal ramjet flies at a height where $T_a = 260$ K and $P_a = 50$ kPa at $M_a = 3.5$. The calorific value of the fuel is $Q = 40$ MJ/kg and the combustion temperature $T_{04} = 2200$ K. Calculate the pressure and the temperature of each point in the thermodynamic cycle, the exhaust velocity $u_e$, the fuel/air ratio $f$, the specific thrust $F/m_a$, and the specific fuel consumption. Assume that the air and the combustion products behave as ideal gases with $\gamma = 1.4$, $C_p = 1000$ J/kg K, and $R = 287$ J/kg · K, that there are no losses and that the nozzle is adapted to ambient pressure.

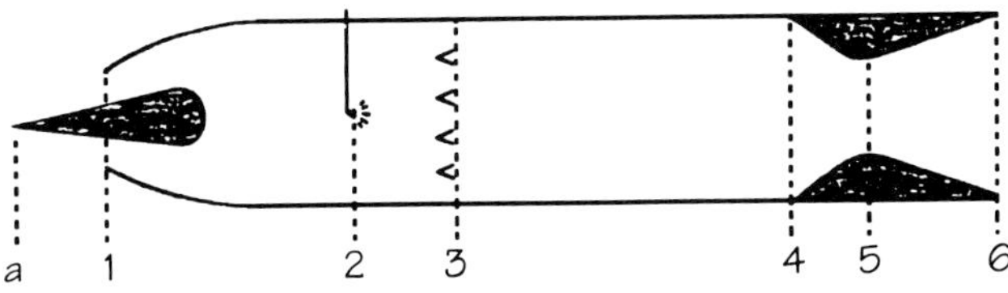


5. A ramjet engine, shown schematically in the sketch, has a converging-diverging inlet. It is designed for flight at $M = 1.9$ at a height of 1200 m ($T_a = 280$ K, $P_a = 86$ kPa). The shock wave is swallowed at design conditions; after that it proceeds towards the diffuser throat, $A_t$, and

reaches optimal operation. The variable geometry exhaust nozzle is and remains adapted to flight conditions. The combustor cross-section is fixed ($A_2 = A_3 = A_4$). The only stagnation pressure losses considered are those caused by the passage through a normal shock and by heat addition in the combustor. Assume that the fluid has constant properties ($\bar{c}, \bar{m}, \gamma$) equal to those of air.

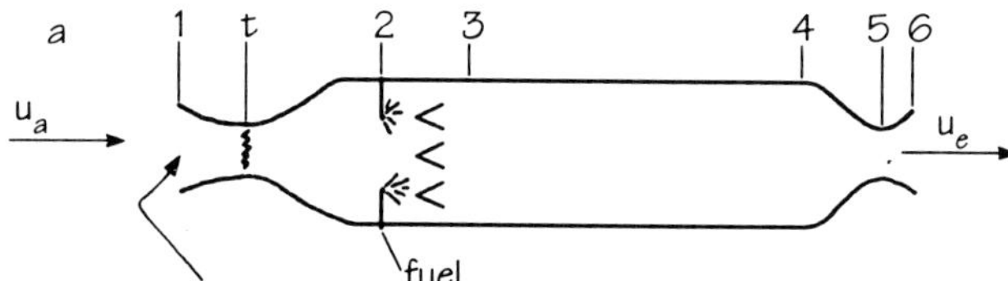


The shock wave stands on the inlet entrance before being swallowed. At optimum flight conditions one has Thrust $F = 50,000$ N; $T_{04} = 1900$ K; $M_4 = 0.55$; $Q = 10,000$ Kcal/kg fuel.

Fluid properties: $\bar{c}_p = 0.24$ Kcal/Kg . K, $\gamma = 1.4$, $m = 29$ kg/kilomole. Calculate the following for optimum operation:

a) Stagnation pressures $P_{0a}, P_{02}, P_{04}$

b) Mach numbers $M_t$, $M_2$ and $M_6$

c) The fuel/air ratio $f$

d) The air mass flow $m_a$

6. A subsonic combustion ramjet flies at a height where $T_a = 225$ K; the temperature in the combustor is 2100 K. The losses in stagnation pressure are given by $r_d = 0.90$, $r_c = 0.93$, $r_n = 0.96$. Assuming that $f \ll 1$, what are the minimum and maximum flight Mach numbers for which positive thrust is obtained? Give a physical explanation for the existence of a maximum flight Mach number.

7. Calculate the performance of a ramjet engine as function of its flight Mach number. The engine is designed to fly at a height of 9.15 km, where $T_a = 230$ K and $P_a = 30$ kPa. The fuel used has a calorific value of 42 MJ/kg and the maximum temperature allowed is 2200 K. According to the manufacturer data the the inlet loss can be calculated from the relation

$$P_{02}/P_{0a} = 1 - 0.1\,(M_a - 1)^{1.5} \qquad \text{for} \quad M_a \geq 1$$

The stagnation pressure ratio across the flameholders is 0.96 and in the combustor it is 0.90. The combustor efficiency $\eta_b$ is 0.98 and the nozzle efficiency $\eta_n = 0.95$. Assume that the velocity in the combustor (between stations 2 and 4) can be neglected. Assume also an adapted nozzle and ideal gases with $\gamma_{0-2} = 1.4$; $\gamma_{3-6} = 1.3$; $c_{p_{0-2}} = 1000$ J/Kg · K and $c_{p_{3-6}} = 1200$ J/Kg · K.

Plot the fuel/air ratio, $f$, the specific thrust $F/m_a$, and the specific fuel consumption as a function of flight Mach number for $1 \leq M_a \leq 5$.

8. Assume that in the ramrocket, shown schematically in the figure, the rocket plume mixes with the incoming air in an ideal constant area duct before expanding through the nozzle. In the ramjet portion the fuel is added to the air stream between stations 2 and 3. Assuming $p_a = p_e$, $m_f/m \ll 1$, $\gamma = \gamma_R$ and $c_p = c_{pR}$, derive the following expressions:

a) $F/m$ and $U_a$ in terms of $T_{03}/T_0$, $p_{04}/p_{03}$, $T_{04}/T_{03}$, $M_a$, $\beta = m_R/m$, and $\gamma$.

b) $T_{04}/T_{03}$ in terms of $\beta$; also, $T_{03}/T_0$, and $T_R/T_0$ in terms of $\beta$.

Here $T_R$ is the adiabatic flame temperature in the rocket chamber.

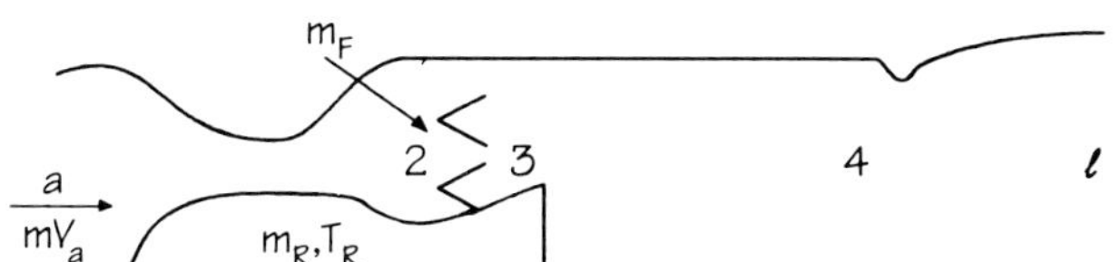


9. Describe the main ramjet applications in the US, France, and the former USSR.

# Chapter 3.

1. Find the cruise range of a scramjet flying at a height of 20 km at Mach 12. It is known that the vehicle has an L/D of 6, uses a fuel having a specific impulse of 1500 sec, the fuel/air ratio is 0.05, and the final mass of the vehicle is 1/3 the initial mass. Assume an adapted nozzle.

2. Calculate the supersonic efficiency of a supersonic inlet having a kinetic energy efficiency of 0.97, if the velocity diminishes by 10% in the inlet.

3. Derive the equations of motion to be used for scramjet nozzle flow applications.

   *Hint*: Use the 2D Reynolds-averaged Navier-Stokes equations for compressible, turbulent flow, supplemented by conservation of species and an equation of state, remembering the nonstationary nature of the process.

4. What is usually meant by hypersonic combustion? How does it differ from supersonic combustion? Which primitive variables can be used to describe it?

5. Explain the calculation methods used by Billig and his group to calculate the combustion processes in hypersonic flow.

6. In what Mach number range does the ramjet-scramjet transition occur? Describe the experimental and theoretical efforts performed in the United States, France, and Russia on this subject (make use of the references cited in Section 5.3.1).

7. Describe the main aerospace plane programs being currently pursued. Compare them and try to evaluate their chances of success.

## Chapter 4

1. Describe and sketch the different types of pulsating combustors discussed in the text.

2. List the major components of a pulse combustor and explain their function.

3. What are the main advances and drawbacks of pulse combustors? Where do you envisage they have a good chance of utilization?

4. Explain the mechanism driving the pulsations in a pulse combustor. Use diagrams to illustrate your explanation.

5. Write down the one-dimensional, unsteady equations of gas dynamics used for a pulse combustor. Explain how they are solved and list the additional issues that must be treated in order to obtain a realistic model.

6. Describe the pulsejet based on a valveless twin combustor developed by Libis and Goldman and discuss the results obtained with this device.

## Chapter 5

1. What are the main components that must be tested during the development of a ramrocket?

2. What facilities are used for the task mentioned in Problem 1? State the main objectives to be attained in such a program.

3. What are (i) air connected test facilities, and (ii) free-jet test facilities? Give examples, using sketches.

4. Enumerate the different types of hypersonic test facilities that you know of and classify them according to their objectives.

5. Describe various types of hypersonic wind tunnels and compare them.

6. Explain what an expansion tube is and how it operates (use sketches).

7. Explain the principle of the reflected shock tunnel (use sketches). Compare the facilities available in Australia (T3 and T4), in the USA (T5 and RFHYL), in France (TCM-2), and Germany (HEG). To answer this question you should consult the references given in Section 5.4.3.

8. Describe the Hypulse facility, stating what test conditions can be attained by its use.

## Chapter 6

1. Define combustion effectiveness.

2. Identify the critical issues involved in the design of a hypersonic combustor.

3. Give some examples of the combined application of computation and experiments in the study of hypersonic combustors.

## Chapter 7

1. Describe the main trends in the development of modern civil aircraft and their powerplants.

2. Discuss the main programs taking place in the 1990s in the field of military aviation.

## Chapter 8

1. Why is it of primary importance to make efficient use of available energy in the design of aircraft powerplants? How is this reflected in recent U.S. programs?

2. Compare the cycle and the performance characteristics of GE and P&W production engines of the late seventies with their E3 engines.

3. Explain the importance of emission levels for civil engines; describe the improvements that can be attained in the E3 program, comparing the results obtained by GE and P&W.

## Chapter 9

1. Classify powerplants for subsonic civil aircraft according to their size and their architecture.

2. Describe the unique features of the Rolls-Royce RB211 engines as compared to the GE CF6 and the P&W JT9D families.

3. Compare the following two large engines which are under development in the 1990s: the RR Trent and the GE90; compare them from all points of view (size, architecture, modularity, introduction of modern techniques, etc.).

4. Discuss similarities and differences between the following medium-sized engines: CFM56, V-2500, RR211-535, and PS-90A.

5. Explain what a propfan is, describe its potential advantages, and compare the types under development.

6. Describe the Garrett ATF3-6 engine, with particular attention to the airflow, the gas flow, and the exhaust system.

7. What is the advantage of the reverse flow combustor of the Garrett TFE731? Describe the peculiar features of

the compressor arrangement in this engine (see Jane's and Bent-McKinley, Ch. 21).

8. List the main applications of small jet engines in the civil and military domains.

9. Describe in detail the Garrett F109 turbofan engine, stating its development philosophy. Explain what the integrity structural program and the accelerated mission test are.

10. Discuss the RTM 322 and MTR 390 programs.

## Chapter 10

1. Compare the three civil supersonic engines that have been built by RR-SNECMA, Kuznetsov, and GE, discussing their advantages and drawbacks.

2. Describe GE double-bypass turbofan, discussing the experimental program carried out for this engine.

3. Describe the P&W VSCE engine, with particular attention to the low emission burner and the low noise coannular nozzle.

4. Describe the turbine bypass engine.

5. Describe the projects for the Mach 3.2 and Mach 5 civil transports carried out by Douglas for the HSCT program, with particular attention to the powerplants.

## Chapter 11

1. What are the main propulsion challenges which the powerplant for a fighter plane must meet?

2. Explain in detail, with the aid of examples, the low observability requirements for military aircraft.

3. What is thrust vector control? Why is it important for military aircraft? Describe some types of TVC.

4. Compare the French M-88 and the Eurojet Turbo EJ200, two modern European military powerplants.

## Chapter 12

1. What are the main options available for vectored thrust engines?

2. Compare the performance of an advanced V/STOL military aircraft with that of a conventionally powered one.

3. Describe the different versions of the Rolls-Royce Pegasus engines.

4. Explain the idea of plenum chamber burning. Discuss the work done by RR in this connection and express your opinion on its potential.

5. Discuss the Yak 36, Yak 38, and Yak 141 with particular emphasis on the Yak 141.

## Chapter 13

1. List the basic requirements common to all combustion chambers.

2. What are the main pollutants emitted from a jet engine?

3. Define the EPA engine classes and the EPA parameter.

4. Discuss the global, environmental concerns associated with engine emissions from aircraft operating at high altitudes.

5. Compare the different techniques that can be used to cool a combustion chamber; discuss in particular the use of lamilloy liners.

6. Explain the concept of prevaporization and give some examples.

7. What is a catalytic combustor? Describe the components of such a device, the processes taking place in it, and the different types.

8. Compare the three following low emission combustors: hot wall, recuperative cooling, and catalytic convector chamber.

9. What type of measurements are used for combustion diagnostics?

10. In obtaining experimental test data, what are the advantages of nonintrusive techniques? Describe in detail at least two of them (use references given in Section 13.6).

## Chapter 14

1. Explain what are acoustic monopoles, dipoles, and quadrupoles.

2. Define the following: Sound pressure level, decibel, and EPNdB.

3. What are the main sources of noise in an aircraft?

4. How is aircraft noise measured near an airport according to FAA and ICAO rules? What is FAR part 36?

5. Discuss methods whereby aircraft noise can be attenuated. In particular, describe methods of noise attenuation used for large turbofans.

6. What is the ANOPP code? What are its uses?

## Chapter 15

1. Describe various types of advanced materials employed in modern powerplants.

2. Explain the following terms: CMC, MMC, PEEK, PMC, SEP-Carbinox, carbon-carbon, hot isostatic pressing, prepregs, and superplastic forming.

3. Discuss the different methods used for turbine blade cooling.

4. What modern materials are used in the Rafale fighter and where?

5. What types of advanced materials will be probably used in the NASP vehicle?

## Chapter 16

1. What is the goal of maintenance standards? How are they achieved?

2. Explain the concept of condition monitoring, using an example.

3. How are waterjets utilized in removing jet engine coatings?

4. What recent events have stressed the importance of maintenance and reliability?

# References


Abarbanel, S. and Kumar, A. (1958). Compact high-order schemes for the Euler equations. *J. of Scient. Computing, 3.* 275–288.

Anderson, G. Y. (1975). Hypersonic propulsion. In: *NASA SP381*, pp. 459–479, Washington, D.C.

Anderson, G. Y., Kumar, A., and Erdos, J. (1990). Progress in hypersonic combustion technology with computation and experiment. *AIAA. Pap. 90-5254.*

Bahr, D. W. (1992). Aircraft turbine engine combustors—development status/challenges. pp. 357–374. In: *Combustion Flow Diagnostics*, Durao, Heiter, Whitelaw and Witzer, eds., *NATO ASI, Series—E207.*

Bangher, J. K., Snape, D. M., and Eardly, B. A. (1984). The design and development of a low emission transply combustor. *AGARD*-CP-353 paper 23.

Barber, T. J. and Cox, G. B., Jr. (1989). Hypersonic vehicle propulsion: a computational fluid dynamics application case study, *J. Propul. Pwr 5*, 492–501.

Barr, P. K., Dwyer, H. A., and Bramlette, T. T. (1988). A one dimensional model of a pulse combustor. *Comb. Sci. Techn., 58*, 315–336.

Barr, P. K. and Dwyer, H. (1991) Pulse combustor dynamics: a numerical study, pp. 673–710. In: Oran, E. S. and Boris, J. P., eds., *Numerical Approaches in Combustion Modeling. Progress in Astronautics and Aeronautics*, Vol. 135, AIAA, Washington, D.C.

Barrère, M. (1988). La combustion supersonique—problèmes posés pour son development. *L'Aér. et l'Astr. 128*, 1988-1, 43–56.

Bauer, C. J. (1982). Next generation trainer (NGT) engine requirements—an application of lessons learned. *AAIA Pap. 82-1084.*

Bayle-Labouré, G. (1991). Pollution émise par les moteurs d'avions. *L'Aér. et L'Astr. 148–149*, 1991 3/4, 125–133.

Beach, H. L. (1979). Hypersonic propulsion. In: *NASA CP 2092*, pp. 387–390, Washington, D.C.

Beckwith, J., Chen, F., Wilkinson, S., Malik, M., and Tuttle, D. (1990). Design and operational features of low disturbance wind tunnels at NASA Langley for Mach numbers 3.5 to 18. *AIAA Pap. 90-1391.*

Ben-Arosh, R. and Gany, A. (1989). Similarity and scale effects in solid fuel ramjet combustors. In: *9th Int. Symp. on Air Breathing Engines*, pp. 127–139, AIAA, Washington, D.C.

Bent, R. D. and McKinley, J. L. (1985). *Aircraft powerplants* McGraw Hill.

Bertin, J. H. et al. (1954). *Aerodynamic Valve*, U.S. Patent No. 2,670,011.

Billig, F. S. (1988). Combustion processes in supersonic flow, *J. Propul. Pwr 4*, 209–216.

Billig, F. S. and Dugger, G. L. (1969). The interaction of shock waves and heat addition in the design of supersonic combustors. In: *12th Symp. (Intern.) on Combustion*, pp. 1125–1134.

Birch, S. (1989). Sänger update. Aerospace Eng. *9*, Aug. 1989, 81–82.

Boury, D., Beaurain A., Lasalmonie A., et Honnorat, Y. (1990). Matériaux et systèmes de matériaux pour la propulsion combinée. Paper 38, In: *AGARD CP-479.*

Braybrook, R. (1991). YAK-141 freestyle. Exclusive details of the World's first supersonic V/STOL fighter. *Air International 41*, 64–67.

Brown, A. C. (1991). Low observable propulsion design. In: *10th Int. Symp. on Air Breathing Engines*, pp. 54–61, AIAA, Washington, D.C.

Brown, A. S. (1989). Revolution in thermoset composites, *Aerospace America 27*(7), 18–23.

Bueteflisch, K. A., ed. (1989) *Proc. 13th Int. Congress on Instrumentation in Aerospace Simulation Facilities.* DLR Res. Center, Goettingen (ICIASF 89).

Calmon, J. (1988). Le turboréacteur à hélices rapides. Pourquoi et comment. *L'Aér. et L'Astr. 132*, 1988-5, 3–15.

Calmon, J. (1989). Les réacteurs militaires: berceau de la technologie. *L'Aér. et L'Astr. 138*, 1989-5, 41–52.

Carpenter, M. H. (1988). Three-dimensional extensions to the SPARK combustion code. *NASP CP-5029. Paper 15.*

Catchpole, B. G. and Runacres, A. (1974). Constant volume gas turbine experiments with gaseous fuel in a rotating pocket combustor, *Note ARL/ME 353*, Aeronaut. Research lab., Dep. of Defence, Melbourne, Australia.

Cattolica, R. J. and Vosen, S. R. (1986). Two dimensional fluorescence imaging of a flame vortex interaction. *Combust. Sci. Technol., 48*, 77–87.

Chant, C. (1989). *Air Defence Systems and Weapons*, p. 66, Braney's Defence Publishers.

Cheng, S. I. (1989). Hypersonic propulsion. *Prog. Energy Combust. Sci. 15*, 183–202.

Chinzei, N., Komura, T., Kudou, K. Murakami, A., Tani, K., Masuya, G., and Wakamatsu, Y. (1991). Effects of injector geometry on scramjet combustor performance. In: *10th Int. Symp. on Air Breathing Engines.* pp. 1219–1227. AIAA, Washington, D.C.

Chiu, H. H. and Croke, E. J. (1980). Combustion performance and noise emission characteristics of pulse combustion, in: *Proc. Symp. on Pulse-Combustion Technology for Heating Applications*, J. M. Clinch, ed., pp. 129–143.

Chopping, D. (1979). Pegasus uprating continues. *Interavia. 34*, 459–60.

Ciepluch, C. C., Davis, D. Y., and Gray, D. E. (1987). Results of NASA's energy efficient engine program. *J. Propul. Pwr. 3*, 560–568.

Colucci, F. (1991). The hybrid solution. *Space, 7*(5), 11–20.

Contensou, P., Marguet, R., and Huet, C. (1972). Etude théorique et experimentale d'un statoréacteur à combustion mixte (domain de vol Mach 3.5/7), *ONERA TP 1121.*

Covault, C., Kandebo, S. W., and Lenorowitz, J. M. (1992). Russian scramjet flight tests. *AWST* 3/2/92, pp. 18–19.

Craigen, J. G. (1976). Mathematical Model of a Pulsating Combustor. Ph.D. Thesis, Univ. of Durham, England.

Cumming, A. (1991). Maintenance standards. *Z. F. Flugwissensch. 15*, 43–48.

Davenport, M., Helmsing, T., Henke, M., and Lyula, G. (1991). V/STOL vs. CTOL—penalties and payoffs for naval aviation. *AIAA Pap. 91-3122.*

Dec, J. E., and Keller, J. O. (1989). Pulse combustor tail-pipe heat-transfer dependence on frequency, amplitude, and mean flow rate, *Combustion and Flame, 77*, 359–374.

Dibble, R. W., and Hollenbach, R. E. (1981). Laser Rayleigh thermometry in turbulent flames. In: *18th Symp. (Inter.) on Combustion*, pp. 1489–1499, The Combustion Institute, Pittsburgh, Pa.

Douglas Aircraft Co. (1989). Study of high-speed civil transport. *NASA CR4235.*

Dreher, A., Bell, R., Flaska, T., and Lozano, M. (1990). Flow field analysis of slush mixing. *AIAA Pap. 90-5265.*

Drummond, J. P. (1979). Numerical solution for perpendicular sonic hydrogen injection into a ducted supersonic airstream, *AIAA J. 17*, 531–533.

Drummond, J. P., Rogers, R. C., and Hussaini, M. Y. (1986). A detailed numerical model of a supersonic reacting mixing layer. *AIAA Pap. 86-1427.*

Dunsworth, L. C. and Reed, G. J. (1979). Ramjet engine testing and simulation techniques. *J. Spacecraft Rockets, 16*, 382–388.

Esgar, J. B., Colladay, R. S., and Kaufman, A. (1970). An analysis of the capabilities and limits of turbine air cooling methods. *NASA* TN-D-5992.

Essman, D. J., Vogel, R. E., Tomlinson, J. G., and Novick, A. S. (1983). TF41/Lamilloy accelerated mission test. *J. Aircraft, 20*, 70–75.

Feiler, C. E. et al. (1975). Propulsion system noise reduction. In: *NASA SP381*, pp. 1–64.

Feiler, C. E., Groeneweg, J. F., Montegani, F. J., Raney, J. P., Rice, E. J. and Stone, J. R. (1979). Noise Reduction. In: *NASA CP2092*, pp. 85–128.

Ferri, A. (1964). Review of problems in application of supersonic combustion. *J. R. Aeronaut. Soc. 68*, 575–597.

Ferri, A. (1968). Review of scramjet propulsion technology. *J. Aircraft, 5*, 3–10.

Ferri, A. (1972). Better marks on pollution for the SST. *Astr. and Aer. 10*(7), 37–41.

Ferri, A. (1973). Reduction of NO formation by premixing, NYU School of Eng. Sci.: *Rep. NYU-AA-73-02.*

Ferri, A. (1973). Mixing controlled supersonic combustion. *Annual Review of Fluid Mech., 5*, 301–308.

Fickeisen, F. C. (1991). Continuing airworthiness requirements and substantiation. In: *ICAS Proc.* 1991, pp. 1844–1851.

Fishbach, L. H., Stitt, L. E., Stone, J. R., and Whitlow, J. B. Jr. (1982). NASA research in supersonic propulsion, a decade of progress, *NASA* TM 82862 (also *AIAA Pap. 82-1048*).

Flithe, K. J., Estey, P. N., and Kniffen, R. J. (1991). The Aquila launch vehicle. A hybrid propulsion space booster. *IAF-91-201*.

Foa, J. V. (1960). *Elements of Flight Propulsion*, John Wiley and Sons, Inc. New York.

Franciscus, L. C. (1981). The turbine bypass engine—a new supersonic cruise research propulsion concept *NASA TM 82608* (also *AIAA Pap 81-1596*).

Fristrom, R. M. and Westenberg, A. A. (1965). *Flame Structure*, McGraw-Hill, N.Y.

Gal-Or, B. (1989). *Vectored Propulsion, Supermaneuverability and Robot Aircraft*, Springer.

Gany, A., and Netzer, D. W. (1986). Combustion studies of metallized fuels for solid fuel ramjets. *J. Propul. Pwr, 2*, 423–427.

Gaydon, A. G. (1974). *Flames—Their Structure, Radiation and Temperature*, 4th edn., Chapman and Hall, London.

Geidel, H. A. (1987). Improved agility for modern fighter aircraft. In: *8th Int. Symp. on Air Breathing Engines*, pp. 47–54, *AIAA*, Washington, D.C.

Gielda, T. and McRae, D. (1986). An accurate, stable, explicit, parabolized Navier-Stokes solver for high-speed flows. *AIAA Pap. 86-1116*.

Gleason, C. C. and Bahr, D. W. (1979). The experimental clean combustion program—Phase III, GE, Cincinnati, Ohio, *R 79 AEG410*.

Goldman, C. (1991). Fuel Flow Rate Modulation in a Ram-Rocket Engine by Means of a Secondary Swirling Flow Injection. M.Sc. Thesis, Technion.

Goldman, C., and Gany, A. (1992). Analysis of thrust modulation of ram-rockets by a vortex valve. *1992 Collection of Papers, 32nd Israel Annual Conf. on Aviation and Astronautics*, pp. 171–177.

Goldman, Y. and Timnat, Y. M. (1983). A study of the pulsating driving mechanism in pulsating combustors. *14th Int. Symp. on Shock Tubes and Waves*, Sydney.

Goldman, Y. and Xieu, D. (1981). Double chamber pulsating combustion system, *Israel J. Techn., 19*, 71–74.

Gosslau, F. (1957). Development of V-1 pulse jet. *AGARDograph No. 20: History of German Guided Missiles Development*, Th. Benecke and A. W. Quick, eds. Appelhaus, Salzhitter-Lebenstedt, West Germany.

Green, W., Tanrikut, S. T., and Sokolowski, D. F. (1983). Development and qualitative characteristics of an advanced two stage combustor. *J. Energy 7*, 345–360.

Greenberg, J. and Wolff, H. (1975). Cold flow evaluation of parameters influencing thrust modulation by a fluidic vortex valve. *Israel J. Techn. 13*, 75–81.

Grobman, J., Anderson, D. N., Diehl, L. A. and Niedzwiecki, P. W. (1975). Combustion and emission technology. In: *NASA SP-381*, pp. 99–138.

Hanby, V. I. and Brown, D. J. (1971). Noise and other problems in the operation of pulsating combustors. *Proc. 1st. Int. Sym. on Pulsating Combustion*, D. J. Brown, ed. Univ. of Sheffield, England, Pap. 16.

Hartree, D. R. (1958). *Numerical Analysis*. Oxford University Press.

Haas, W., Rodi, W., and Schoenung, B. (1988). Film Kuehlung von Turbinenschaufel durch Ausblasung an einer Lochreihe. Z. Flugwissensch. 12, 159–172.

Hauser, C. H., Haas, J. F., Reid, L. and Stepke, F. S. (1979). Turbomachinery technology. In: *NASA CP 2092*, pp. 231–272.

Hedland, E., Higgins, C., Rozanski, C., Fehring, N. and Krueger, D. (1990). The new high Reynolds Mach 8 capability in the NSWC hypervelocity wind tunnel. *AIAA Pap. 90-1379*.

Herbst, W. B. (1987). Thrust vectoring—why and how? In: *Symp. on Air Breathing Engines*, pp. 41–46, *AIAA*, Washington, D.C.

Hill, P. G. and Peterson, C. P. (1965; second edition, 1992). *Mechanics and Thermodynamics of Propulsion*. Addison-Wesley.

Hooper, W. G. (1989). Mach 5 and Mach 8 hypersonic test facility. *AIAA Pap. 89-2297*.

Hornung, H. G. and Belanger J. (1990). Role and techniques of ground testing for simulation of flows up to orbital speed. *AIAA Pap. 90-1377*.

Horsley, W., Kroenke, I., and Chaneller, F. (1990). Slush hydrogen density gage operation in extreme environments. *AIAA Pap. 90-5235*.

Ikawa, H. (1991). Rapid methodology for design and performance predictions of integrated supersonic combustion ramjet engine. *J. Propul. Pwr. 7*, 437–444.

Jacobson, N. S. (1992). High temperature durability considerations for HSCT combustor. *NASA TP 3162*.

Jones, H. P. and Launder, B. E. (1972). The prediction of laminarization with a two equation model of turbulence. *Int. J. Heat Mass Transfer 15*, 301.

Kamath, H. (1989). Parabolized Navier-Stokes algorithms for chemically reacting flows. *AIAA Pap. 89-0386*.

Kanda, T., Masuga, G., and Wakamatsu, Y., Chinzei, N. and Kanmuri, A. (1989). A comparison of scramjet engine performance among various cycles. *AIAA Pap. 89-2676.*

Kanda, T., Masuga, G., Wakamatsu, Y. (1991). Propellant feed system of a regeneratively cooled scramjet. *J. Propul. Pwr. 7,* 299–304.

Kandebo, S. W. (1991). NASP team evaluates interim systems to contain early X-30 program costs. *Aviation Week Space Techn.,* Sept. 2, pp. 63–64.

Kentfield, A. C. (1980). Pressure-gain combustion, a review of recent progress, *Proc. Symp. on Pulse-Combustion Technology for Heating Applications,* J. M. Clinch, ed. pp. 22–45.

King, W. S. (1967). On swirling nozzle flows. *J. Spacecraft Rockets 4,* 1404–5.

Kniffen, R. J. and Flithe, K. J. (1991). The development of a 200,000 lbf thrust hybrid rocket booster. *IAF-91-265.*

Koelle, D. (1990). Saenger advanced STS—Progress report 1990. *AIAA Pap. 90-5200.*

Kors, D. L. (1988). Combined cycle propulsion for supersonic flight. *Acta Astronaut 18,* 191–200.

Korting, P. A. O. G. and Schöyer, H. E. R. (1985). Determination of the regression rate in solid fuel ramjet by means of the ultrasonic pulse echo method. *Proc. Heat Transfer in Fire and Combustion Systems.* C. K. Law, ed. HTD, Vol. 45, New York.

Korting, P. A. O. G., van der Geld, C. W. M., and Timnat, Y. M. (1986). Investigation of a solid fuel combustion chamber. In: *28th Israel A. Conf. on Aviation and Astronautics,* 60–68

Korting, P. A. O. G. and Reitsma, H. J. (1988). The solid fuel combustion chamber. In: *TNO Prins Maurits Laboratory. Diverse and Dynamic,* pp. 163–171, TNO. Rijswijk.

Krieger, K. N., Batson, J. D., Martins, H. F. and Steele, M. A. (1988). Design and development of the Garrett F109 turbofan engine. *Canadian Aeronaut. Space J. 34,* 170–177.

Kuechemann, D. (1965). Hypersonic aircraft and their aerodynamic problems. *Prog. Aeronaut. Sci. 6,* 271–353.

Kuentzmann, P. and Lengellé, G. (1977). Recent research activity at ONERA in combustion instability and erosive burning. TP 1977-30.

Lane, R. J. (1968). Choice of engine cycle for high performance military aircraft. Paper 2 in *AGARD—Advanced Components for Turbojet Engines.*

Launder, B. E. and Spalding, D. B. (1972). *Mathematical Models of Turbulence,* Academic Press, London-New York.

Laruelle, G. (1985). Synthesis of aerodynamic studies of air intakes of a highly maneuvering missile at high Mach numbers. In: *7th Int. Symp. on Air Breathing Engines,* pp. 125–134, AIAA, Washington, D.C.

Laruelle, G. (1987). Air intakes for supersonic missiles: design criteria and development. In: *8th Int. Symp. on Air Breathing Engines,* pp. 84–93, AIAA, Washington, D.C.

Lenorowitz, J. M. (1991). Next generation French fighter. *Aviation Week Space Techn. 134,* No. 21, 20–22.

Levy, Y. and Lockwood, F. C. (1981). Velocity measurements in particle laden turbulent free jet. *Combustion and Flame, 40,* 333–339.

Lewis, W. J. (1991). Propulsion for supersonic STOVL aircraft. In: *Supplementary Papers from the 10th Int. Symp. on Air Breathing Engines.* Rolls-Royce, Bristol.

Lewis, W. J. and Simpkin, P. (1981). Multimission V/STOL with vectored thrust engines. *AIAA Pap. 81-1363.*

Libis, N. and Goldman, Y. (1992). Study of a pulse ramjet based on twin valveless combustors coupled to operate in antiphase. *32nd Israel Conf. on Aviation and Astronautics,* 163–170.

Lockwood, F. C. and Odidi, A. O. O. (1975). Measurement of mean and fluctuating temperature and of ion concentration in round free-jet turbulent diffusion and premixed flames. In: *15th Symp. (Intern.) on Combustion,* pp. 561–571. The Combustion Institute, Pittsburgh, Pa.

Lockwood, R. M. (1964). Pulse-reactor low cost lift-propulsion engines, *AIAA Pap. 64-172.*

Logan, P. (1987). Simultaneous measurements of temperature, density and mass flux in supersonic turbulence. *Progress in Astronautics and Aeronautics.* Vol. 112, pp. 116–140, AIAA. N.Y.

Mady, C. J., Hickley, P. J., and Netzer, D. W. (1978). Combustion behavior of solid fuel ramjet combustors. *J. Spacecraft Rockets 15,* 131–132.

Mager, A. (1961). Approximate solution of isentropic swirling flow through a nozzle. *ARS J. 31,* 1140–1148.

Maita, M. (1991). Space plane program in Japan. In: *10th Int. Symp. on Air Breathing Engines,* pp. 62–70. AIAA, Washington, D.C.

Mangiarotty, R. A., Marsh, A. H., and Feder, F. (1968). Ducted lining materials and concepts. In: *NASA SP-189,* pp. 29–52.

Marguet, R. (1989). Ramjet research and applications in France. In: *9th Int. Symp. on Air Breathing Engines.* pp. 49–57, AIAA, Washington, D.C.

Marguet, R. and Cazin, Ph. (1985). Ramjet research in France: Realities and perspectives. In: *7th Int. Symp. on*

*Air Breathing Engines*, pp. 215–224, A I A A, Washington, D.C.

Marquardt (1968). Test facilities engineering operations. Van Nuys, CA.

McClinton, C. R. (1990). CFD Support of NASP Design. *AIAA Pap. 90-3252.*

McCormack, F. N. (1969). The effect of viscosity on hyper-velocity impact cratering. *AIAA. Pap. 69-354.*

Melis, M. and Gladden, H. (1990). A unique high heat flux facility for testing hypersonic engine components. *AIAA Pap. 90-5228.*

Mestre, A. and Lagain, G. (1984). Conception d'un foyer a' flux inverse pour petites turbomachines. *AGARD* CP 353, Paper 10.

Metcalfe, M. T., Eaton, R. A., and Snape, D. M. (1991). The impact of air transport on the environment. In: *10th Int. Symp. on Air Breathing Engines*, pp. 221–228, A I A A, Washington, D.C.

Minoda, M., Sakata, K., Tamaki, T., Saitoh, T. and Yasuda, A. (1991). Feasibility study of air-breathing turboengines for horizontal takeoff and landing space planes. *J. Propul. Pwr. 7*, 821–828.

Morgan, R. G. and Stalker, R. J. (1989). Hypersonic combustion of hydrogen in a shock tunnel. In: *9th Int. Symp. on Air Breathing Engines*, pp. 577–584, A I A A, Washington, D.C.

Morris, S. J., Geiselhart, K. A., and Coen, P. G. (1989). Performance potential of an advanced technology Mach 3 turbojet engine installed in a conceptual high-speed civil transport. *NASA TM 4144.*

Mularz, E. J., Gleason, C. C., and Dodds, W. J. (1979). Combustor concepts for aircraft gas turbine low power emission reduction. *J. Energy 3*, 55–61.

Muylaert, J., Voiron, T., Sagnier, P., Lourme, D., Papirnyk, O., Hannemann, K., Buetefisch, K., and Koppenwallner, G. (1991). Review of the European wind tunnel performance and simulation requirements. In: *ESA SP 318*, pp. 559–574.

Natan, B., Gany, A., and Wolff, H. (1982). Thust modulation by injection of secondary swirling flow and secondary combustion. *Acta Astron. 9*, 703–711.

Netzer, D. W. (1977). Modeling solid-fuel ramjet combustion. *J. Spacecraft Rockets, 14*, 762–766.

Netzer, D. W. (1981). Primitive variable model application to solid fuel ramjet combustion. *J. Spacecraft Rockets, 18*, 127–132.

Netzer, A. and Gany, A. (1991). Burning and flameholding characteristics of a miniature solid fuel ramjet combustor. *J. Propul. Pwr 7*, 357–363.

Nielsen, A. E. (1992). NASP means more than scramjets. *Space News 3*, No. 21, p. 15.

Nonweiler, T. (1963). Delta wings of shapes amenable to shock wave theory. *J. R. Aeronaut. Soc. 67*, 39–40.

Nored, D. L., Dugan, J. T. Jr., Saunders, N. T. and Zemanski, J. A. (1979). Aircraft energy efficiency (ACEE) status report. In: *NASA CP2092*, pp. 1–58.

Northam, G. B. and Anderson, G. Y. (1986). Supersonic combustion ramjet research at Langley. *AIAA Pap. 86-0159.*

Norton, D. J., Farquhar, B. W., and Hoffman, J. P. (1969). An analytical experimental investigation of swirling flows in nozzles. *AIAA J. 10*, 1992–2000.

Parkinson, R. and Conchie, P. (1990). Hotol, *AIAA Pap. 90-5201.*

Petrash, D. A. Diehl, L. A., Jones, R. F., and Mularz, E. J. (1979). Emission reduction. In: *NASA CP2092*, pp. 59–84.

Picard, C. A. (1988). Use of new materials and new technologies in modern aircraft structures. *L'Aér. et L'Astr. 130*, 1988-3, 4–29.

Pierson, J. (1987). Les technologies nouvelles et le développement des programmes aéronautiques civils actuels. *L'Aér. et L'Astr. 123–124*, 1987 2–3, 4–9.

Ponizy, B. and Wojcicki, S. (1984). On modeling of pulse combustors. *20th Symp. (Int.) on Combustion*. The Combustion Inst., Pittsburgh, pp. 2019–2024.

Porter, C. D. (1958). Valveless gas-turbine-combustors with pressure gain, *ASME Pap. 58-GTP-11.*

Preisser, S. J., Golab, R. A., Seiner, J. M., and Powell, C. A. (1990). Supersonic jet noise: its generation, prediction, and effect on people and structures. *SAE TP 901927.*

Puster, R. L., Rebush, D. E., and Kelly, H. N. (1987). Modification to the Langley 88 high temperature tunnel for hypersonic propulsion testing. *AIAA Pap. 87-1887.*

Putnam, A. A., and Brown, D. J. (1974). Combustion noise: problems and potentials. In: *Combustion Technology—Some Modern Developments*, H. B. Palmer and J. M. Beer, eds. Academic Press.

Putnam, A. (1980). A review of pulse combustor technology, in: *Proceedings of the Symposium on Pulse-Combustion Technology for Heating Applications*, J. M. Clinch, ed, pp. 1–21.

Ramohalli, K. and Yi, J. (1990). Hybrids revisited. *AIAA 90-1962.*

Richards, E. J. and Meads, D. J., eds. (1968). *Noise and Acoustic Fatigue in Aeronautics*, John Wiley and Sons, Ltd.

Roberts, R., Fiorentino, A., and Greene, W. (1977). Experimental clean combustor program, Phase III, *NASA CR-135253*.

Roffe, G. and Ferri, A. (1975). Prevaporisation and premixing to obtain low oxides of nitrogen in gas turbine combustors. *NASA CR-2495*.

Rogers, R. C., Drummond, J. P., and Weidner, E. H. (1987). Numerical analyses of shock tunnel data for hydrogen injection into supersonic air flows. Paper presented at the 24th JANNAF Combustion Meeting, Monterey, California.

Ronald, T. M. (1990). Materials for hypersonic engines. Paper 36 in *AGARD CP-479*.

Rose, A. and Gupta, R. (1984). Combustion diagnostics by photo-deflection spectroscopy. In: *20th Symp. (Intern.) on Combustion*, pp. 1339–1345, The Combustion Institute, Pittsburgh, Pa.

Rosfjord, T. J. (1976). Catalytic combustion for gas turbine engines. *AIAA Pap. 76-46*.

Ruffles, P. C. (1984). Small engine technology, *Aer. J. 88*, 10–16.

Ruhmann, D. C., Stoyack, J. III, and Martin, R. (1992a). Missiles and space systems with a material difference. *Aerospace America 30*, no. 9, 42–43.

Ruhmann, D. C., Bates, W. F. Jr., Dexter, H. B., and June, R. R. (1992b). New materials drive high performance aircraft. *Aerospace America 30*, no. 9, 46–49.

Sanborn, J. W., Lenerz, J. E., and Johnson, J. D. (1989). Advanced turbofan engine combustion system design and test verification. *J. Propul. Pwr. 5*, 502–509.

Sayles, D. C. (1975). Development of test motors for advanced controllable propellants. *J. Spacecraft Rockets, 12*, 174–178.

Schaber, R., Schwab, R. R., and Voss, H. (1991). Preliminary design of a ramjet powered supersonic anti-ship missile. *AIAA Pap. 91-3127*.

Schadow, K. C. (1969). Experimental investigation of Boron combustion in air-augmented rockets. *AIAA J. 7*, 1070–1076.

Schadow, K. C. (1981). Boron combustion in ducted rockets. *AGARD-CP-307*.

Schadow, K. C., Cords, H. F., and Chiese, D. J. (1978). Experimental studies of combustion processes in a tubular combustor with fuel addition along the wall, *Comb. Sci. Technol. 19*, 51–57.

Schoedl, R. (1980). Laser-two-focus (L2F) velocimeter for automatic flow vector measurements in the rotating components of turbomachines, *ASME J. Fluid Engng 102*, 412–419.

Schöyer, H. F. R., and Korting, P. A. O. G. (1984). Window on science visit. U.S.A. March 21–April 22, *Report PML 1984-C25*.

Schulte, G., Pein, R., and Högl, A. (1987). Temperature and concentration measurements in a solid fuel ramjet combustion chamber. *J. Propul. Pwr. 3*, 114–120.

Shepard, C. E. and Warshawsky, I. (1952). Electrical techniques for compensation of thermal time lag of thermocouples and resistance thermometer elements. *NASA-TN-2703*.

Smith, C. J. (1990). The promise of advanced materials for a 21th Century UBE. *AIAA Pap. 90-2396*.

Sokolowski, D. E. and Rhode, J. E. (1981). The $E^3$ combustors status and challenge, *NASA TM 82684* also *AIAA Pap. 81-1353*.

Soltheran, A. (1971). Some combustion aspects of plenum chamber burning, pp. 58–63, in: *Combustion and Heat Transfer in Gas Turbines*, Norster, E. R., ed., Pergamon.

Sosounov, V. A. (1974). Some problems concerning optimal ducted rocket engines with secondary burning. *Proc. 2nd. Int. Symp. on Air Breathing Engines*, Sheffield.

Spalding, D. B., Launder, B. E., Morse, A. P., and Maples, G. (1974). Combustion of hydrogen-air jets in local chemical equilibrium. *NASA CR-2407*.

Stalker, R. J. (1972). Development of a hypervelocity wind tunnel, *Aeronaut. J. 76*, 374–384.

Stalker, R. J. and Morgan, R. G. (1987). Scramjet testing in impulse facilities. In: *8th Int. Symp. on Air Breathing Engines*, pp. 66–74, AIAA, Washington, D.C.

Stalker, R. J. (1989). Thermodynamic and wave processes in high Mach number propulsion ducts. *AIAA Pap. 89-0261*.

Stevenson, C. A. and Netzer, D. W. (1981). Primitive-variable model application to solid-fuel ramjet combustion. *J. Spacecraft 18*, 89–94.

Stoltz, L. and Graff, J. (1981). Boron-aluminum blade advances. *AAIA Pap. 81-1479*.

Stull, F. D. Craig, R. R., and Hojnacki, J. T. (1974). Dump combustor parametric investigations. *ASME Fluid Mechanics of Combustion, Joint Fluids Eng. and CSMG Conf.*, pp. 135–142.

Stull, F. D., Craig, R. R., Streby, G. D., and Vanka, S. P. (1985). Investigation of a dual inlet side dump combustor using liquid fuel injection. *J. Propuls. Pwr. 1*, 83–88.

Swithenbank, J. (1966). Hypersonic air-breathing propulsion, *Progr. Aeronaut. Sci. 8*, 229–294.

Swithenbank, J., Eames, I., Chin, S., Ewan, B., Yang, Z. Cao, J., and Zhao, X. (1989). Turbulent mixing in supersonic combustion systems. *AIAA Pap. 89-0260*.

Tamagno, J., Bakos, R., Pulsonetti, M., and Erdos, J. (1990). Hypervelocity real gas capabilities of GASL's expansion tube (HYPULSE) facility. *AIAA Pap. 90-1390*.

Tanrikut, S., Marshall, R. L., and Sokolowski, D. E. (1981). Improved combustor durability, segmented approach with advanced cooling techniques, *AIAA Pap. 81-1354*.

Taylor, J. W. R., ed. (1978). Olympus 593. In: *Jane's All the World's Aircraft*, pp. 694–695.

Tharrat, C. E. (1965). The propulsive duct, *Aircraft Engineering, 37*, Part I, 327–337, Part II, 359–371.

Tharrat, C. E. (1966). The propulsive duct, *Aircraft Engineering, 38*, Part III, 23–25.

Thring, M. W., ed. (1961). *The Collected Works of F. H. Reynst*, Pergamon Press.

Timnat, Y. M. (1987). *Advanced Chemical Rocket Propulsion*. Academic Press.

Tsujimoto, Y. and Machii, N. (1986). Numerical analysis of a pulse combustor, *21th Symp. (Int.) on Combustion*. The Combustion Inst., Pittsburgh, Pa. pp. 539–546.

Vanka, S. P., Stull, F. D., and Craig, R. R. (1983). Analytical characterization of flow fields in side-inlet dump combustors, *AIAA Pap. 83-0199*.

Vanka, S. P., Craig, R. R., and Stull, F. D. (1986). Mixing, chemical reaction and flowfield development in ducted rockets, *J. Propul. Pwr. 2*, 331–338.

Vedikhin, G. V., Gafarov, A. S., Dolgikh, E. K., and Kandalintseva, M. V. (1979). Burning of fuel in the pulsating combustion regime, *Izv. VUZ. Aviatsionnaya Tekhn. 22*(3), 75–77.

Venishi, K., Rogers, R. C., and Northam, G. B. (1987). Three-dimensional computation of transverse hydrogen jet combustion in a supersonic airstream. *AIAA Pap. 87-0089*.

Vinogradov, V., Grachko, V., Petrov, M., and Sheechman, J. (1990). Experimental investigation of a 2-D dual model scramjet with hydrogen fuel at Mach 4–6. *AIAA Pap. 90-5269*.

Vogelsang, L. B. and Gunnink, J. W. (1985). An evaluation of new materials for the next generation of transport aircraft structures. Paper presented at Symp. on The next generation of aircraft, Delft.

Vos, J. B. (1987). The Calculation of Turbulent Reacting Flows with a Combustion Model Based on Finite Chemical Kinetics. Ph.D. Thesis, Delft University.

Waldman, B. and Harsha, P. (1990). The NASP Program. *AIAA Pap. 90-5252*.

Waltrup, P. J. (1987). Liquid-fueled supersonic combustion ramjets: a research perspective. *J. Propul. Pwr. 3*, 515–524.

Waltrup, P. J., Billig, F. S., and Stockbridge, R. D. (1979). Procedure for optimizing the design of scramjet engines. *J. Spacecraft Rockets 16*, 163–171.

Waltrup, P. J., Billig, F. S., and Stockbridge, R. D. (1982). Engine sizing and integration requirements for hypersonic missile applications. In: *Symp. Ramjets and Ramrockets in Military Applications. AGARD-CP-307*.

Wanstall, B. (1984). Integral rocket ramjets for long legs at supersonic speeds, *Interavia 39*, 331–334.

Watson, J. D. and Sisson, S. (1992). Waterjets for removing engine coatings. *Aerospace Eng. 12*, No.4, pp. 13–14.

Wear, J. D., Trout, A. M., Smith, J. M., and Jones, R. E. (1978). Design and preliminary results of a semitranspiration cooled (Lamilloy) liner for a high-pressure, high-temperature combustor. *AIAA Pap. 78-997*.

Weber, R. J. and MacKay, J. S. (1958). An analysis of ramjet engines using supersonic combustion. *NACA TN* 4386.

Wesoky, H. L., Prather, M. J., and Kayten, G. G. (1990). NASA's high speed research program; an introduction and status report. *SAE TP 90-1923*.

Wilson, D. R. (1990). Development and calibration of a continuous flow arc heated hypersonic wind tunnel. *AIAA Pap. 90-1381*.

Yackle, A. R. (1983). V/STOL, a practical weapon system. *Astr. and Aer. 21,*(1), 18–26.

Yamanaka, T. (1990). The Japanese space planes. R&D Overview. *AIAA* Pap. 90-5223.

Zhongqin, Z., Zhenpeng, T., Jinfu, T., and Wenlan, F. (1986). Experimental investigation of combustion efficiency of air-augmented rockets. *J. Propul. Pwr. 2*, 305–310.

Zucrow, M. J. (1958). *Aircraft and Missile Propulsion*, Chp. 9, Wiley and Sons. N.Y.

Zvuloni, R., Levy, Y., and Gany, A. (1989). Investigation of a small solid fuel ramjet. *J. Propul. Pwr. 5*, 146–152.

# Index